Advances in Sustainable Bioenergy Production and Biomass Waste Reutilization

Advances in Sustainable Bioenergy Production and Biomass Waste Reutilization

Editors

Steven Lim
Shuit Siew Hoong
Pang Yean Ling
Santi Chuetor

MDPI

Basel • Beijing • Wuhan • Barcelona • Belgrade • Novi Sad • Cluj • Manchester

Editors

Steven Lim
Chemical Engineering
Universiti Tunku Abdul Rahman
Kajang
Malaysia

Shuit Siew Hoong
Chemical Engineering
Universiti Tunku Abdul Rahman
Kajang
Malaysia

Pang Yean Ling
Chemical Engineering
Universiti Tunku Abdul Rahman
Kajang
Malaysia

Santi Chuetor
Chemical Engineering
King Mongkut's University of Technology North Bangkok
Bangkok
Thailand

Editorial Office
MDPI
St. Alban-Anlage 66
4052 Basel, Switzerland

This is a reprint of articles from the Special Issue published online in the open access journal *Sustainability* (ISSN 2071-1050) (available at: www.mdpi.com/journal/sustainability/special_issues/Bioenergy_Production_Biomass_Waste_Reutilization).

For citation purposes, cite each article independently as indicated on the article page online and as indicated below:

Lastname, A.A.; Lastname, B.B. Article Title. *Journal Name* **Year**, *Volume Number*, Page Range.

ISBN 978-3-7258-0624-9 (Hbk)
ISBN 978-3-7258-0623-2 (PDF)
doi.org/10.3390/books978-3-7258-0623-2

Contents

About the Editors . . . . **vii**

Preface . . . . **ix**

Yasirah Yusoff, Ee Sann Tan and Firas Basim Ismail
A Comparison of Feedstock from Agricultural Biomass and Face Masks for the Production of Biochar through Co-Pyrolysis
Reprinted from: *Sustainability* **2023**, *15*, 16000, doi:10.3390/su152216000 . . . . **1**

Hui Wun Tan, Yean Ling Pang, Steven Lim, Woon Chan Chong, Chin Wei Lai and Ahmad Zuhairi Abdullah
Exploring the Potential of Utilizing Aquatic Macrophytes for Enhanced Phytoremediation of Zinc in Artificial Wastewater: Characteristics and Parameter Studies
Reprinted from: *Sustainability* **2023**, *15*, 15170, doi:10.3390/su152015170 . . . . **17**

Yoonah Jeong, Jae-Sung Kim, Ye-Eun Lee, Dong-Chul Shin, Kwang-Ho Ahn and Jinhong Jung et al.
Investigation and Optimization of Co-Combustion Efficiency of Food Waste Biochar and Coal
Reprinted from: *Sustainability* **2023**, *15*, 14596, doi:10.3390/su151914596 . . . . **39**

Mooktzeng Lim and Ee Sann Tan
Techno-Economic Feasibility Study for Organic and Plastic Waste Pyrolysis Pilot Plant in Malaysia
Reprinted from: *Sustainability* **2023**, *15*, 14280, doi:10.3390/su151914280 . . . . **51**

Segundo Rojas-Flores, Magaly De La Cruz-Noriega, Luis Cabanillas-Chirinos, Santiago M. Benites, Renny Nazario-Naveda and Daniel Delfín-Narciso et al.
Green Energy Generated in Single-Chamber Microbial Fuel Cells Using Tomato Waste
Reprinted from: *Sustainability* **2023**, *15*, 10461, doi:10.3390/su151310461 . . . . **67**

Ashvinder Singh Gill, Kam Huei Wong, Steven Lim, Yean Ling Pang, Lloyd Ling and Sie Yon Lau
Investigation of Microwave Irradiation and Ethanol Pre-Treatment toward Bioproducts Fractionation from Oil Palm Empty Fruit Bunch
Reprinted from: *Sustainability* **2024**, *16*, 1275, doi:10.3390/su16031275 . . . . **79**

Shamala Gowri Krishnan, Fei Ling Pua and Zhang Fan
Effect of Chemical Pre-Treatment on the Catalytic Performance of Oil Palm EFB Fibre Supported Magnetic Acid Catalyst
Reprinted from: *Sustainability* **2023**, *15*, 8637, doi:10.3390/su15118637 . . . . **99**

Małgorzata Hawrot-Paw and Aleksander Stańczuk
From Waste Biomass to Cellulosic Ethanol by Separate Hydrolysis and Fermentation (SHF) with *Trichoderma viride*
Reprinted from: *Sustainability* **2022**, *15*, 168, doi:10.3390/su15010168 . . . . **113**

Yean Ling Pang, Yen Ying Quek, Steven Lim and Siew Hoong Shuit
Review on Phytoremediation Potential of Floating Aquatic Plants for Heavy Metals: A Promising Approach
Reprinted from: *Sustainability* **2023**, *15*, 1290, doi:10.3390/su15021290 . . . . **123**

Dave Mangindaan, Emil Robert Kaburuan and Bayu Meindrawan
Black Soldier Fly Larvae (*Hermetia illucens*) for Biodiesel and/or Animal Feed as a Solution for Waste-Food-Energy Nexus: Bibliometric Analysis
Reprinted from: *Sustainability* **2022**, *14*, 13993, doi:10.3390/su142113993 . . . . **146**

Jia Ying Tan, Wah Yen Tey, Joongjai Panpranot, Steven Lim and Kiat Moon Lee
Valorization of Oil Palm Empty Fruit Bunch for Cellulose Fibers: A Reinforcement Material in Polyvinyl Alcohol Biocomposites for Its Application as Detergent Capsules
Reprinted from: *Sustainability* **2022**, *14*, 11446, doi:10.3390/su141811446 . . . . **164**

Besek Mariam Mohamad Jahis, Zul Ilham, Sugenendran Supramani, Mohamad Nor Azzimi Sohedein, Mohamad Faizal Ibrahim and Suraini Abd-Aziz et al.
Ganodiesel: A New Biodiesel Feedstock from Biomass of the Mushroom *Ganoderma lucidum*
Reprinted from: *Sustainability* **2022**, *14*, 10764, doi:10.3390/su141710764 . . . . **184**

About the Editors

Steven Lim

Steven Lim is an Associate Professor at the Department of Chemical Engineering, Lee Kong Chian Faculty of Engineering and Science, Universiti Tunku Abdul Rahman, Malaysia. He is also the chairperson of the Centre for Advanced and Sustainable Materials Research. His research interests revolve around devising sustainable and cost-effective methods to synthesize renewable energy (biodiesel, bioethanol, and syngas) and other value-added products (bioplastic, cellulose nanofiber, and triacetin) from renewable sources such as biomass. He has published much of his research in high-impact journals, such as Bioresource Technology, the Journal of Hazardous Material, and the Journal of Applied Energy.

Shuit Siew Hoong

Dr. Shuit Siew Hoong is an assistant professor in the Department of Chemical Engineering, Universiti Tunku Abdul Rahman. He is actively involved in research regarding renewable energy, membrane teachnology, nanocatalysis and water or wastewater treatment.

Pang Yean Ling

Pang YL completed her Bachelor's and Doctoral Degrees in Chemical Engineering in 2009 and 2013, respectively, from Universiti Sains Malaysia (USM). After a year of post-doctoral research at the University of Malaya (UM), she joined Universiti Tunku Abdul Rahman (UTAR) in December 2014 and has been there until the present point. She is currently an Associate Professor in the Department of Chemical Engineering of UTAR. Currently, she is working on research projects related to green chemistry principles, such as the use of biomass relative to valuable components and the use of any green alternative methods to synthesize highly efficient nanomaterials for environmental remediation and biofuel production.

Santi Chuetor

Santi Chuetor is focused on the development of an integrative biorefinery, environmentally ecofriendly process for lignocellulose pretreatment, green chemistry, dry fractionation of lignocellulosic biomass for bio-based chemicals and energy applications, as well as integration of techno-economic assessment and environmental evaluation of lignocellulosic valorization.

Preface

As the global community grapples with the urgent need for sustainable energy sources and environmental conservation, the pivotal role of bioenergy production and biomass waste reutilization has come to the forefront of scientific inquiry and technological innovation. This Special Issue collection, "Advances in Sustainable Bioenergy Production and Biomass Waste Reutilization", seeks to illuminate the cutting-edge research and transformative developments in these critical fields. The quest for sustainable energy solutions has never been more imperative, given the escalating challenges posed by climate change, dwindling fossil fuel reserves, and the environmental impact of conventional energy sources. Bioenergy, derived from renewable biological resources, emerges as a promising alternative, offering the potential to mitigate greenhouse gas emissions and contribute to a more sustainable and resilient energy landscape.

This collection assembles a diverse array of contributions from esteemed researchers, scientists, and engineers who have dedicated their expertise to addressing the multifaceted aspects of sustainable bioenergy production. From advancements in biomass feedstock selection and cultivation to breakthroughs in conversion technologies, the papers within this collection span the entire bioenergy production chain. Equally significant is the focus on biomass waste reutilization, as sustainable practices demand not only efficient energy production but also the responsible management of byproducts.

The interdisciplinary nature of the research presented in this Issue reflects the collaborative efforts required to address the complex challenges associated with sustainable bioenergy. Contributions encompass fields such as biology, chemistry, engineering, agronomy, and environmental science, demonstrating the interconnectedness of diverse disciplines in achieving comprehensive and viable solutions. As we embark on this journey through the pages of "Advances in Sustainable Bioenergy Production and Biomass Waste Reutilization", it is our hope that the insights and discoveries shared herein will inspire further research, innovation, and policy developments. By disseminating knowledge and fostering dialogue, this collection aims to contribute significantly to the ongoing global efforts towards a more sustainable, resilient, and environmentally conscious energy future.

We extend our gratitude to the dedicated researchers who have contributed their work to this collection, as well as to the readers who share our commitment to advancing the frontiers of sustainable bioenergy production and biomass waste reutilization.

May this compilation serve as a catalyst for positive change and propel us towards a future where bioenergy plays a pivotal role in meeting our energy needs while preserving the delicate balance of our planet's ecosystems.

Steven Lim, Shuit Siew Hoong, Pang Yean Ling, and Santi Chuetor
Editors

sustainability

MDPI

Article

A Comparison of Feedstock from Agricultural Biomass and Face Masks for the Production of Biochar through Co-Pyrolysis

Yasirah Yusoff [1], Ee Sann Tan [2,*] and Firas Basim Ismail [3]

[1] Institute of Sustainable Energy, Universiti Tenaga Nasional, Kajang 43000, Malaysia; yasirahyusoff@uniten.edu.my
[2] Department of Mechanical Engineering, Universiti Tenaga Nasional, Kajang 43000, Malaysia
[3] Institute of Power Engineering, Universiti Tenaga Nasional, Kajang 43000, Malaysia; firas@uniten.edu.my
* Correspondence: eesann@uniten.edu.my

Abstract: This study explores the pyrolysis of disposable face masks to produce chemicals suitable for use as fuel, addressing the environmental concern posed by single-use face masks. Co-pyrolysis of biomass with face mask plastic waste offers a promising solution. The research focuses on the co-pyrolysis of biomass and face masks, aiming to characterise the properties for analysis and optimisation. Selected agricultural biomass and face mask plastic waste were subjected to temperatures from 250 °C to 400 °C for co-pyrolysis. Slow pyrolysis was chosen because face masks cannot be converted into useful bioproducts at temperatures exceeding 400 °C. The samples were tested in four different ratios and the study was conducted under inert conditions to ensure analysis accuracy and reliability. The results indicate that face masks exhibit a remarkable calorific value of 9310 kcal/kg. Face masks show a two-fold increase in calorific value compared with biomass alone. Additionally, the low moisture content of face masks (0.10%) reduces the heating value needed to remove moisture, enhancing their combustion efficiency. This study demonstrates the potential of co-pyrolysis with face masks as a means of generating valuable chemicals for fuel production, contributing to environmental sustainability.

Keywords: pyrolysis; face mask; biomass; sustainability; EFB

check for updates

Citation: Yusoff, Y.; Tan, E.S.; Ismail, F.B. A Comparison of Feedstock from Agricultural Biomass and Face Masks for the Production of Biochar through Co-Pyrolysis. *Sustainability* **2023**, *15*, 16000. https://doi.org/10.3390/su152216000

Academic Editor: Maurizio Volpe

Received: 22 September 2023
Revised: 23 October 2023
Accepted: 27 October 2023
Published: 16 November 2023

1. Introduction

The World Health Organisation declared a pandemic in March 2020 in response to the global outbreak of COVID-19 [1]. This outbreak has led to the loss of millions of lives and directly impacted over 180 million individuals worldwide [2]. In quick response to this crisis, many nations mandated the use of personal protective equipment, particularly face masks [3–5]. With an estimated global production of over 120 billion face masks in 2020, these masks became critical tools for controlling the spread of the disease [6].

Since the onset of the COVID-19 outbreak, the disposal of these face masks has become an ongoing issue, from waste management to recycling challenges. In a local context, Malaysia often resorts to incineration and landfilling for waste disposal. Compounding the problem, no proper guidelines are set for the disposal of used face masks, leading to a surge in mismanaged face mask waste and raising genuine environmental concerns [7]. Regarding biomass waste, Malaysia is blessed with agricultural abundance, making the utilisation of waste from the agricultural industry, such as empty fruit bunch (EFB) and rice husk, a common approach for energy production. Biomass encompasses the by-products of crop cultivation and farming, including stems, husks, barks, grains, and leaves, which are typically overlooked. This biomass serves as a unique feedstock rich in carbon that can be repeatedly used without contributing to increased carbon dioxide emissions, thus aiding in climate protection. Converting biomass into useful materials, through heat and chemical reactions, plays a dual role by creating materials for renewable energy and promoting clean energy production [8].

Biochar is the solid product resulting from pyrolysis. It has a structure with hollow spaces between the particles and contains a considerable amount of carbon [9]. Biochar serves various purposes, such as extracting pollutants from water solutions [9,10], improving and healing soil [11], and crafting bio-based composites [9,12], among other uses. Therefore, the potential of biochar depends on the characteristics of the materials or feedstocks used [13]. Numerous studies have consistently shown that specific organic materials, such as EFBs, rice husks, and leaves, can substantially increase the carbon content and provide a substantial calorific value [14]. Commonly used agricultural biomass for biochar or bio-oil production include EFB, rice straw, rice husk and food waste [2,9,10]. These chosen agricultural biomass sources offer great potential due to their composition and organic content. However, information and data on the synergy between face masks and agricultural waste are lacking, which this study seeks to explore.

The most commonly used face masks, such as three-ply disposable face masks, are composed of mixtures of metallic compounds and various polymer types, including polyacrylonitrile, polycarbonate, polyurethane, polyester, polystyrene, polyethylene, and primarily polypropylene [15–17]. Disposable face masks have now become a major source of microplastic waste generation, potentially causing harmful effects [18,19]. Therefore, proper disposal of these masks must be ensured, as they may pose a major waste issue, especially for marine life [8,20]. This research aims to address this growing waste issue by repurposing both biomass and face mask waste for energy production. This approach can reduce the carbon footprint and help manage greenhouse gas emissions. It not only improves waste management but also contributes to renewable energy generation, promoting a more sustainable environment. This approach offers a promising solution to waste management challenges whilst simultaneously reducing greenhouse gas emissions in Malaysia.

The incineration of waste is a major source of greenhouse gas emissions, as it often releases carbon dioxide and other harmful pollutants into the atmosphere [21]. Additionally, improper disposal of used face masks can contribute to pollution and worsen the environmental waste problem. Co-pyrolysis aims to reduce greenhouse gas emissions whilst repurposing waste materials through the co-pyrolysis of biomass and face masks. This process can result in the production of biochar, a solid substance rich in carbon, which can be used as a solid fuel, contributing to more environmentally friendly and sustainable energy production. The carbon stored in biochar can help mitigate carbon emissions, and has the potential to increase char production during combustion due to alterations in the organic material's composition and subsequent energy release. In essence, these specific organic sources have proven effective in enhancing char production through combustion thanks to their unique characteristics, including their considerable carbon content and moderate calorific value.

Furthermore, biochar produces fewer emissions when used as a solid fuel compared with conventional fossil fuels, resulting in an overall reduction in the carbon footprint. In addition to contributing to proper waste management, this process also mitigates greenhouse gas emissions into the environment by using the potential of waste materials to create a valuable resource. This research direction aligns with Malaysia's efforts to reduce carbon emissions and achieve numerous environmental goals. The concurrent treatment of waste management and greenhouse gas emissions through the co-pyrolysis of face masks and biomass offers a more sustainable approach to addressing these issues in Malaysia. It represents a step towards developing an energy- and environmentally friendly solution that supports the nation's sustainability initiatives and reduces its carbon impact.

This novel approach not only holds the potential to enhance waste management practices but could also bolster the generation of renewable energy. The concept of sustainability, achieved through the conversion of waste into energy, provides a promising solution for a nation dealing with persistent waste management challenges, encompassing both medical and municipal waste. Furthermore, research on the combination of biomass is lacking, especially regarding rice husk and EFB, combined with disposable face masks using the co-pyrolysis method to produce solid fuel for energy production. Studies have explored the option of converting disposable face masks into biochar through pyrolysis [21–23]. However, unlike previous research, this study suggests innovative ways to manage medical waste by combining used face masks with biomass to create biochar and bio-oil.

A low-temperature reactor with a retort heating system was used for the slow co-pyrolysis to extract energy from the waste. The choice of slow pyrolysis is due to its low rate of heating and low heating temperature. A determined maximum temperature of 400 °C with a heating rate of 5 °C and a nitrogen flowrate of 2 L/min makes slow pyrolysis suitable for the type of feedstocks being used. The small size of face mask particles at 0.5 mm makes slow pyrolysis the preferred method, as the heat capacity of ground face masks at slow pyrolysis settings helps extract more bio-products, leaving a substantial amount of biochar and bio-oil from the various feedstocks used. The ratio of feedstocks is also a crucial factor, as this greatly influences the quality and yield of the biochar due to the synergistic effects each feedstock can contribute [24]. These criteria are essential to carefully determine to ensure the optimum temperature, heating rate, and ratio are established.

In summary, this study explores the conversion of disposable face masks into biochar through slow pyrolysis, which was chosen for its efficiency in extracting a better yield of biochar and bio-oil. The selection of feedstocks and their ratios plays a critical role in determining the biochar yield. The products from the co-pyrolysis undergo comprehensive analysis to identify the most promising sample for further optimisation.

2. Selection of Feedstock Samples

The recycling of face masks poses a significant challenge, primarily due to the risk of potential COVID-19 infections associated with their intricate structure. Improper disposal of used masks is a common occurrence, exacerbated by the large quantity of waste generated, which places additional stress on waste management systems [25]. This issue is even more critical in developing nations, where waste management often receives minimal attention, leading to uncontrolled disposal of face masks into aquatic environments, drainage systems, and soil. Reports suggest that improper handling of solid waste, including face masks, can increase the risk of COVID-19 transmission [26].

Directly incinerating masks is considered an easier disposal method as it effectively eliminates germs through high-temperature combustion [27]. However, this approach comes with numerous environmental challenges, including toxic gas emissions and air pollution [9]. Hence, it is essential to develop an innovative method to address this escalating waste problem. Thermochemical conversion is a highly promising approach for breaking down and transforming disposable masks effectively [28].

To achieve successful waste conversion, specific optimal temperatures are required. Combining face masks with other materials for co-pyrolysis offers the advantage of obtaining bio-products at much lower temperatures [29]. Among various source materials, the combination of face masks with biomass for conversion is viewed as a productive way to create more valuable end products [30]. Agricultural waste is often chosen for pyrolysis experiments due to its abundance. The high production of palm oil leads to substantial EFB waste [31], and such valuable products should not go to waste. This applies for rice husk and leaves as well. Although leaves can naturally decompose, optimising them as a valuable resource is a more sustainable approach due to their rich content.

Table 1 presents various feedstocks commonly used for pyrolysis within the biomass group, including those considered for this research. Sugarcane, wood chips, wheat straw, and bamboo [32–35] are compared to the feedstocks selected for this study, taking into account their calorific value and carbon content. In the case of face masks, their high calorific value may be attributed to the presence of several polymer materials. However, owing to the waste management challenges associated with face masks, they were chosen as a feedstock for this research.

Table 1. Feedstock Comparison.

Feedstocks	Calorific Value (kcal/kg)	Carbon Content (%)
Face Mask	9310	73.8
EFB	4113	43.3
Rice Husk	3411	35.8
Leaves	2955	31.8
Sugarcane	1969	58.28
Woodchip	3629	53.3
Wheat Straw	1808	46.2
Bamboo Leaves	4732	46.98

As for EFB and rice husk, these two materials are readily available in Malaysia, where the project is based, making them suitable choices. Malaysia is one of the largest producers of palm oil and rice, ensuring easy accessibility and abundance of these feedstocks. Leaves, on the other hand, are a common agricultural waste readily available worldwide. Despite their lower calorific value compared with face masks, EFB, wood chips, and bamboo, the choice is motivated by waste management issues. Selecting leaves aims to highlight their potential as an energy source due to their widespread availability and slow degradation. Bamboo is a high-calorific biomass source rich in carbon, but its limited abundance and waste management challenges, particularly in Malaysia, make it a less suitable choice. Wood chips have alternative applications, such as furniture production (e.g., plywood or pellets), which is why they are not the focus here. Environmental concerns, including deforestation for wood production, further discourage the use of wood chips. Throughout this project, substantial consideration is given to both environmental issues and economic aspects to guide the research effectively.

2.1. Feedstocks

Feedstocks are chosen in this study for several reasons. Considering environmental and economic factors, the selection is based on high availability and energy content.

2.1.1. Face Masks

Face masks have become a major waste issue during the COVID-19 outbreak. Even before the pandemic, disposable face masks were among the most highly accumulated medical waste. Recycling medical waste requires an efficient disposal method, and thermal transformation processes are the most effective approach after considering the potential constraints [36]. Co-pyrolysis offers a potential solution for the proper disposal of used face masks, preventing them from ending up in landfills or being improperly discarded. By converting face mask waste into valuable products through co-pyrolysis, environmental impacts can be minimised, promoting sustainable waste management practices.

The uniqueness of face masks lies in the composition of the materials. They are mostly made up of several plastics, such as polypropylene, often with fillers like polyester, polyurethane, polyamide, polyethylene, polystyrene, polyacrylonitrile, polycarbonate, and viscose fibre, with polypropylene having a high carbon content [37]. Referring to Table 1, their high carbon and calorific value make face masks a suitable feedstock. The calorific value can be converted into energy. With low moisture and oxygen content, face masks require much less heat energy for the combustion process, making them an attractive option.

Although processing face masks can be complicated, the initiative to use face masks is necessary due to the high waste volume despite COVID-19 being less prevalent nowadays [37]. Face masks still constitute a large portion of medical waste, and co-pyrolysis of face masks can lead to the recovery of valuable resources such as biochar, bio-oil, and bio-gas. These resources can be utilised for various purposes, including soil improvement, renewable energy generation, and biofuel or biochemical production.

2.1.2. Empty Fruit Bunch

Malaysia is a major producer of palm oil, and as a by-product of the palm oil industry, EFB is readily available in copious quantities. The abundance of EFB makes it a convenient and cost-effective feedstock for co-pyrolysis processes in the country. EFB has a high-energy content due to its lignocellulosic composition [38]. As such, it is a favourable biomass feedstock for co-pyrolysis, as it can contribute to the production of biochar, bio-oil, and syngas with significant energy potential. The utilisation of EFB in co-pyrolysis processes aligns with Malaysia's objective of producing more renewable energy sources and reducing its dependence on fossil fuels.

2.1.3. Rice Husk

Malaysia is a major producer of rice, and as a result, rice husk is available in massive quantities as an agricultural residue. This abundant supply makes rice husk a viable and cost-effective feedstock for co-pyrolysis processes in the country. Rice husk has a high silica content, which makes it a suitable candidate for co-pyrolysis [39]. The presence of silica acts as a catalyst or support material during the pyrolysis process, enhancing the production of valuable products such as biochar and bio-oil. Additionally, rice husk has a high calorific value, making it a desirable biomass feedstock for energy production in co-pyrolysis applications.

2.1.4. Leaves

Leaves are readily available in Malaysia, particularly in tropical regions with dense vegetation. As a common organic waste material, leaves can be easily collected, making them a convenient and accessible feedstock for co-pyrolysis processes. Leaves also pose a major waste management challenge when not properly disposed of. The challenges of using leaves are commonly due to their state of accumulating dirt and having to undergo thorough cleaning and grinding. The advantage of using leaves as feedstock is that they have high carbon and calorific value, which may contribute to potential energy [40]. Co-pyrolysis offers a sustainable solution by converting leaves into valuable by-products such as biochar, bio-oil, and bio-gas. This process not only reduces the environmental burden of leaf waste but also promotes resource recovery and contributes to sustainable waste management practices in Malaysia [9].

3. Slow Co-Pyrolysis as a Medium for Energy Production from Disposal of Face Masks and Agricultural Waste

Pyrolysis entails the controlled heating of biomass feedstocks, such as wood, agricultural residues, or energy crops, in an oxygen-free environment, resulting in the production of biochar, bio-oil, and syngas. Numerous research findings consistently demonstrate the effectiveness of biochar as a soil amendment. Its porous structure enhances soil fertility by improving water retention, reducing nutrient leaching, and promoting microbial activity. Additionally, the carbon sequestration potential of biochar contributes to long-term climate change mitigation [41].

The selection of these feedstocks for slow co-pyrolysis primarily stems from their compatibility with lower heating rates and temperature ranges. It was observed that subjecting the samples to higher temperatures above 450 °C [1] tends to result in ash formation. Therefore, slow pyrolysis is the most suitable method for processing these feedstock samples. This study focuses on biochar; therefore, maintaining the slow co-pyrolysis setting is desirable, as it enhances both the quantity and quality of biochar production whilst also yielding a significant amount of bio-oil [42]. The slow co-pyrolysis settings employed in this research are as follows: a maximum temperature of 400 °C, a heating rate of 5 °C per minute, and a nitrogen flowrate of 2 L/min. The average sample volume is 10–11 g and is manually mixed inside the crucible. The sieve size during grinding is set at 0.50 mm and the predetermined residence time is 30 min after reaching the target temperature. Residence time is crucial for stabilising the char [43]. After numerous tests, a residence time of 30 min was found to be optimal, as extending it beyond this time leads to the conversion of char into ashes.

Within the framework of co-pyrolysis, biochar is the solid, carbon-rich material produced when agricultural waste and used face masks are heated together in a low-oxygen atmosphere. This co-pyrolysis addresses two critical issues: maximising the utilisation of agricultural biomass waste and managing the increasing use of face masks due to the COVID-19 pandemic. In the context of the specific feedstocks used in this study, the primary emphasis is placed on the production of biochar. This choice is supported by strong evidence indicating that during slow pyrolysis, the production of char significantly surpasses that of bio-oil and bio-gas in terms of quantity. Additionally, biochar maintains a notable calorific value post-combustion, making it an attractive and sustainable resource for various energy applications.

The aim of co-pyrolysing these feedstocks is to produce biochar, a carbon-rich compound with potential uses in improving soil quality and sequestering carbon for use as a solid fuel. Biochar is a sustainable choice, as it repurposes waste materials to create a valuable end product that addresses various environmental and agricultural challenges. Biochar offers remarkable potential as a solid fuel due to its high-energy content, sustainability, and positive environmental impact. It efficiently generates heat for cooking and heating whilst also storing carbon, making it a carbon-negative option. Biochar combustion results in fewer emissions [44], making it an environmentally friendly choice. This versatile solid fuel can be used in various devices, and its local sourcing reduces transportation costs. Furthermore, the residual ash from biochar combustion enhances soil quality and agricultural output. Although the adoption of biochar as a solid fuel may vary, it aligns with environmentally friendly and sustainable energy practices.

The potential of bio-oil as a renewable fuel source is already well known and common in the industry. Its high energy density makes it suitable for heating and electricity generation, offering a sustainable alternative to conventional fossil fuels [45]. Moreover, the further refining of bio-oil will allow for the production of biofuels and biochemicals, addressing the need for cleaner and more sustainable alternatives. Bio-oil from the pyrolysis of various plastic waste, such as polyethylene or polypropylene, is reported to have calorific values of between 43% and 53%, respectively, whereas face masks, known for their higher calorific value, yield between 43% and 80% of bio-oil [46].

Bio-gas, a valuable by-product of pyrolysis, has been extensively studied for its versatility as a fuel source. It has shown promise in power generation through combustion in gas turbines or internal combustion engines. Additionally, bio-gas serves as a valuable feedstock for chemical synthesis, enabling the production of a diverse range of chemicals and fuels. The yield and compound identification in the sample are influenced by the heating temperature, heating rate, and mixture ratio [47]. However, co-pyrolysis produces fewer nitrogen oxides and sulphur oxides due to the inert environment in which the process takes place. This results in better quality and yield in the solid biochar and bio-oil processes [48].

The chosen method for slow co-pyrolysis involves the utilisation of the Carbolite Gero TG2 12/125/425, produced by Carbolite (Hope Valley, UK), which is a horizontal tube furnace, as depicted in Figure 1. This specific apparatus is selected due to its ability to operate over a broad temperature range, extending from as low as 30 °C to as high as 1200 °C [49]. This wide temperature range enables the execution of numerous tests aimed at determining the optimal temperature for feedstock combustion. The sample is placed within the horizontal tube, and the combustion testing is carried out according to predetermined settings. Upon completion of the testing, the apparatus is allowed to cool down until it reaches a temperature below 100 °C, at which point the biochar can be manually collected directly from the crucible. Subsequently, the biochar undergoes proximate and TGA analysis as a second step to ascertain the char yield, facilitating a comparison with raw samples. This comparative analysis is of paramount importance in the research, aiming to identify the most promising sample among the feedstocks and the ideal ratio with the highest char yield.

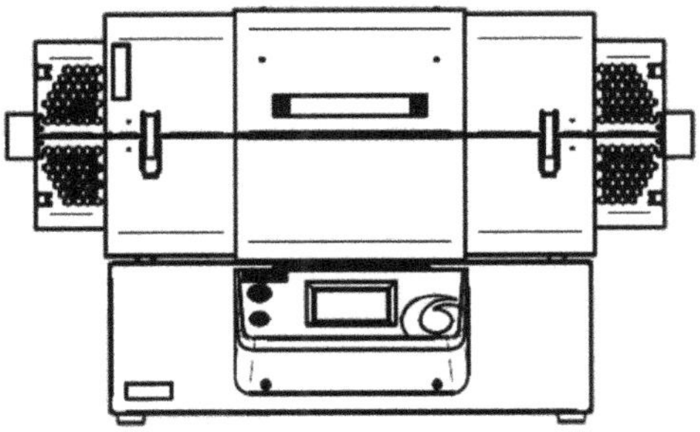

Figure 1. Carbolite Gero TG2 12/125/425 [49].

4. Design of Experiment

Figure 2 illustrates the methodology designed for this experiment. The process begins with the preparation of feedstocks, which involves blending biomass and face masks within a reactor. The feedstocks are subjected to grinding using a 0.5 mm sieve, and the average sample mass used for testing falls within the range of 10–11 g. Following this, the mixture is heated under an inert atmosphere within a temperature range of 250–400 °C [50]. The sample ratios are categorised as 25%, 50%, and 75% of face mask to 75%, 50% and 25% of biomass, as well as an exclusive 100% of all three types of biomass and 100% face mask [24]. The heating rate is set at 5 °C/min, with a constant nitrogen flow of 2 L/min to maintain an inert condition within the combustion chamber [51]. A dwelling duration of 30 min is observed after reaching the target temperature, as experiments have shown that a dwelling duration exceeding 30 min may result in the conversion of biochar into ash. The residence time plays a critical role as it aids in completing and further enhancing co-pyrolysis combustion [43].

The experimental properties are determined based on references from various research studies [24] and numerous trial-and-error runs conducted to identify the optimum temperature and heating rate for each sample. The processed samples remain within the chamber until they attain the appropriate operating temperature before they are collected and processed. This approach mitigates the risk of sudden temperature drops within the combustion tube. The co-pyrolysis experiments entail variations in the process parameters, including temperature, heating rate, and residence time, aimed at determining the optimal conditions for the thermal degradation of the feedstocks.

The subsequent steps involve subjecting the by-products to characterisation. The biochar produced in the experiments undergoes proximate and TGA analysis to determine the calorific value and yield of the samples. Characterisation also encompasses the analysis of the composition, heating value, viscosity, and density of the bio-oil. The experimental data are subject to statistical analysis to identify one final potential sample. When no significant improvements are noted in the yield, a re-evaluation of the parameters and a round of optimisation become necessary. In summary, the methodology for studying the co-pyrolysis of biomass with face mask waste entails a combination of experimental, analytical, and statistical techniques to optimise process parameters, evaluate product quality and yield, and assess the environmental and economic implications of the process.

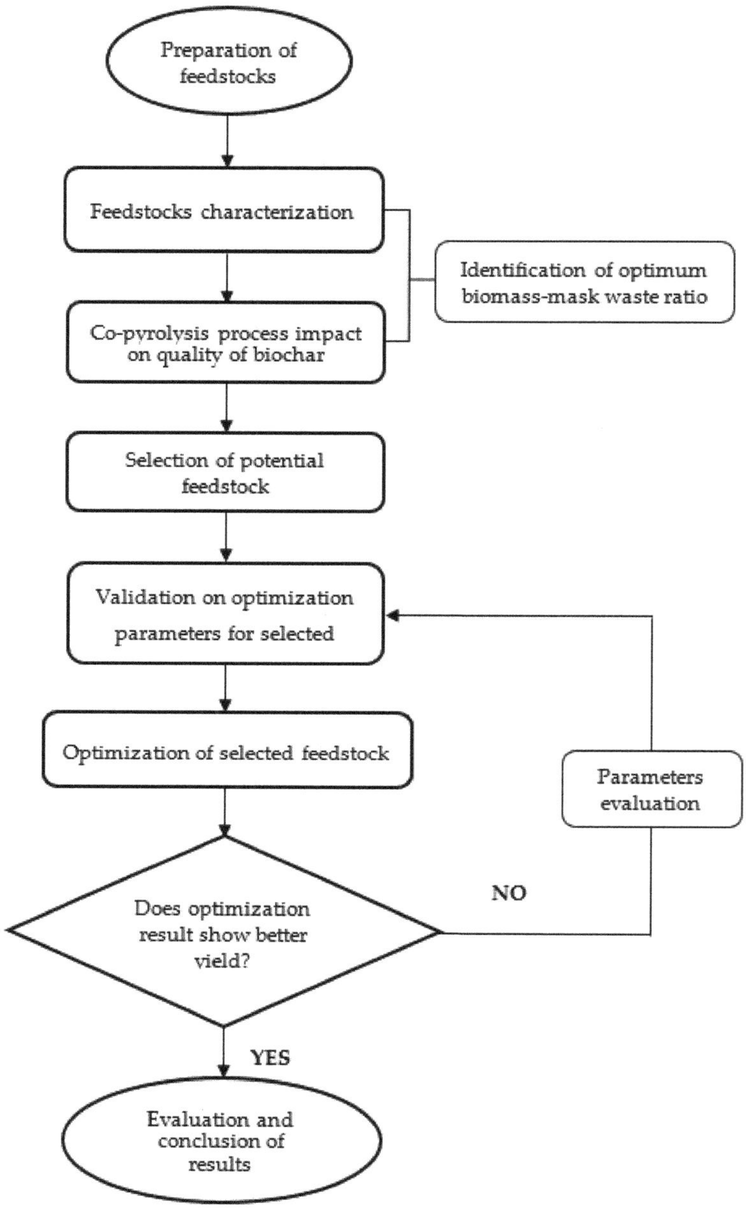

Figure 2. Flowchart of feedstock preparation.

5. Results

5.1. Ultimate and Proximate Analysis

Ultimate and proximate analysis are two crucial methods employed to characterise the composition of co-pyrolysis products. Ultimate analysis entails the determination of the elemental composition of the co-pyrolysis products, typically achieved through the combustion of a sample followed by a measurement of the resulting gases. This process furnishes information concerning the quantities of carbon, hydrogen, nitrogen, sulphur, and oxygen within the products. Proximate analysis provides data on moisture, ash, volatile matter, and fixed carbon in a material. These data are instrumental in evaluating the energy content and potential applications of co-pyrolysis products. Face mask analysis adheres to the ASTM E872-82 [52] and E1755-01 [53] standards [51]. For EFB, the ASTM E 1756-01 [54] standard is utilised [55]. In the case of rice husk and leaves, ASTM-D7582-15 [56] is used [57,58]. Table 2 presents the comprehensive results of the ultimate and proximate analysis conducted to ascertain the precise elemental content within the feedstocks.

Table 2. Ultimate, proximate analysis and calorific value results.

Analysis	Face Masks	EFB	Rice Husk	Leaves
Moisture Content (%)	0.10	11.9	11.7	31.7
Ash Content (%)	8.62	1.36	16.3	12.0
Volatile Matter (%)	91.3	72.2	58.0	45.7
Fixed Carbon (%)	<0.1	14.4	13.9	10.6
Carbon (%)	73.8	43.3	35.8	31.8
Hydrogen (%)	11.8	5.98	5.33	6.35
Nitrogen (%)	0.23	0.60	0.60	0.72
Oxygen (%)	14.2	50.1	58.2	61.0
Sulphur (%)	<0.01	0.03	0.05	0.21
Gross Calorific Value (kcal/kg)	9310	4113	3411	2955

5.1.1. Moisture Content

The moisture content of the sample is a crucial factor significantly influencing the material's strength properties. A low moisture content facilitates thorough combustion by reducing energy consumption during the combustion process and often lowering the heating rate during pyrolysis [59]. This absorbed moisture plays a role in creating binding forces between the particles by occupying available spaces between the biomass particles. Figure 3 illustrates that leaves have the highest moisture content at 31.7%, whereas EFB and rice husk have moisture contents of 11.9% and 11.7%, respectively, owing to their origins as agricultural residues. Face masks, due to their component materials, are expected to exhibit the lowest moisture content. Although the initial moisture content of the biomass particles was maintained at an optimised level, the samples remained prone to absorbing moisture during feedstock preparation.

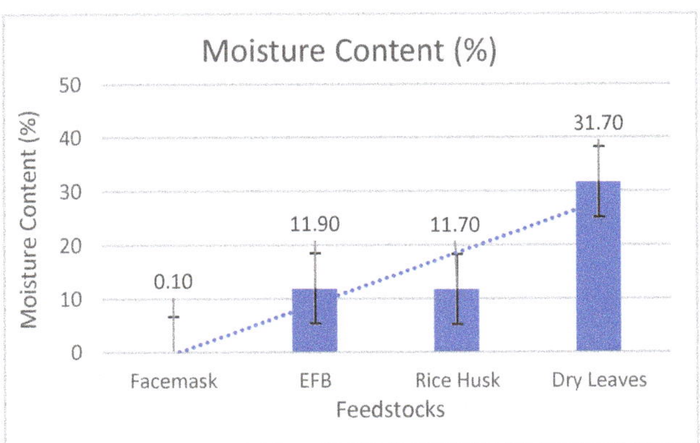

Figure 3. Moisture content.

5.1.2. Volatile Matter

Face masks contain synthetic materials that are highly flammable, enabling them to release numerous gases when subjected to heat. As depicted in Figure 4, face masks exhibit a volatile matter content of 91.3%. EFB demonstrates a substantial amount at 72.2%, whereas rice husk possesses 58%, and leaves contain 45.7% of combustible materials, rendering them suitable for applications such as biomass combustion or bioenergy production. Volatile matter refers to the flammable substances within a material that can transform into gases when exposed to heat. This percentage indicates how much of a material's weight will transition into gas during heating [60,61]. Materials with higher volatile matter content hold greater energy potential and can serve as valuable sources of renewable energy, whereas those with lower volatile matter content may find diverse uses, such as composting or mulching.

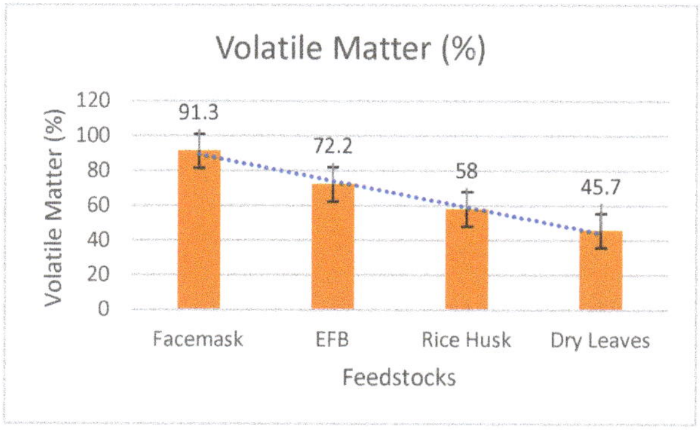

Figure 4. Volatile matter.

5.1.3. Ash Content

Ash content, the least desirable by-product arising from combustion, represents a non-combustible residue left behind after the combustion process. This residue, known as ash, comprises inorganic elements such as phosphorus, silicon, calcium, potassium, and chlorine [61]. After burning the face masks, approximately 8.62% of their weight remains as inorganic ash. This higher percentage may be attributed to the presence of additives or synthetic materials in the masks. EFB leaves behind only 1.36% ash after combustion, indicating a lower quantity of inorganic residue. This characteristic makes it suitable for biomass energy production, as it leads to cleaner combustion. Rice husk exhibits an ash content of 16.3%, whereas leaves exhibit 12.0% ash after combustion.

5.1.4. Fixed Carbon

For fixed carbon, face masks contain an insignificant amount of solid carbon (<0.1%), as they are primarily composed of non-carbon materials such as synthetic polymers. EFB contains a considerable portion of solid carbon (14.4%), whereas rice husk also contains a significant amount of solid carbon (13.9%). Leaves exhibit a fixed carbon content of 10.6%. Fixed carbon content is a critical factor in evaluating the energy potential and applicability of these materials in different industries. Fixed carbon refers to the stable carbonaceous material remaining within a substance after the removal of volatile matter and ash during combustion [62]. It represents the carbon component that remains in solid form during heating. Higher fixed carbon content indicates better energy yields during combustion or conversion processes, whereas lower levels may lead to alternative uses, such as in agriculture or composting.

5.2. Ultimate Analysis

Figure 5 shows that face masks possess a high carbon content of 73.8%. This observation suggests that a significant proportion of the mask's weight comprises carbon, commonly found in synthetic materials like polypropylene or polyethylene. EFB contains a carbon content of 43.3%, whereas rice husk contains 35.8% carbon, with leaves exhibiting the lowest carbon content of 31.8%. Carbon content is a critical factor for assessing energy potential. Materials with higher carbon content typically yield more energy during combustion [21]. The hydrogen, nitrogen, and sulphur content in these samples is relatively low. The low hydrogen content implies that the samples produce less heat energy, but this can be balanced with oxygen levels [23].

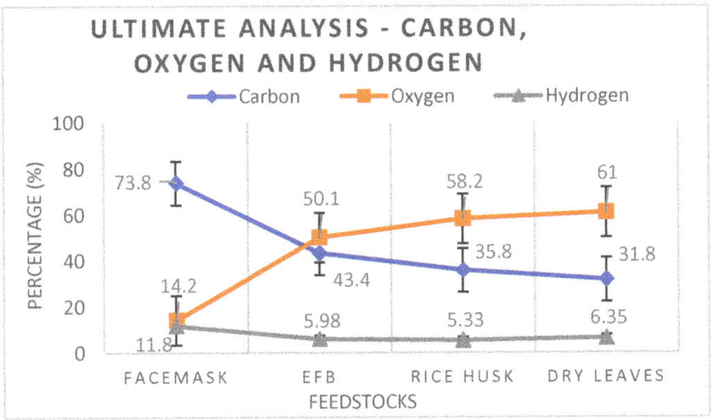

Figure 5. Carbon, Oxygen, and Hydrogen.

Materials with low nitrogen and sulphur content have only minimal amounts of these elements, which is advantageous for energy production as it reduces harmful emissions and environmental impacts. Concerning oxygen levels, face masks exhibit the lowest value at 14.2%, whereas EFB has a high oxygen content of 50.1%, followed by rice husk at 58.2%, and leaves at 61.0%. The higher oxygen content in the biomass samples may be attributed to their origin from agricultural residue, which often contains cellulose and hemicellulose. Having high oxygen content in biomass materials is beneficial, as it leads to clean combustion and reduces harmful emissions like carbon monoxide and soot [63].

5.3. Calorific Value

Calorific value, also known as the heating value, denotes the amount of heat energy released when a specific quantity of a substance undergoes complete combustion. It is measured in kcal/kg and plays an important role in evaluating the energy content of unconventional materials. As indicated in Figure 6, face masks exhibit a high calorific value of 9310 kcal/kg, indicating a substantial energy content, making them a potentially valuable energy source. EFB demonstrates a moderate calorific value of 4113 kcal/kg, rendering it a suitable option for energy generation, especially given its abundant availability as a by-product of the palm oil industry.

The calorific value of rice husk is 3411 kcal/kg, whereas leaves have a calorific value of 2955 kcal/kg, offering a reasonable energy content and making them a practical choice as a renewable energy source through combustion or gasification techniques. Higher calorific values in materials render them more favourable for energy production, as they provide greater heat energy per unit [62]. This knowledge facilitates the optimised selection and utilisation of materials for specific energy applications, promoting efficient resource management. A higher calorific value results in a stronger energy release during combustion [63]. For example, the face mask boasts the highest calorific value at 9310 kcal/kg, surpassing EFB, rice husk, and leaves, thereby potentially offering the most significant energy output.

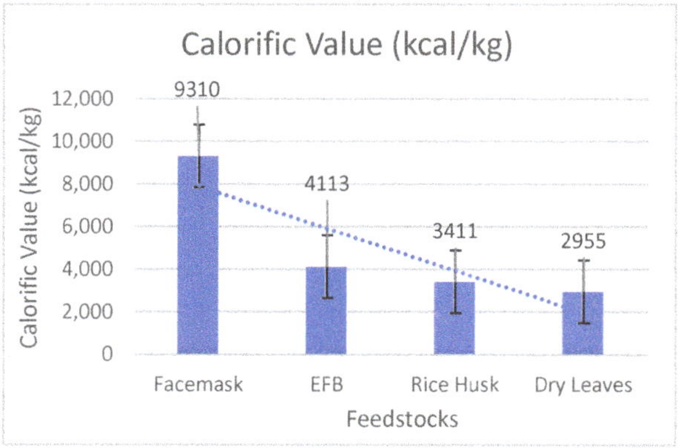

Figure 6. Calorific value.

5.4. Biochar Yield Results

Biochar yield, which is the amount of biochar produced through the pyrolysis process, is a critical parameter often associated with the carbonisation of organic materials. Biochar production is usually quantified in terms of weight or volume. Factors such as the type of biomass used, pyrolysis conditions, and the equipment employed can influence the actual quantity generated. Analysing biochar yield is pivotal for assessing the effectiveness and success of biochar production. Biochar yield is computed using the formula:

$$\text{Biochar yield } (\%) = \left[\frac{\text{Weight of biochar}}{\text{Weight of total feedstocks}}\right] \times 100$$

Table 3 presents the results of biochar yield from samples subjected to co-pyrolysis. The outcome for leaves with a face mask biochar is noteworthy, indicating the potential effectiveness of the co-pyrolysis method in achieving a substantial biochar yield. The absence of results for EFB with a face mask and rice husk with a face mask is a matter for future research. Table 3 underscores how well the process transforms the starting materials into biochar, demonstrating its practical and efficient use in biomass.

Table 3. Biochar yield results.

Feedstocks	Biochar Yield (%)
Face Mask	27.08 [64]
EFB	34.27 [65]
EFB + Face Mask	-
Rice Husk	33.07 [66]
Rice Husk + Face Mask	-
Leaves	32.87 [9]
Leaves + Face Mask	40.20 [9]

Table 4 outlines the roadmap for the future phases of this study, delineating the anticipated outcomes. These projections are based on insights from prior research in the field, supported by logical analytical calculations and sound assumptions. The expected values were derived from the lowest and highest yield results obtained from various research efforts. This table serves as a guide for the upcoming work, providing direction towards an in-depth understanding of the subject matter.

Table 4. Expected biochar yield results.

Feedstocks	Biochar Yield (%)
Face Mask	25–35
EFB	28–38
Rice Husk	30–43
Leaves	28–38
Leaves + Face Mask	30–45

6. Conclusions

The results were meticulously examined, and a comprehensive evaluation of the selected samples for both biomass and plastic materials was conducted. This study primarily focused on the comparison of feedstock samples for the production of biochar through the co-pyrolysis process, both individually and in combination. The paper draws upon existing research and proposes an alternative approach to generate solid fuel for energy from waste management processes. Given the considerable amount of agricultural waste globally, co-pyrolysis holds significant potential for further development and could make a substantial impact in addressing the issue of environmental waste pollution.

The findings have unveiled the elemental composition of certain samples in terms of percentage, assisting in determining the most suitable composition for use in the co-pyrolysis process. Regarding ultimate carbon analysis, face masks exhibit over 70% carbon content when compared with EFB. Notably, face masks also boast a calorific value exceeding 126% when compared with the second highest, which was EFB. The pivotal benchmark here is the calorific value, as face masks display a calorific value twice as high as other feedstocks. This implies that face mask raw samples can potentially contribute more energy during combustion and may result in a higher energy content in biochar, a crucial factor for achieving an enhanced biochar yield.

In addition to their low moisture content, face masks offer significant potential as a feedstock for energy compared with other feedstocks. With the right ratios of feedstocks and heating rates, they ensure a higher yield of char. Furthermore, the synergy achieved by combining these two distinct materials may lead to an improved quality and yield of oil and char. The biochar yield results demonstrate that the hybrid biochar of leaves with a face mask exhibits a higher yield at 40.20% compared with the second-highest yield of EFB at 34.27%. This finding underscores the potential for an enhanced biochar yield through the synergy of combining face masks with agricultural waste. However, much research needs to be conducted because of the lack of results for rice husk with a face mask and EFB with a face mask.

Currently, scientific laboratories worldwide are actively engaged in developing innovative technologies and conducting studies to further enhance the quality of bio-products. This paper also discusses pilot-scale studies previously reported in the academic literature, emphasising the need for additional research to achieve the best results and optimal outcomes. There is a crucial need to delve deeper into the dynamics of the reactions, as they play a pivotal role in obtaining the desired output and ensuring the efficient operation of the co-pyrolysis process.

Author Contributions: Conceptualisation, Y.Y. and E.S.T.; methodology, Y.Y.; validation, Y.Y. and E.S.T.; resources, Y.Y. and E.S.T.; data curation, Y.Y.; writing—original draft preparation, Y.Y.; writing—review and editing, Y.Y. and E.S.T.; visualisation, Y.Y.; supervision, E.S.T. and F.B.I.; project administration, E.S.T. and F.B.I.; funding acquisition, E.S.T. All authors have read and agreed to the published version of the manuscript.

Funding: This research was funded by the AAIBE Chair of Renewable Energy at UNITEN with project code 202001KETTHA. The authors would like to thank MDPI for providing the funding to publish in this journal.

Institutional Review Board Statement: Not applicable.

Informed Consent Statement: Not applicable.

Data Availability Statement: No new data were created.

Acknowledgments: The authors would like to acknowledge the direct and indirect contributors to this project and research.

Conflicts of Interest: The authors declare no conflict of interest.

References

1. Li, C.; Yuan, X.; Sun, Z.; Suvarna, M.; Hu, X.; Wang, X.; Ok, Y.S. Pyrolysis of waste surgical masks into liquid fuel and its life-cycle assessment. *Bioresour. Technol.* **2022**, *346*, 126582. [CrossRef] [PubMed]
2. Hou, Y.; Feng, Z.; He, Y.; Gao, Q.; Ni, L.; Su, M.; Ren, H.; Liu, Z.; Hu, W. Co-pyrolysis characteristics and synergistic interaction of bamboo residues and disposable face mask. *Renew. Energy* **2022**, *194*, 415–425. [CrossRef]
3. Lam, S.C.; Suen, L.K.P.; Cheung, T.C.C. Global risk to the community and clinical setting: Flocking of fake masks and protective gears during the COVID-19 pandemic. *Am. J. Infect. Control* **2020**, *48*, 964–965. [CrossRef] [PubMed]
4. Su, Y.; Zhang, Z.; Zhu, J.; Shi, J.; Wei, H.; Xie, B.; Shi, H. Microplastics act as vectors for antibiotic resistance genes in landfill leachate: The enhanced roles of the long-term aging process. *Environ. Pollut.* **2021**, *270*, 116278. [CrossRef]
5. Wang, J.; Shen, J.; Ye, D.; Yan, X.; Zhang, Y.; Yang, W.; Li, X.; Wang, J.; Zhang, L.; Pan, L. Disinfection technology of hospital wastes and wastewater: Suggestions for disinfection strategy during coronavirus Disease 2019 (COVID-19) pandemic in China. *Environ. Pollut.* **2020**, *262*, 114665. [CrossRef]
6. Yudell, M.; Roberts, D.; DeSalle, R.; Tishkoff, S. NIH must confront the use of race in science. *Science* **2020**, *369*, 1314–1315. [CrossRef]
7. Maulana, N.B.; Mapa, M.T.; Tan, P. THE Facemask Disposal Circumstance During Pandemic COVID-19. *Malays. J. Soc. Sci. Humanit.* **2022**, *7*, e001536. [CrossRef]
8. Adelodun, B.; Ajibade, F.O.; Ibrahim, R.G.; Bakare, H.O.; Choi, K.-S. Snowballing transmission of COVID-19 (SARS-CoV-2) through wastewater: Any sustainable preventive measures to curtail the scourge in low-income countries? *Sci. Total. Environ.* **2020**, *742*, 140680. [CrossRef]
9. Emenike, E.C.; Iwuozor, K.O.; Agbana, S.A.; Otoikhian, K.S.; Adeniyi, A.G. Efficient recycling of disposable face masks via co-carbonization with waste biomass: A pathway to a cleaner environment. *Clean. Environ. Syst.* **2022**, *6*, 100094. [CrossRef]
10. Iwuozor, K.O.; Emenike, E.C.; Ighalo, J.O.; Omoarukhe, F.O.; Omuku, P.E.; Adeniyi, A.G. A Review on the thermochemical conversion of sugarcane bagasse into biochar. *Clean. Mater.* **2022**, *6*, 100162. [CrossRef]
11. He, Y.; Gu, F.; Xu, C.; Wang, Y. Assessing of the influence of organic and inorganic amendments on the physical-chemical properties of a red soil (Ultisol) quality. *Catena* **2019**, *183*, 104231. [CrossRef]
12. Onifade, D.V.; Ighalo, J.O.; Adeniyi, A.G.; Hameed, K. Morphological and Thermal Properties of Polystyrene Composite Reinforced with Biochar from Plantain Stalk Fibre. *Mater. Int.* **2020**, *2*, 150–156. [CrossRef]
13. Ighalo, J.O.; Adeniyi, A.G. Thermodynamic modelling and temperature sensitivity analysis of banana (*Musa* spp.) waste pyrolysis. *SN Appl. Sci.* **2019**, *1*, 1086. [CrossRef]
14. Wee, M.X.; Chin, B.L.; Saptoro, A.; Yiin, C.L.; Chew, J.J.; Sunarso, J.; Yusup, S.; Sharma, A. A review on co-pyrolysis of agriculture biomass and disposable medical face mask waste for green fuel production: Recent advances and thermo-kinetic models. *Front. Chem. Sci. Eng.* **2023**, *17*, 1141–1161. [CrossRef]
15. Rahman, M.Z.; Hoque, M.E.; Alam, M.R.; Rouf, M.A.; Khan, S.I.; Xu, H.; Ramakrishna, S. Face Masks to Combat Coronavirus (COVID-19)—Processing roles, requirements, efficacy, risk and Sustainability. *Polymers* **2022**, *14*, 1296. [CrossRef]
16. Shaw, K.A.; Zello, G.A.; Butcher, S.J.; Ko, J.B.; Bertrand, L.; Chilibeck, P.D. The impact of face masks on performance and physiological outcomes during exercise: A systematic review and meta-analysis. *Appl. Physiol. Nutr. Metab.* **2021**, *46*, 693–703. [CrossRef]
17. Aragaw, T.A. Surgical face masks as a potential source for microplastic pollution in the COVID-19 scenario. *Mar. Pollut. Bull.* **2020**, *159*, 111517. [CrossRef]
18. Canning-Clode, J.; Sepúlveda, P.; Almeida, S.; Monteiro, J. Will COVID-19 Containment and Treatment Measures Drive Shifts in Marine Litter Pollution? *Front. Mar. Sci.* **2020**, *7*, 691. [CrossRef]
19. Dharmaraj, S.; Ashokkumar, V.; Hariharan, S.; Manibharathi, A.; Show, P.L.; Chong, C.T.; Ngamcharussrivichai, C. The COVID-19 pandemic face mask waste: A blooming threat to the marine environment. *Chemosphere* **2021**, *272*, 129601. [CrossRef]
20. Anastopoulos, I.; Pashalidis, I. Single-use surgical face masks, as a potential source of microplastics: Do they act as pollutant carriers? *J. Mol. Liq.* **2021**, *326*, 115247. [CrossRef]
21. Jung, S.; Lee, S.; Dou, X.; Kwon, E.E. Valorization of disposable COVID-19 mask through the thermo-chemical process. *Chem. Eng. J.* **2021**, *405*, 126658. [CrossRef] [PubMed]
22. Sun, S.; Yuan, Y.; Chen, R.; Xu, X.; Zhang, D. Kinetic, thermodynamic and chemical reaction analyses of typical surgical face mask waste pyrolysis. *Therm. Sci. Eng. Prog.* **2021**, *26*, 101135. [CrossRef]
23. Yousef, S.; Eimontas, J.; Striūgas, N.; Abdelnaby, M.A. Pyrolysis kinetic behaviour and TG-FTIR-GC–MS analysis of Coronavirus Face Masks. *J. Anal. Appl. Pyrolysis* **2021**, *156*, 105118. [CrossRef]

24. Rosas-Medellín, D.; Martínez-Urbizu, S.; Barbosa, R.; Alonso-Lemus, I.; Escobar, B. Co-pyrolysis of two environmental issues: Face mask and Sargassum spp. for efficacious solid waste management and its AEMFC applications. *Int. J. Hydrog. Energy* 2023, *in press*. [CrossRef]
25. Park, C.; Choi, H.; Lin, K.-Y.A.; Kwon, E.E.; Lee, J. COVID-19 mask waste to energy via thermochemical pathway: Effect of Co-Feeding food waste. *Energy* **2021**, *230*, 120876. [CrossRef] [PubMed]
26. Nzediegwu, C.; Chang, S.X. Improper solid waste management increases potential for COVID-19 spread in developing countries. *Resour. Conserv. Recycl.* **2020**, *161*, 104947. [CrossRef]
27. Geyer, R.; Jambeck, J.R.; Law, K.L. Production, use, and fate of all plastics ever made. *Sci. Adv.* **2017**, *3*, e1700782. [CrossRef]
28. Chen, R.; Zhang, D.; Xu, X.; Yuan, Y. Pyrolysis characteristics, kinetics, thermodynamics and volatile products of waste medical surgical mask rope by thermogravimetry and online thermogravimetry-Fourier transform infrared-mass spectrometry analysis. *Fuel* **2021**, *295*, 120632. [CrossRef]
29. Torres, F.G.; De-La-Torre, G.E. Historical microplastic records in marine sediments: Current progress and methodological evaluation. *Reg. Stud. Mar. Sci.* **2021**, *46*, 101868. [CrossRef]
30. Xu, E.G.; Ren, Z.J. Preventing masks from becoming the next plastic problem. *Front. Environ. Sci. Eng.* **2021**, *15*, 6–8. [CrossRef]
31. Sukiran, M.A.; Loh, S.K.; Bakar, N.A. Conversion of pre-treated oil palm empty fruit bunches into bio-oil and bio-char via fast pyrolysis. *J. Oil Palm Res.* **2018**, *30*, 121–129. [CrossRef]
32. Kimura, L.M.; Santos, L.C.; Vieira, P.F.; Parreira, P.M.; Henrique, H.M. Biomass Pyrolysis: Use of Some Agricultural Wastes for Alternative Fuel Production. *Mater. Sci. Forum* **2010**, *660–661*, 259–264. [CrossRef]
33. Pedišius, N.; Praspaliauskas, M.; Pedišius, J.; Dzenajavičienė, E.F. Analysis of Wood Chip Characteristics for Energy Production in Lithuania. *Energies* **2021**, *14*, 3931. [CrossRef]
34. Sedmihradská, A.; Pohořelý, M.; Jevič, P.; Skoblia, S.; Beňo, Z.; Farták, J.; Čech, B.; Hartman, M. Pyrolysis of wheat and barley straw. *Res. Agric. Eng.* **2020**, *66*, 8–17. [CrossRef]
35. Sahoo, S.S.; Vijay, V.K.; Chandra, R.; Kumar, H. Production and characterization of biochar produced from slow pyrolysis of pigeon pea stalk and bamboo. *Clean. Eng. Technol.* **2021**, *3*, 100101. [CrossRef]
36. Skrzyniarz, M.; Sajdak, M.; Zajemska, M.; Iwaszko, J.; Biniek-Poskart, A.; Skibiński, A.; Morel, S.; Niegodajew, P. Plastic Waste Management towards Energy Recovery during the COVID-19 Pandemic: The Example of Protective Face Mask Pyrolysis. *Energies* **2022**, *15*, 2629. [CrossRef]
37. De-La-Torre, G.E.; Aragaw, T.A. What we need to know about PPE associated with the COVID-19 pandemic in the marine environment. *Mar. Pollut. Bull.* **2021**, *163*, 111879. [CrossRef] [PubMed]
38. Wee, M.; Jie, X. The Potentials of Co-Pyrolysis of Empty Fruit Bunch (EFB) and Disposable Face Mask Wastes. 2022. Available online: https://www.icheme.org/media/18941/2021-research-melvin-wee-xin-jie.pdf (accessed on 26 August 2023).
39. Tt, A.K.; Mech, N.; Ramesh, S.; Gandhimathi, R. Evaluation of composite briquettes from dry leaves in energy applications for agrarian communities in India. *J. Clean. Prod.* **2022**, *350*, 131312. [CrossRef]
40. Kumar, M.; Upadhyay, S.N.; Mishra, P.K. Pyrolysis of Sugarcane (*Saccharum officinarum* L.) Leaves and Characterization of Products. *ACS Omega* **2022**, *7*, 28052–28064. [CrossRef]
41. Collard, F.-X.; Blin, J. A review on pyrolysis of biomass constituents: Mechanisms and composition of the products obtained from the conversion of cellulose, hemicelluloses and lignin. *Renew. Sustain. Energy Rev.* **2014**, *38*, 594–608. [CrossRef]
42. Tan, H.; Lee, C.T.; Ong, P.Y.; Wong, K.Y.; Bong, C.P.C.; Li, C.; Gao, Y. Pyrolysis and Fast Pyrolysis on The Quality of Lignocellulosic and Lignin-Based Biochar A Review on The Comparison Between Slow Pyrolysis and Fast Pyrolysis on the Quality of Lignocellulosic and Lignin- Based Biochar. In *IOP Conference Series: Materials Science and Engineering*; IOP Publishing: Bristol, UK, 2021. [CrossRef]
43. Sun, J.; He, F.; Pan, Y.; Zhang, Z. Effects of pyrolysis temperature and residence time on physicochemical properties of different biochar types. *Acta Agric. Scand. Sect. B Soil Plant Sci.* **2017**, *67*, 12–22. [CrossRef]
44. Srivatsav, P.; Bhargav, B.S.; Shanmugasundaram, V.; Arun, J.; Gopinath, K.P.; Bhatnagar, A. Biochar as an Eco-Friendly and Economical Adsorbent for the Removal of Colorants (Dyes) from Aqueous Environment: A Review. *Water* **2020**, *12*, 3561. [CrossRef]
45. Diao, R.; Wang, C.; Luo, Z.; Zhu, X. The valorization of co-pyrolysis bio-oil derived from bio-oil distillation residue and walnut shell via coupling fractional condensation and lyophilization. *J. Clean. Prod.* **2021**, *294*, 126263. [CrossRef]
46. Luo, Z.; Zhu, X.; Deng, J.; Gong, K.; Zhu, X. High-value utilization of mask and heavy fraction of bio-oil: From hazardous waste to biochar, bio-oil, and graphene films. *J. Hazard. Mater.* **2021**, *420*, 126570. [CrossRef]
47. Özsin, G.; Pütün, A.E. Insights into pyrolysis and co-pyrolysis of biomass and polystyrene: Thermochemical behaviors, kinetics and evolved gas analysis. *Energy Convers. Manag.* **2017**, *149*, 675–685. [CrossRef]
48. Jouhara, H.; Ahmad, D.; van den Boogaert, I.; Katsou, E.; Simons, S.; Spencer, N. Pyrolysis of domestic based feedstock at temperatures up to 300 °C. *Therm. Sci. Eng. Prog.* **2018**, *5*, 117–143. [CrossRef]
49. Carbolite Gero. *C. Tg2 12/125/425. Operation Manual*; Carbolite Gero Ltd.: Hope Valley, UK, 2021; 74p.
50. Manić, N.; Janković, B.; Stojiljković, D.; Angelopoulos, P.; Radojević, M. Thermal characteristics and combustion reactivity of coronavirus face masks using TG-DTG-MS analysis. *J. Therm. Anal. Calorim.* **2022**, *147*, 10131–10143. [CrossRef]
51. Yousef, S.; Eimontas, J.; Zakarauskas, K.; Striūgas, N.; Mohamed, A. A new strategy for using lint-microfibers generated from clothes dryer as a sustainable source of renewable energy. *Sci. Total Environ.* **2020**, *762*, 143107. [CrossRef]

52. *ASTM E872-82;* Standard Test Method for Volatile Matter in the Analysis of Particulate Wood Fuels. ASTM: West Conshohocken, PA, USA, 2019.
53. *ASTM E1755-01;* Standard Test Method for Ash in Biomass. ASTM: West Conshohocken, PA, USA, 2001.
54. *ASTM E1756-01;* Standard Test Method for Determination of Total Solids in Biomass. ASTM: West Conshohocken, PA, USA, 2001.
55. Park, J.-W.; Heo, J.; Ly, H.V.; Kim, J.; Lim, H.; Kim, S.-S. Fast pyrolysis of acid-washed oil palm empty fruit bunch for bio-oil production in a bubbling fluidized-bed reactor. *Energy* **2019**, *179*, 517–527. [CrossRef]
56. *ASTM D7582-15;* Standard Test Methods for Proximate Analysis of Coal and Coke by Macro Thermogravimetric Analysis. ASTM: West Conshohocken, PA, USA, 2015.
57. Wakatuntu, J.; Olupot, P.W.; Jjagwe, J.; Menya, E.; Okure, M. Optimization of pyrolysis conditions for production of rice husk-based bio-oil as an energy carrier. *Results Eng.* **2023**, *17*, 100947. [CrossRef]
58. Lay, M.; Rusli, A.; Abdullah, M.K.; Hamid, Z.A.A.; Shuib, R.K. Converting dead leaf biomass into activated carbon as a potential replacement for carbon black filler in rubber composites. *Compos. Part B Eng.* **2020**, *201*, 108366. [CrossRef]
59. Foffi, R.; Savuto, E.; Stante, M.; Mancini, R.; Gallucci, K. Study of Energy Valorization of Disposable Masks via Simulation Approach. *Energies* **2022**, *15*, 2103. [CrossRef]
60. Oginni, O.; Singh, K. Influence of high carbonization temperatures on microstructural and physicochemical characteristics of herbaceous biomass derived biochars. *J. Environ. Chem. Eng.* **2020**, *8*, 104169. [CrossRef]
61. Dhyani, V.; Bhaskar, T. A comprehensive review on the pyrolysis of lignocellulosic biomass. *Renew. Energy* **2018**, *129*, 695–716. [CrossRef]
62. Kai, X.; Yang, T.; Shen, S.; Li, R. TG-FTIR-MS study of synergistic effects during co-pyrolysis of corn stalk and high-density polyethylene (HDPE). *Energy Convers. Manag.* **2019**, *181*, 202–213. [CrossRef]
63. Asadi, A.; Zhang, Y.; Mohammadi, H.; Khorand, H.; Rui, Z.; Doranehgard, M.H.; Bozorg, M.V. Combustion and emission characteristics of biomass derived biofuel, premixed in a diesel engine: A CFD study. *Renew. Energy* **2019**, *138*, 79–89. [CrossRef]
64. Ortega, F.; Pula, H.J.; Calero, M. Characterization of the Products of the Catalytic Pyrolysis of Discarded COVID-19 Masks over Sepiolite. *Appl. Sci.* **2023**, *13*, 3188. [CrossRef]
65. Lee, X.J.; Lee, L.Y.; Hiew, B.Y.Z.; Gan, S.; Thangalazhy-Gopakumar, S.; Ng, H.K. Valorisation of oil palm wastes into high yield and energy content biochars via slow pyrolysis: Multivariate process optimisation and combustion kinetic studies. *Mater. Sci. Energy Technol.* **2020**, *3*, 601–610. [CrossRef]
66. Zhang, Y.; Ma, Z.; Zhang, Q.; Wang, J.; Ma, Q.; Yang, Y.; Luo, X.; Zhang, W. Comparison of the Physicochemical Characteristics of Bio-char Pyrolyzed from Moso Bamboo and Rice Husk with Different Pyrolysis Temperatures. *BioResources* **2017**, *12*, 4652–4669. [CrossRef]

sustainability

MDPI

Article

Exploring the Potential of Utilizing Aquatic Macrophytes for Enhanced Phytoremediation of Zinc in Artificial Wastewater: Characteristics and Parameter Studies

Hui Wun Tan [1], Yean Ling Pang [1,2,*], Steven Lim [1,2], Woon Chan Chong [1,2], Chin Wei Lai [3] and Ahmad Zuhairi Abdullah [4]

1 Department of Chemical Engineering, Lee Kong Chian Faculty of Engineering and Science, Universiti Tunku Abdul Rahman, Kajang 43000, Selangor, Malaysia; huiwun95@gmail.com (H.W.T.); stevenlim@utar.edu.my (S.L.); chongwchan@utar.edu.my (W.C.C.)
2 Centre for Photonics and Advanced Materials Research, Universiti Tunku Abdul Rahman, Kajang 43000, Selangor, Malaysia
3 Nanotechnology & Catalysis Research Centre (NANOCAT), Institute for Advanced Studies, University of Malaya, Kuala Lumpur 50603, Malaysia; cwlai@um.edu.my
4 School of Chemical Engineering, Universiti Sains Malaysia, Nibong Tebal 14300, Penang, Malaysia; chzuhairi@usm.my
* Correspondence: pangyl@utar.edu.my or pangyeanling@hotmail.com; Tel.: +60-39-086-0288; Fax: +60-39-019-8868

Abstract: Heavy metal pollution due to industrialization can threaten the surrounding environment and living organisms. Phytoremediation is a green technique that uses hyperaccumulator plants to eliminate or decrease heavy metals in polluted water bodies. The aim of this study was to investigate the changes in morphology of *Pistia stratiotes* (water lettuce) and *Eichhornia crassipes* (water hyacinth) before and after phytoremediation of zinc (Zn) by using scanning electron microscopy (SEM), electron dispersive X-ray spectroscopy (EDX) and Fourier transform infrared spectroscopy (FTIR). The SEM images showed the formation of small granular aggregates on the surfaces of the leaf and root. EDX results confirmed the uptake of Zn metal, especially in the plant roots. The FTIR spectra showed the Zn metal binding with several characteristic functional groups (O-H, C-H and C=O bonds). Different parameters were also studied to optimize the Zn uptake rate. Water lettuce achieved 80.1% phytoremediation of Zn after 5 days at optimum conditions (10 ppm of Zn, 6 ppm of sodium chloride and natural solution pH). Meanwhile, water hyacinth reached up to 88% when increasing the sodium chloride up to 9 ppm. In conclusion, Zn phytoremediation using both plants can be a potential remediation method for improving the quality of water.

Keywords: phytoremediation; zinc; water lettuce; water hyacinth; characteristics; parameter studies

check for updates

Citation: Tan, H.W.; Pang, Y.L.; Lim, S.; Chong, W.C.; Lai, C.W.; Abdullah, A.Z. Exploring the Potential of Utilizing Aquatic Macrophytes for Enhanced Phytoremediation of Zinc in Artificial Wastewater: Characteristics and Parameter Studies. *Sustainability* **2023**, *15*, 15170. https://doi.org/10.3390/su152015170

Academic Editor: Dino Musmarra

Received: 20 August 2023
Revised: 27 September 2023
Accepted: 10 October 2023
Published: 23 October 2023

1. Introduction

Environmental pollution is slowly escalating to become a worldwide threat which requires proper mitigation and prevention. To date, fast-paced industrialization and urbanization have contributed to the pollution of water, forcing around 40% of the global population to face water scarcity [1]. According to the United States Environmental Protection Agency [2], metals or metalloids with a density of 5 g/cm^3 and above are classified as heavy metals, which includes aluminum (Al), arsenic (As), cadmium (Cd), chromium (Cr), copper (Cu), lead (Pb), selenium (Se), mercury (Hg) and zinc (Zn). These heavy metals are capable of leaching into drinking water from household plumbing pipelines, natural mineral deposits, municipal waste disposal facilities and industrial activities which include mining operations, electronic manufacturing, petroleum refineries and the pharmaceutical industry. Once the polluted water is consumed by humans, even at a low concentration, the health problems imposed will be fatal. For instance, Al can cause neurotoxic effects, Pb can cause impaired mental and physical development among children, Cr can potentially damage the liver and cause vomiting, while Cd can be carcinogenic and cause multiple diseases such as endocrine disruption and osteoporosis [3].

Environmental monitoring agencies in various countries have developed approved limits for the heavy metal levels in consumable water in lieu of their detrimental effects on the surrounding environment [4]. In general, industrial wastewater treatment technologies can be divided into physical, chemical, physiochemical and biological methods, each with their own pros and cons [1]. Physical treatment methods include membrane filtration, screening and sedimentation and are generally cheaper and easier to operate but have a low treatment efficiency. Chemical methods such as chemical precipitation, ion exchange and solvent extraction can be metal-selective with a high treatment capacity but suffer from high maintenance cost and the production of unwanted sludge. Meanwhile, physiochemical methods include electrodialysis, coagulation and flocculation, as well as photocatalysis; they can also be metal-selective and highly efficient, but the drawbacks are the relatively expensive operation cost and that it is labor-intensive [1,5,6].

As a green alternative, biological treatment, which can extract harmful heavy metals in a more eco-friendly manner, is strongly recommended by researchers [4]. In this context, carrying out phytoremediation of heavy-metals-polluted water by using green plants has become one of the research hotspots in the area of green remediation technology [7]. Phytoremediation is a technique that directly use various types of plants to adsorb/absorb, accumulate or detoxify, reduce harmful effects and minimize heavy metal contamination in any water sources or soils through biological, chemical and physical processes. It is a cost-effective, long-term sustainable method that is less destructive to the surrounding environment and highly suitable for under-developed and developing countries [8]. This process uses green plants known as hyperaccumulator plants to remove heavy metal pollutants from the polluted environment through different techniques such as phytovolatilization, rhizofiltration, phytostabilization and phytoextraction [4]. Phytovolatilization refers to the conversion of toxic metals into less harmful volatile forms before their release into the atmosphere via the foliage system. On the other hand, rhizofiltration involves the removal of pollutants from contaminated water through processes such as adsorption onto roots or absorption by roots. Phytostabilization aims to immobilize heavy metals, reducing their bioavailability within the food chain. Meanwhile, phytoextraction entails the uptake, translocation and accumulation of contaminants in the aerial parts of plants [9]. According to An et al. [9], a hyperaccumulator plant is a plant species which can accumulate significant levels of heavy metals within their aerial parts without exhibiting phytotoxic symptoms. In contrast to non-hyperaccumulator plants, hyperaccumulator plants can bioaccumulate up to 100 times the amount of heavy metals under similar conditions. In recent years, studies have shown that the research focus has been slowly shifting from the discovery of hyperaccumulator plant species to the internal growth and metabolisms of the plants. This is because a research focus on uncovering the underlaying characteristics of the plants can help us understand and maximize the efficacy of phytoremediation as a solid and greener choice among all the conventional wastewater treatment methods.

However, in order for the phytoremediation process to be as efficient as possible, the selection of suitable plant species is imperative. Favorable characteristics of suitable phytoremediation plants include having a high biomass yield, the ability to absorb and tolerate copious amounts of heavy metals during its life cycle by transporting the absorbed heavy metals to the aerial parts of the plant and a fast growth rate [4]. Apart from favorable characteristics for phytoremediation purposes, external factors such as climate, temperature, pH, light irradiation, salinity and nutrient availability can also influence the growth of plants and phytoremediation performance. *Pistia stratiotes* (water lettuce) and *Eichhornia crassipes* (water hyacinth) are two of the most common hydrophytes used for the phytoremediation treatment of heavy-metals-laden wastewater due to their widely spread habitat, ease of maintenance and economical nature. Water lettuce, also known as water cabbage or shellflower, belongs to the Araceae and is mostly found in lakes, ponds and streams. It has a rapid growth rate and is also capable of surviving under high metal stress by accumulating metals in its roots and leaves, making it suitable for phytoextraction purpose [10]. Meanwhile, water hyacinth is closely related to the Liliaceae and can be easily found in large quantities throughout the year due to its inundation. The root of water hyacinth is unique in that it can accumulate and extract large amounts of heavy metals such as Cd, Cu, Pb, Zn and Hg. Similar to water lettuce, water hyacinth is highly recommended for the phytoremediation treatment of industrial effluent containing heavy metals due to its high biomass production rate and tolerance towards extreme surrounding environments, as well as its high metal uptake rate [11]. Water lettuce and water hyacinth were chosen for the Zn phytoremediation study due to their high growth rate, short growth cycles, low energy demands and high availability in Malaysia. To date, reports on the change in the surface characteristics of these two plants before and after performing phytoremediation when treating Zn-containing wastewater is rarely found in the literature. Although work with a similar idea has been conducted by Zheng et al. [12], in which the competitive interaction between the Cu and Cd metal sorption on water hyacinth plant roots was studied, the whole experiment was conducted based on the adsorption–desorption of metals on dried plant roots, rather than the phytoremediation process. In addition, it was reported in the literature that the phytoremediation period required to remove heavy metals was usually about 1–4 months [13–16]. Thus, the novelty of this study is the focus on the differences in the surface characteristics before and after phytoremediation, as well as the comparison of phytoremediation performance between water lettuce and water hyacinth in a rather short period. The aim of the present study was to analyze the important parameters (phytoremediation duration, Zn concentration, pH and salinity) affecting the phytoremediation performance of both plants with regard to the Zn-contaminated wastewater through the one-factor-at-a-time (OFAT) optimization method.

2. Materials and Methods

2.1. Chemicals and Materials

Fresh water lettuce and water hyacinth were purchased from the same local supplier to ensure uniformity. Reagent-grade zinc nitrate hexahydrate ($Zn(NO_3)^2 \cdot 6H_2O$, 98%) was obtained from Sigma-Aldrich (St. Louis, MO, USA) as the zinc source. It has a molecular weight of 297.49 g/mol and is highly soluble in water. Hydrochloric acid (HCl, 37%), sodium hydroxide (NaOH, 99%) and sodium chloride (NaCl, 99%) were obtained from Merck (Darmstadt, Germany). All the chemicals were used as received and distilled water was used throughout the study.

2.2. Preparation of Plants

The fresh plants purchased were selected to have a weight of between 20 and 30 g each with green young leaves. The plants were first washed with running tap water to remove residual soil particles, dirt and insect larvae that had grown on them. These plants were then transferred to water tanks containing only distilled water with light exposure in a 10 h light/14 h dark cycle to allow them to acclimatize to their new surroundings. After one week of acclimatization, the plants were harvested and patted dry using clean paper towel, before being used in the subsequent experiment.

2.3. Preparation of Zn Stock Solution and Sodium Chloride Stock Solution

A total of 1000 ppm stock solution of Zn and sodium chloride was prepared by dissolving the zinc salt and sodium chloride, respectively, in distilled water. The stock solutions prepared were stored at room temperature and shielded from light exposure to prevent any degradation in concentration.

2.4. Phytoremediation of Zn

The experiment was carried out in a glass beaker filled with 1 L of synthetic Zn-containing wastewater solution under natural light irradiation. The solution was stirred by using a magnetic stir bar on a hotplate prior to the phytoremediation experiment to ensure a uniform dispersion of the Zn content. Next, the acclimatized plants were transferred to beakers filled with 1 L solution containing 5 ppm of Zn each in order to study the effect of the duration of phytoremediation. Afterwards, 1 L solutions with different concentrations of Zn (5, 10, 15, 20 and 25 ppm) were prepared from the as-prepared Zn stock solution. This range of Zn concentrations was selected based on the US EPA standard's regulated maximum contaminant level of Zn in drinking water and the Malaysia national water quality standard, as well as the limits for sewage and industrial effluent regulated under the Environmental Quality Act 1974 [17–19]. After that, 1 M of hydrochloric acid and 1 M of sodium hydroxide solutions were prepared and used to adjust the pH of the solution with specific Zn concentrations. Finally, 1 L solutions of specific Zn concentrations with specific pH and different salinity (3, 6, 9, 12 and 15 ppm) were prepared by using the as-prepared sodium chloride stock solution. At least 3 replicates were obtained to calculate the errors shown in the figures provided for the parameter studies. At the end of the experiment, the treated plants were harvested and dried in an oven at 100 °C overnight, cut into small pieces and stored in properly labeled polythene bags for further characterization analysis.

2.5. Plants Characterization

Both fresh and treated plants were washed and dried in an oven at 90 °C overnight. The dried samples were cut and separated into leaf and root parts. The separated parts were then finely cut into tiny pieces and passed through a No. 40 mesh sieve to obtain uniformly sized samples for further characterization [20]. All characterization studies were conducted using three replicates of both fresh and treated plant parts. The scanning electron microscopy (SEM) images of the fresh and treated dried plants were studied by using a Hitachi SEM S-3400N scanning electron microscope that was operated at 15 kV. The samples were mounted on the aluminum holder with carbon-conductive tape. The electron dispersive X-ray spectroscopy (EDX) analysis was conducted using Ametek EDAX software. The Fourier transform infrared spectroscopy (FTIR) spectra of the fresh and treated dried plants were analyzed by using the Thermo attenuated total reflection-FTIR with a scan number of 64 through a range of wavenumbers between 400 and 4000 cm^{-1}.

2.6. Analysis of Liquid Samples

After phytoremediation experiment, a certain volume of liquid sample was withdrawn from the solution and a syringe filter with a pore size of 0.45 μm was used to filter out the root hairs and other solid particles from the extracted liquid samples. The concentrations of Zn in the liquid samples were determined by using Perkin Elmer inductively coupled plasma–optical emission spectrometry (ICP-OES), model Optima 7000 DV. The removal efficiency of Zn by the plants was calculated as follows:

$$\text{Removal efficiency (\%)} = (C_0 - C_t)/C_o \times 100\% \qquad (1)$$

where C_0 is the initial concentration of Zn before phytoremediation and C_t is the concentration of Zn at any reaction time t (days).

3. Results and Discussion

3.1. Characterization of Water Lettuce and Water Hyacinth before and after Phytoremediation of Zn

The functional groups of water lettuce and water hyacinth at their respective leaves and roots before and after phytoremediation of Zn were studied by using FTIR analysis. The results are shown in Figures 1 and 2, respectively.

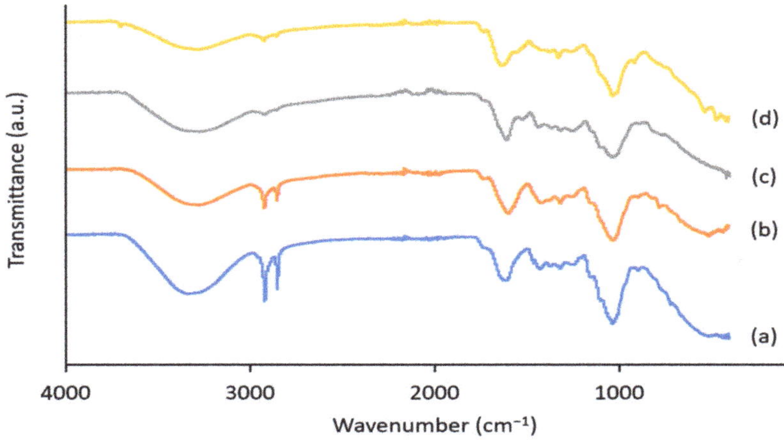

Figure 1. FTIR spectra for (a) leaf without Zn, (b) leaf with Zn, (c) roots without Zn and (d) roots with Zn for water lettuce.

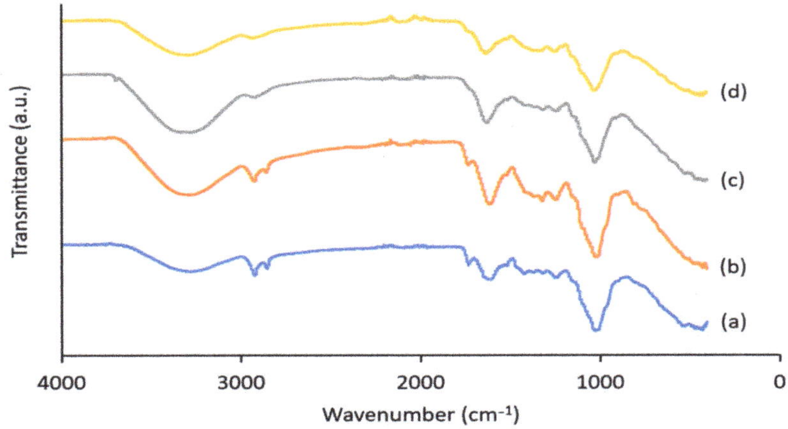

Figure 2. FTIR spectra for (a) leaf without Zn, (b) leaf with Zn, (c) roots without Zn and (d) roots with Zn for water hyacinth.

Figure 1 shows the FTIR spectra of both the water lettuce leaf and roots. As illustrated in Figure 1 (a) to (d), a wide and strong absorption band at 3340 to 3290 cm^{-1} could be observed, which corresponded to the existence of stretching vibrations of O-H bonds due to the presence of water remaining within the plant itself. As Zn was absorbed by the water lettuce, it could be observed that the band shifted from 3340 to 3290 cm^{-1} and from 3300 to 3290 cm^{-1} for the leaf and roots, respectively. The shifting of the band could arise from the cationic interaction between Zn^{2+} ions with the hydroxyl group for metal oxygen binding [14]. Meanwhile, for the leaf of water lettuce before and after Zn uptake, two sharp peaks could be observed at 2910 and 2850 cm^{-1}, which were associated with the stretching vibrations of C-H bonds within the cellulose structure of the water lettuce leaf. According to [15], these bands can be attributed to the presence of asymmetric methylene (CH_2) stretching around 2930 to 2910 cm^{-1} and symmetric CH_2 stretching around 2860 to 2840 cm^{-1}, which suggested the presence of aliphatic materials within the leaf cuticle such as wax and cutin. In contrast, the presence of C-H bonds around 2910 and 2850 cm^{-1} was very minimal for the water lettuce roots, as evidenced by the insignificant intensity of both peaks for Figure 1 (c) and (d). Additionally, a large absorption peak at 1600 to 1620 cm^{-1}

could also be observed in Figure 1 (a) to (d), which corresponded to the presence of C=O bonds. As shown in Figure 1, upon Zn uptake, the FTIR spectra displayed alterations, particularly in the bands associated with O-H, C-H and C=O. These changes suggested the participation of cationic elements such as Na^+ and K^+ in the adsorption of Zn^{2+} ions onto the plant surfaces through proton and metal exchange processes [16]. According to [21], the interaction between metals with functional groups of water lettuce acting as active sites could result in the alteration of shape, shifting of position and change in intensity of the characteristic peaks.

Figure 2 illustrates the FTIR spectra for both leaf and roots of water hyacinths. Similar to the water lettuce, a broad absorption band at around 3300 cm^{-1} could be observed in all the leaf and roots of water hyacinth before and after Zn uptake, which suggested the presence of O-H stretching vibration mode. Next, two smaller absorption bands at 2920 and 2850 cm^{-1} could only be found in the leaf of the water hyacinth which corresponded to the stretching vibration of C-H bonds. As mentioned beforehand, the presence of these aliphatic contents (symmetric and asymmetric CH_2 stretching band) was exclusive to the structure of the cuticles [15]. On the other hand, two absorption peaks were also found at 1730 and 1600 cm^{-1} for the water hyacinth leaf before and after Zn uptake, which were attributed to the stretching vibration of C=O bonds. However, the characteristic peak for C=O bonding could only be observed at around 1620 cm^{-1} for the roots of water hyacinth before and after phytoremediation of Zn. It could be observed that the intensity of the characteristic peak for C=O bonding changed as the water hyacinth underwent Zn phytoremediation, indicating possible interactions between the characteristic functional group and Zn metals [14].

In short, the FTIR spectra showed that both water lettuce and water hyacinth contained characteristic functional groups for most polysaccharides such as O-H, C-H and C=O bonds. According to [20], the presence of these diverse functional groups could potentially act as a binder for Zn ions. In this context, the binding interactions between the characteristic functional groups and Zn metals during phytoremediation, which resulted in the alteration of these characteristic peaks, were in good agreement with the results obtained by [14].

The surface morphology of the leaf and roots of water lettuce and water hyacinth before and after phytoremediation of Zn were investigated by using SEM analysis. The results are shown in Figures 3 and 4. Figure 3a,b show the SEM images of the leaf of water lettuce before and after the uptake of Zn, respectively. The rod-like structure with a folded shape of the leaf represented the trichomes of the plant, which were able to trap tiny air bubbles and keep the water lettuce floating on the water surface. After the phytoremediation of Zn, a cluster of small granules was observed around the voids between the trichomes as indicated by the red arrow shown in the figure, which were absent in the leaf prior to Zn uptake. According to [22], this could be associated with the formation of Zn metal complexes that subsequently immobilized the complexes within the voids. On the other hand, Figure 3c,d represent the SEM images of the water lettuce roots before and after Zn phytoremediation, respectively. Prior to the absorption of Zn, the surface of the plant roots was smooth and linear. Upon the Zn uptake, small aggregates and bulges alongside a creased surface morphology could be observed on the plant roots. The altered appearance of the plant roots could be attributed to the interactions between functional groups present on the roots' surface and the cationic Zn metal, leading to the formation of stable metal-chelating complexes [23].

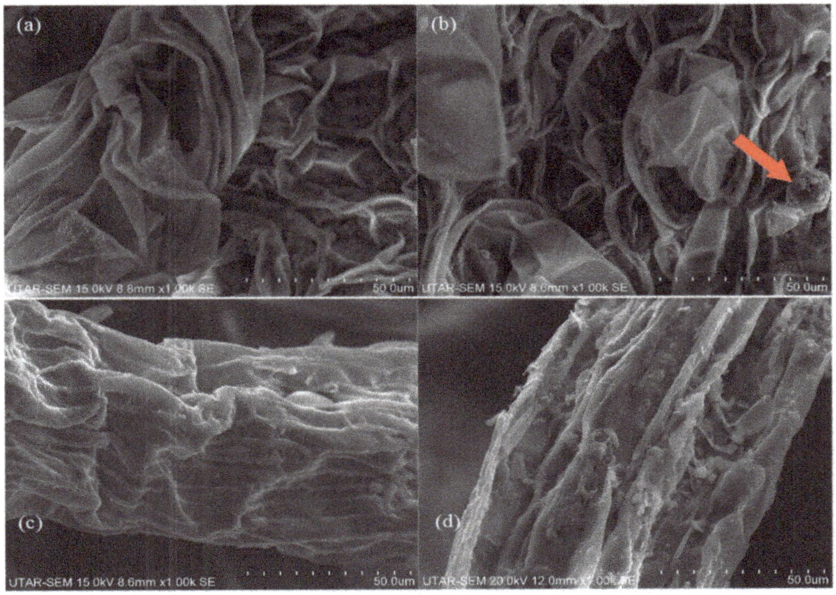

Figure 3. SEM images for (**a**) leaf without Zn uptake, (**b**) leaf with Zn, (**c**) roots without Zn and (**d**) roots with Zn for water lettuce.

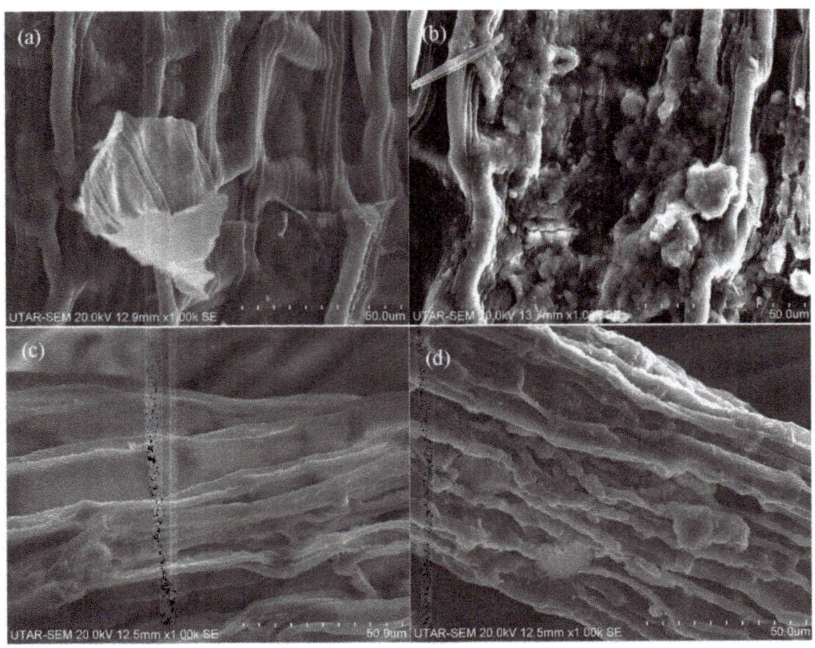

Figure 4. SEM images for (**a**) leaf without Zn uptake, (**b**) leaf with Zn, (**c**) roots without Zn and (**d**) roots with Zn for water hyacinth.

Figure 4a,b present the SEM images of the water hyacinth leaf before and after Zn uptake. In general, the surface morphology of the leaf was linear and covered with a copious amount of evenly distributed stomata. After the water hyacinth was subjected to a solution containing Zn metal, the stomata on the epidermis were saturated with aggregates and the surface had swollen. The presence of aggregates suggested the binding between the Zn metal and the carboxylic groups and other functional groups present to form stable metal chelates [21]. On the contrary, Figure 4c,d show the SEM images for the water lettuce roots before and after absorption of Zn. The epidermal cells of the water hyacinth roots were originally oblong and closely packed. When Zn was absorbed, the oblong-shaped roots broke into tiny fibrous roots which were also covered with small individual granules and agglomerates. The dissociation of roots and formation of tiny granules on the surface is in accordance with the result obtained by [22], where different features of plant surfaces such as carboxylic groups, esters and lignin were highly involved in the absorption of metals via the chelation of metal complexes.

Figures 5 and 6 represent the EDX mapping of water lettuce and water hyacinth leaf and roots after Zn phytoremediation. Figure 5 confirms the presence of Zn in both the leaf and roots of water lettuce, as indicated by its distribution across the plant tissues. Figure 5a shows a lower density of Zn distribution within the leaf of the water lettuce compared with its root counterparts that are shown in Figure 5b. Similarly, the distribution pattern of Zn within both the leaf and roots of water hyacinth, as shown in Figure 6a,b, followed the same trend, with a higher density of Zn distribution in the roots compared with the aerial part of the plant.

The EDX results for all the plant samples are shown in Table 1. The EDX analysis of the plant leaf and roots for both water lettuce and water hyacinth before Zn uptake consist of carbon (C), oxygen (O), sodium (Na), calcium (Ca) and potassium (K) elements. The Na and Ca were only observed in the roots and leaves of both plants, respectively, before the Zn uptake. Meanwhile, prior to the uptake of Zn, K was detected in the leaf and roots of water lettuce and water hyacinth. Ca and K were sources of nutrients for the growth and sustainability of the plants. For instance, Ca was responsible for the development of cell walls to resist diseases, and K was used to strengthen the plants during early growth and for the retention of water. On the other hand, Na was not a necessary nutrient for the growth and development of the plants, and a high amount of Na can induce salt stress, which is unbeneficial to the plants [24]. The atomic percentage of K in the leaves of water hyacinth was lower than its roots counterpart, which is in accordance with the findings reported by [25]. The presence of these mineral elements (Na, Ca and K) was in part due to the intake of nutrients in the previous surrounding environment during the growth of the water lettuce and water hyacinth prior to the application of Zn metal for phytoremediation.

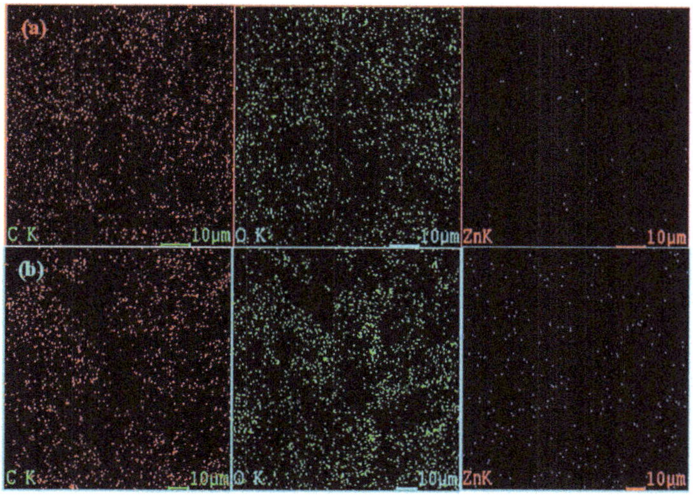

Figure 5. EDX mapping of (**a**) leaf and (**b**) roots after Zn uptake by water lettuce.

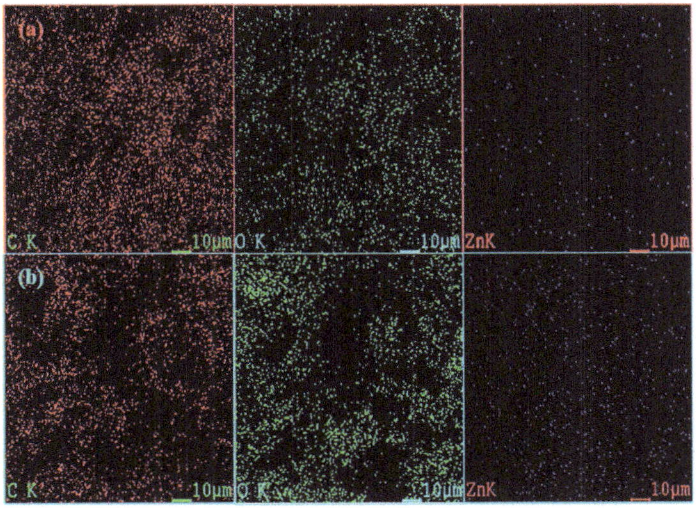

Figure 6. EDX mapping of (**a**) leaf and (**b**) roots after Zn uptake by water hyacinth.

Table 1. Elemental atomic percentage of all samples extracted from EDX analysis.

Plant Sample		Elemental Atomic Percent (%)					
		C	O	Zn	Na	Ca	K
Water lettuce	Leaf (without Zn)	43.3	33.7	-	-	16.2	6.8
	Leaf (with Zn)	57.1	41.5	1.4	-	-	-
	Roots (without Zn)	44.4	35.7	-	4.1	7.6	8.2
	Roots (with Zn)	40.6	52.1	7.3	-	-	-
Water hyacinth	Leaf (without Zn)	36.2	37.2	-	-	18.4	8.2
	Leaf (with Zn)	77.1	20.7	2.2	-	-	-
	Roots (without Zn)	48.7	46.8	-	1.6	-	2.9
	Roots (with Zn)	46.8	43.8	9.4	-	-	-

It was observed that as the plants were subjected to the solution containing Zn metal, the atomic percentage of Zn was increased. For both types of plant, the atomic percentage of Zn was found to be higher in the roots compared with their respective leaf counterparts. Not only that, the atomic percentage of Zn in the leaf and roots of water hyacinth after phytoremediation of Zn was also greater than that of the water lettuce. This finding could be related to the higher contaminant reduction capability of water hyacinth, which resulted in a greater amount of Zn being accumulated within the plant biomass [26]. After the uptake of Zn, the mineral elements were not detected through EDX analysis due to their release to the solution as a result of the stronger chelation ability of Zn^{2+} with the negatively charged functional groups found on the plant roots, replacing the originally bounded cationic mineral elements [12]. Meanwhile, the absence of mineral elements in the plant leaves after Zn uptake indicated the utilization of the mineral elements to facilitate cellular activities such as sequestration, compartmentalization and efflux of excess metals to reduce the stress induced by metal [27].

3.2. *Parameter Effect*

3.2.1. Effect of Duration of Phytoremediation

The duration of phytoremediation of Zn by water lettuce and water hyacinth was studied extensively, and the results are illustrated in Figure 7. The study on the duration of phytoremediation of Zn by both plants was conducted over a period of a week (7 days) to determine the time required to achieve maximum Zn removal efficiency by both plants.

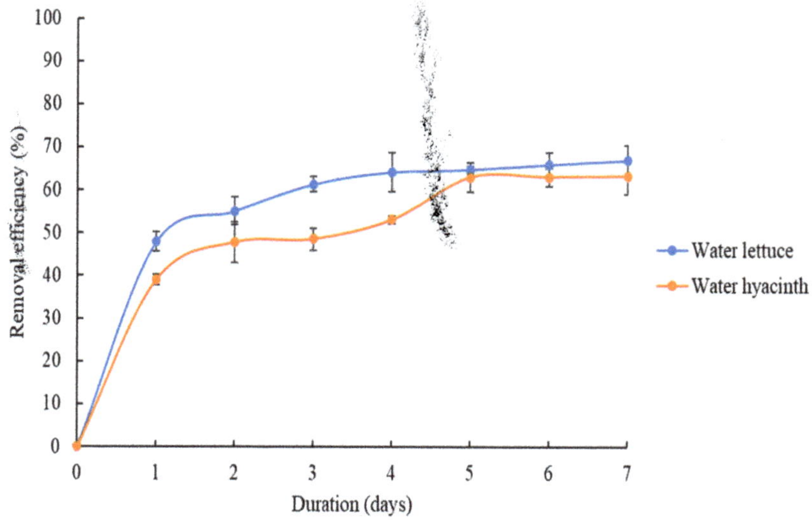

Figure 7. Duration of phytoremediation of Zn by water lettuce and water hyacinth.

Based on Figure 7, both plants exhibited the highest rate of removal efficiency during the first day of phytoremediation, where the water lettuce and water hyacinth recorded a 49% and 39% Zn removal efficiency, respectively. Nevertheless, the efficiency of removal of Zn by the water lettuce showed a continuous uptrend over the first four days of phytoremediation treatment, before reaching a constant removal rate on day 5 and onwards. The removal rate of Zn achieved by using water lettuce in the present study was greater than the one reported by [27].

On the other hand, water hyacinth experienced a steady increase in Zn removal efficiency during the first five days, and therefore took a slightly longer period of time to achieve a constant removal rate on day 6. The slight difference in time taken to achieve a constant rate of Zn removal between the two plants might be due to the faster growth rate of water lettuce, which could potentially absorb more Zn as one of the important nutrients required for plant growth in terms of metabolic and physiological mechanisms [28]. Having said that, both water lettuce and water hyacinth clearly showed a significant Zn removal efficiency within the first five days, which was indicative of the quick attainment of the saturation state. Upon reaching the saturation state, the plants were displaying a certain degree of difficulty with further absorbing Zn from the solution, even though the Zn concentration had reduced with the passage of time. In this context, a similar result was also obtained by [29], where the water hyacinth and water lettuce showed similar performances in terms of removing Zn. In this study, 5 days of phytoremediation duration was selected for subsequent studies, as it was found to be the most optimum timeframe for both plants to achieve their respective highest attainable Zn removal efficiency before saturation.

3.2.2. Effect of Zn Concentration

The phytoremediation of different concentrations of Zn by water lettuce and water hyacinth were studied. The results of the study are shown in Figures 8 and 9. For water lettuce, the Zn removal efficiency at different concentrations showed an uptrend for the first four days of the experiment. As discussed previously, at 5 ppm Zn concentration, the removal efficiency by water lettuce had reached its saturation point, with an almost stagnant removal rate of 65% on day 5 of the study. Meanwhile, the Zn removal efficiency continued to escalate throughout the whole five days of study for 10 ppm and 15 ppm Zn concentrations. The 10 ppm of Zn, in particular, had achieved the highest removal efficiency of 72% on day 5 compared with other concentrations of Zn. The removal efficiency for 15 ppm of Zn was recorded at 59% on the final day of study, which was lower than both the 5 and 10 ppm Zn concentrations. Nonetheless, the rate of removal for the 15 ppm Zn concentration was still increasing even on the fifth day of the study, indicating a possibility of achieving a higher Zn removal efficiency with a longer phytoremediation duration.

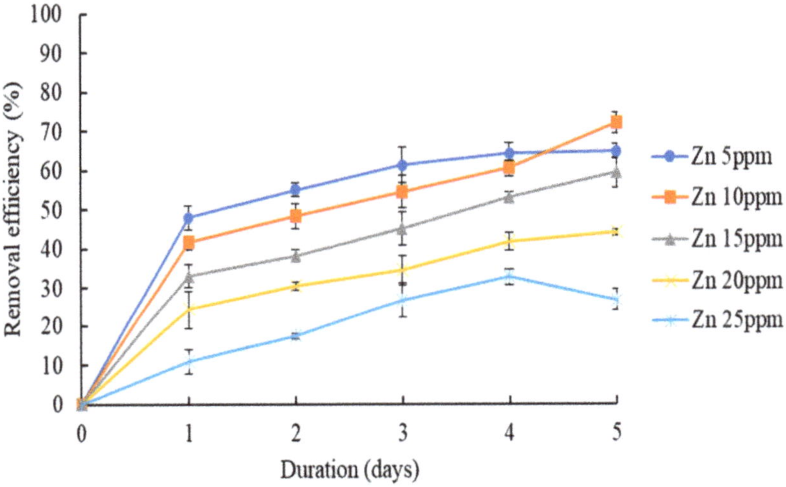

Figure 8. Effect of Zn concentration on the removal efficiency of Zn by water lettuce (duration = 5 days).

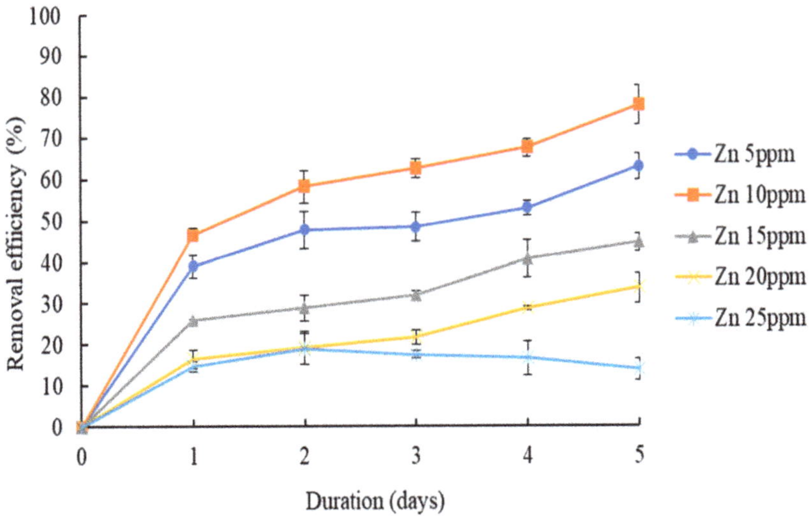

Figure 9. Effect of Zn concentration on the removal efficiency of Zn by water hyacinth (duration = 5 days).

On the other hand, the removal efficiency in higher concentrations of Zn at 20 and 25 ppm by water lettuce did not cross the 50% mark throughout the study period. At a Zn concentration of 20 ppm and above, the rate of removal was increasing slowly and reached a final removal efficiency of 44.1% on the fifth day of the study. Although the removal efficiency at 20 ppm of Zn was increasing at a very slow pace, the obtained data suggested that similar to the cases of Zn concentrations at 10 ppm and 15 ppm, the water lettuce could potentially remove more Zn contaminant if subjected to a longer study period. In the meantime, instead of showing a slowdown in terms of removal rate, the removal efficiency in 25 ppm of Zn on day 5 of the experiment was found to be lower than that on day 4, which indicated a possible re-release of absorbed Zn contaminants by the wilting plant. It was also observed that the plant leaves started to turn yellowish on day 4, which suggested the chlorosis of the plant leaves and overall necrosis of the plant due to exposure to a high concentration of Zn. This finding is in line with the result reported by [30]. In their study, it was observed that a high concentration of Zn had reduced the expression of photosynthetic pigments and interfered with the photosystem, as evidenced by the chlorosis of the fronds.

Moreover, [31] also reported that the growth of plants was retarded when exposed to a higher concentration of Zn due to the toxicity imposed onto the plant itself, subsequently causing the plant to suffer from wilting. Although hyperaccumulator plants such as water lettuce were capable of defending themselves from phytotoxicity induced by toxic heavy metals, excessively high concentrations of heavy metals could still endanger a plant's overall metabolic and physiological mechanisms. According to [32], reactive oxygen species (ROS) could be produced as a consequence of excessive accumulation of toxic heavy metals within the cytosol. The production of ROS could then lead to oxidative damage to the plant cells and damaging of DNA, as well as hampering of the cell antioxidant and homeostasis mechanisms. In addition, ROS could also damage the structure of the plant cells including the cell membrane, chloroplast and photosynthetic pigments, ultimately causing serious retardation to the plant's growth [33].

Based on Figure 9, water hyacinth showed an overall uptrend in terms of removal efficiency of Zn at all Zn concentrations except for 25 ppm of Zn. Similar to water lettuce, the water hyacinth's efficiency of removal of Zn was the highest at 78%, which was achieved when the plant was exposed to 10 ppm of Zn concentration. The second highest removal efficiency of Zn (63%) was attained when the Zn concentration was at 5 ppm. The difference in the removal efficiency with increasing Zn concentration could be associated with the higher number of Zn^{2+} ions available in the solution that can be absorbed by the plant. As the bioavailability of Zn was higher with an increased Zn concentration, the phytoremediation efficiency of Zn by water hyacinth was also increased due to more Zn^{2+} ions being available at the uptake sites. However, as the Zn concentration was further increased to 15, 20 and 25 ppm, the rate of removal of Zn by water hyacinth began to drop in a similar fashion to that of water lettuce. The final removal efficiencies of Zn at these concentrations were 44.7%, 33.7% and 13.9%, respectively, in ascending order. It could be observed that the removal rates of Zn at concentrations of 5, 10, 15 and 20 ppm were showing a continuing pattern even on the final day of the study, although the removal rate at 5 ppm of Zn by water hyacinth was found to be reaching a stagnant condition on day 6 and onwards in our earlier study. Thus, it is highly likely that the Zn removal efficiency by water hyacinth at these concentrations could reach even higher levels if given a longer time to perform phytoremediation. On the other hand, the rate of removal of Zn was almost identical during the first two days of the study when the plant was exposed to 20 and 25 ppm of Zn. Despite that, the similarity in their removal rate began to deviate starting from day 2 onwards, where the removal efficiency at 25 ppm of Zn was declining from its peak value of 18.9% to 13.9%. The reason behind the reducing Zn removal efficiency by water hyacinth at a concentration of 25 ppm was also similar to that of the water lettuce. During the study, the fronds of the water hyacinth were facing chlorosis at the relatively high concentration of Zn, whereby the color of the plant leaves turned from green to yellow. This chlorosis symptom was a direct result of the phytotoxicity experienced by the plant at such a high concentration of Zn. Based on Figures 8 and 9, it can be seen that water lettuce and water hyacinth demonstrated a great absorption and removal efficiency when subjected to 10 ppm of Zn. The high removal rate of Zn by these two plants might be associated with their special characteristics such as broad leaves and rapid growth rate, as well as their highly vascular and fibrous root system. Notwithstanding their special features, the Zn removal efficiency of both plants showed a decline at concentrations of 15 ppm and above, which could be attributed to the saturation of active sites for Zn absorption and the lower tolerance of plants to such high concentrations of Zn [34]. Thus, the concentration of Zn at 10 ppm was selected for subsequent studies, and the treated Zn concentration also falls within the permissible limit of effluent discharge into water bodies as stated by [35].

3.2.3. Effect of pH Solution

The effect of the solution's pH (pH 2, 4, 6, 8 and 10) on the water lettuce's and water hyacinth's removal efficiency of Zn was studied, and the results are shown in Figures 10 and 11, respectively. Based on Figure 8, pH 6, which was the natural solution pH, led to the highest removal efficiency of Zn by water lettuce (75%). No removal was observed at a solution pH of 2, even on the final day of study.

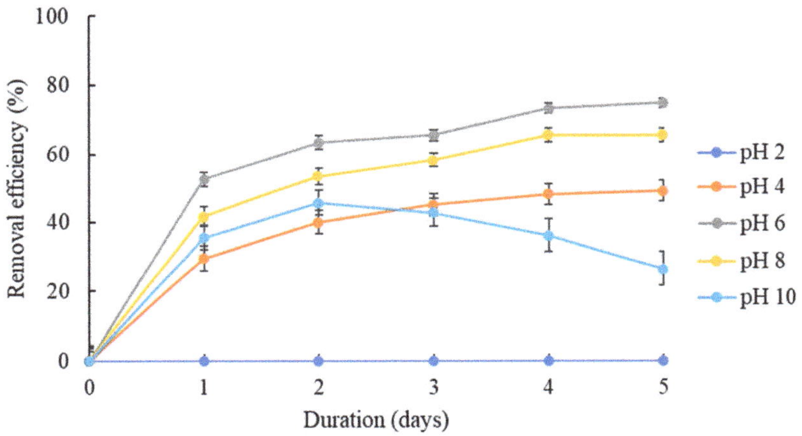

Figure 10. Effect of solution pH on the removal efficiency of Zn by water lettuce (duration = 5 days and Zn concentration = 10 ppm).

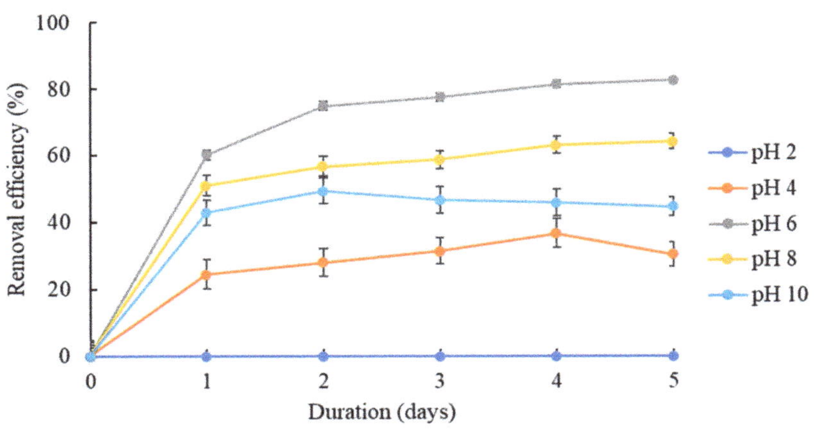

Figure 11. Effect of solution pH on the removal efficiency of Zn by water hyacinth (duration = 5 days and Zn concentration = 10 ppm).

The solution's pH plays a crucial role in the phytoremediation of Zn by accumulator plants. It is one of the external environmental factors which can significantly affect the growth of a plant. According to [36], the pH of a solution is dictated by the concentration or percentage of hydrogen ions (H^+). The solution pH can influence the toxicity of contaminants, and it also governs the form of substance in the solution. Not only that, hydrolysis, precipitation, complexation and redox reactions are also pH-dependent, which can directly affect the availability and speciation of heavy metals for absorption. Consequently, water lettuce was not able to remove Zn in an effective manner at solution pH 2, while at this pH, water lettuce was simultaneously showing symptoms of chlorosis, death and curling of plant leaf tips on the second day of the study.

When the solution pH was increased to pH 4, maintained at its natural solution pH (pH 6) or increased to pH 8, water lettuce was able to absorb Zn metal, showing removal efficiencies of 49.1%, 75% and 65.7%, respectively. At a slightly acidic condition (pH 4), the presence of H^+ ions potentially caused the competition with the Zn^{2+} ions for the available uptake sites of the plant roots. As a result, the amount of Zn metal removed from the solution was lower in an acidic condition. This finding also aligns with [37], where the removal efficiency of Zn by water lettuce was found to be the lowest (49.1%) at pH 4, suggesting the binding of metal ions with inorganic acid which ultimately leads to the formation of complexes and reduces the metal uptake. In this study, hydrochloride acid (HCl) was used to adjust the solution pH to its acidic condition, and thus, the reaction between Zn^{2+} ions and Cl^- ions might produce the chloride complex, which resulted in a lower availability of Zn for plant uptake during the phytoremediation process [38]. On the other hand, when water lettuce was subjected to Zn-containing solution at pH 8, we recorded and overall increment in the rate of removal of Zn throughout the whole study period, with it reaching the saturation state on day 4 of the study.

Nevertheless, when the solution pH was further increased to pH 10, the overall removal efficiency of Zn was poorer than for a solution pH of 8. The reduction in removal efficiency of Zn at increasingly alkaline conditions could be associated with the presence of OH^- ions, which might react with the Zn^{2+} ions, thereby causing the formation of $Zn(OH)_2$ precipitate and subsequently leading to a lower availability of Zn^{2+} ions. As a result, the formation of Zn precipitate at high-pH conditions could reduce the availability of Zn^{2+} ions to be absorbed by the active sites on plant roots during phytoremediation [39]. At a solution pH of 10, the removal efficiency of Zn also started to show a decline after the second day of the study, resulting in a final removal efficiency of only 26.6%. The decrement in the removal efficiency of Zn after its peak value on day 2 (45.6%) at solution pH 10 demonstrated that the re-release of absorbed Zn contaminants was evidenced by the yellowing of green leaves and necrosis of the plant during the second day of the study. According to [28], the ideal pH for healthy growth of water lettuce was neutral, and it was able to tolerate a range of pH values from 4 to 8. Therefore, the decaying water lettuce at both extreme conditions of solution pH (pH 2 and pH 10) observed in this study has verified this result.

Based on Figure 11, water hyacinth exhibited the highest removal efficiency (82.5%) of Zn at a natural solution pH of 6, while the removal efficiency of Zn was the lowest (−30.5%) at a solution pH of 4. Meanwhile, there was no Zn removal by water hyacinth at the solution pH of 2. From the obtained result, it can be observed that a lower removal efficiency of Zn occurred at a higher and lower pH than the natural solution pH (pH 6).

At solution pH 4, the removal efficiency of Zn by water hyacinth was increasing over the first four days of the study, before showing signs of plant necrosis and chlorosis, which ultimately led to a removal efficiency of 30.5% on the final day of the study. The same situation could be observed when water hyacinth was subjected to a Zn-metal-laden solution at a pH of 10. Under this highly alkaline condition, the water hyacinth only managed to survive up until the second day of experiment, before experiencing curling and yellowing of young leaves as well as necrosis, which caused the re-release of absorbed Zn contaminants and the natural decline in the efficiency of removing Zn. The water hyacinth ended the study with a removal efficiency of 44.7% at a solution pH of 10. Nonetheless, unlike solution pH 2, water hyacinth was able to survive better in solution pH 4 and 10 without immediate necrosis and chlorosis upon the beginning of the study. This finding is in accordance with the observation stated by [40], where water hyacinth was capable of surviving within the range of pH 4 to 10 by adjusting the solution pH to its requirements.

Meanwhile, water hyacinth did not experience necrosis and chlorosis when subjected to a solution pH of 8. As in the case of the natural pH solution, the Zn removal rate at a solution pH of 8 was in an overall increasing pattern throughout the study period, with a final Zn removal efficiency of 64%. The lower removal efficiency of Zn at solution pH 8 compared with the natural solution's pH could be ascribed to the reduced availability of Zn^{2+} ions for the plant's uptake, since these reacted with the OH^- ions present in the alkaline solution and subsequently generated the $Zn(OH)_2$ precipitate [39]. In short, both water lettuce and water hyacinth exhibited the highest removal efficiency of Zn at the natural solution pH of 6. As the ideal range for optimal growth of both plants was in between pH 6 and 7, and the pH of the permissible discharge of glove industry effluent was also between 6.5 and 7.5 [35], the natural solution pH of 6 was chosen as the optimum solution pH for the following study.

3.2.4. Effect of Salinity Concentration

Salinity is one of the most important environmental conditions and could significantly affect the metal phytoremediation performance of plants. According to [41], salinity stress could negatively affect the growth, respiration and photosynthesis mechanisms of a plant. In this context, sodium chloride was used as a source to alter the salinity of the solution. The effect of salinity concentrations (3, 6, 9, 12 and 15 ppm) on the removal efficiency of water lettuce and water hyacinth when removing Zn were studied, and the results are shown in Figures 12 and 13, respectively. The plants' efficiency when removing Zn was observed to be highly dependent on the salinity of the solution.

Based on Figure 12, the water lettuce's removal efficiency of Zn was the highest (80.1%) when the salinity concentration was at 6 ppm, while the lowest removal efficiency of Zn (5%) occurred at a salinity concentration of 15 ppm. According to [8], an increasing salinity concentration could lead to a direct and detrimental effect on the growth of a plant. This statement was clearly valid in conjunction with the results obtained from the present study. As the concentration of salinity was increased from 6 ppm to 15 ppm, a noticeable drop in removal efficiency of Zn could be observed. This was due to the excessive presence of sodium and chloride ions under high-salinity conditions, which could impair the growth and development of the plant through cytotoxicity. In addition, the high salinity concentration could also induce the generation of ROS, which leads to oxidative stress, distortion of genomic stability and damage of DNA [42]. The resulting damage imposed onto water lettuce at such high salinity concentrations (12 and 15 ppm) could be observed, as the removal efficiency of Zn started to reduce from its peak value on the second day of the study.

Nevertheless, the overall removal efficiency of Zn was in a constantly increasing pattern when the salinity concentration was increased from 3 to 9 ppm, which hinted at the possibility of achieving a higher removal efficiency with extended phytoremediation duration. In addition, the efficiency of removal of Zn also increased with a salinity concentration increase up to 9 ppm. The positive effects of low salinity concentrations on the water lettuce suggested that the cellular metabolisms and antioxidant activity, as well as a root-to-shoot translocation of nutrients, were not adversely affected [43]. In addition to that, the formation of Zn-chloride salt complexes in the presence of a tolerable amount of sodium chloride could potentially lead to a higher Zn removal rate by water lettuce. This was in accordance with the study conducted by [44], which suggested that the ability of accumulator plants to tolerate a low level of salinity stress could directly translate into an increased translocation rate of salt from the surroundings to the aerial plant parts.

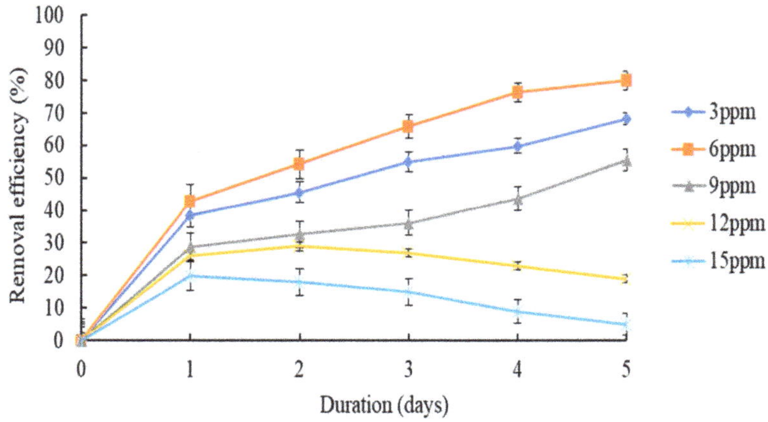

Figure 12. Effect of salinity concentrations on the removal efficiency of Zn by water lettuce (duration = 5 days, Zn concentration = 10 ppm and natural solution pH = pH 6).

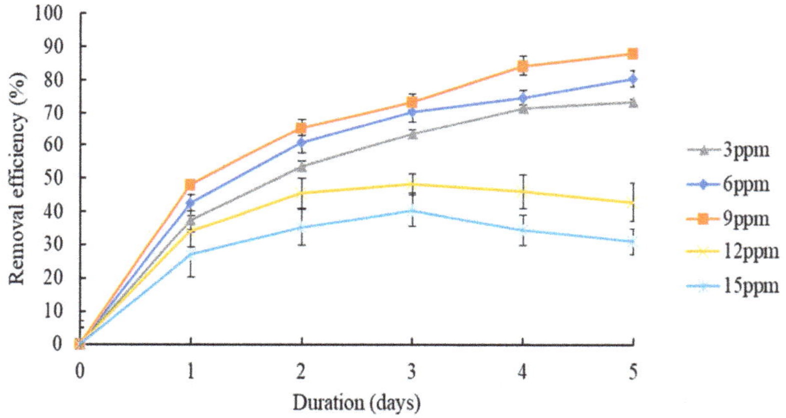

Figure 13. Effects of salinity concentrations on the removal efficiency of Zn by water hyacinth (duration = 5 days, Zn concentration = 10 ppm and natural solution pH = pH 6).

Based on Figure 13, the removal efficiency of Zn by water hyacinth was found to be the highest (88%) at a salinity concentration of 9 ppm, where a slightly higher tolerance towards salinity could be observed compared with water lettuce. On the other hand, identical to water lettuce, water hyacinth also displayed the lowest removal efficiency of Zn (31%) at a salinity concentration of 15 ppm. The overall trend of removal efficiency of Zn across different salinity concentrations for water hyacinth was also highly analogous to that of the water lettuce. Nevertheless, the most noticeable difference between these two plants could be observed at a salinity concentration of 9 ppm. Unlike the water lettuce, water hyacinth clearly showed its higher tolerance towards salinity in terms of the threshold limit for salinity concentration. When the salinity concentration was increased from 3 to 9 ppm, the reaction between positively charged Zn^{2+} ions and the salt-derived anions could form stable Zn-chloride ligands with enhanced mobility. As a result, when the ligands arrived at the root surfaces, Zn could then be absorbed by the plant roots via disassociation from the ligands [45]. However, as the salinity concentration was further increased up to 15 ppm, the effect of the enhanced Zn mobility was quickly outweighed by the negative effects caused by high salinity, such as the closure of stomata, lower transpiration rate, growth inhibition and oxidative damage caused by ROS generation [46]. Symptoms of necrosis and chlorosis could also be observed in water hyacinth at such high salinity concentrations, with the plant suffering from curling leaves, premature fall of leaves and discoloration of green young leaves. In conclusion, the optimum salinity concentration for water lettuce and water hyacinth was selected to be 6 and 9 ppm, respectively.

3.3. Possible Phytoremediation Strategies of Zn

The phytoremediation process of Zn-contaminated solution involves several mechanisms such as rhizofiltration, phytoextraction and phytostabilization, as shown in Figure 14. Rhizofiltration involves the absorption and sequestration of metal contaminants from water bodies. In this context, plants with a long and fibrous rhizosphere can potentially absorb a large amount of water, thereby allowing vast amounts of metal contaminants to be absorbed and adsorbed on the plant roots concurrently as well. According to [47], the rhizofiltration mechanism is more suitable than other available phytoremediation mechanisms when treating heavy metals from contaminated aqueous solutions. Aquatic plants such as water lettuce and water hyacinth are able to take up and translocate metals by utilizing their roots as active sites for adsorption on root surfaces or absorption into and sequestration within the roots prior to translocation to the aerial parts of the plant.

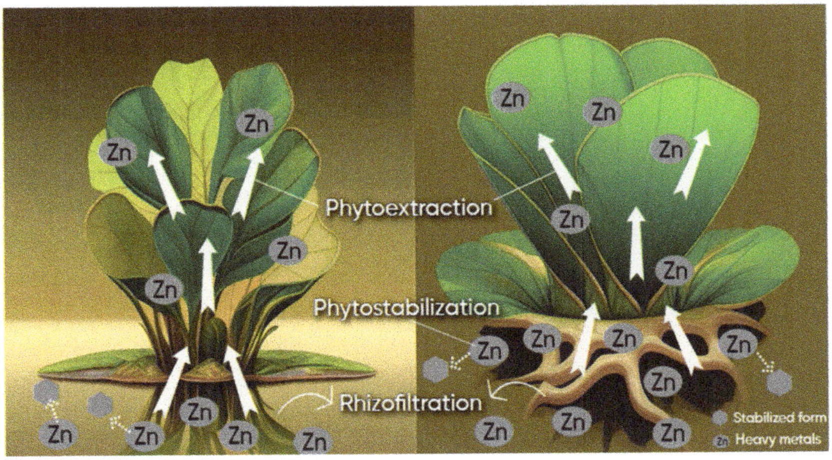

Figure 14. Mechanisms involved in the phytoremediation of Zn-contaminated solution by water hyacinth (**left**) and water lettuce (**right**).

Phytoextraction is defined as the process of accumulating heavy metals from the contaminated surrounding via the roots and transferring them towards the aboveground plant tissues, where the heavy metals are accumulated [48]. Due to the convenience of harvesting just the plant shoots, which are metal-enriched, phytoextraction by cultivating the plants directly on the contaminated sites is deemed the most commercially viable method of phytoremediation. According to [49], the phytoextraction mechanism requires the candidate plants to be tolerant towards metal-induced stress and capable of producing a high quantity of biomass. Apart from that, the in situ phytoextraction treatment also allows subsequent post-treatment handling methods such as composting, compaction and thermal decomposition to further minimize the risk of handling biomass with high levels of metal contaminants [50]. Plants can also establish phytostabilization, which is capable of constraining the movement of metal contaminants within the vadose zone through immobilization or root accumulation. Phytostabilization is technically a process used to stabilize the metal contaminants and reduce their mobility in order to prevent further intrusion into the food chain. In this context, [51] pointed out that both chemical (metal availability) and biological (metal plant uptake, ecotoxicological essays) tests should be carried out to monitor the metal contaminants if phytostabilization is chosen as the method for remediation.

3.4. Comparative Results and Literature Review

Table 2 lists several studies conducted by various researchers on the phytoremediation efficiency of various plants relating to different types of heavy metals. Based on Table 2, it can be noticed that most of the phytoremediation experiments were conducted with at least 30 days of duration. For instance, during the experiment conducted by Shehata et al. [52], both the Kenaf and Flax plants were cultivated for 8 weeks on the contaminated soil, consisting of several heavy metals such as Cr, Co, Cd and Mn, before harvesting. The phytoremediation efficiencies achieved by the Kenaf and Flax were between 12 and 36.5% and 9.4 and 45.2%, respectively. On the other hand, in another experiment conducted by Zhang et al. [53], Ryegrass, which is a hyperaccumulator plant, was incubated in sediment consisting of Cd, Pb and Zn at a temperature of 25 °C and a light/dark cycle of 12 h/12 h for a period of 30 days. After the incubation period, the phytoremediation efficiencies recorded were 47.8% for Cd, 37.2% for Pb and 42.5% for Zn. Meanwhile, Spiny pigweed was chosen by Njoku and Nwani [54] to be planted in pots containing heavy-metal-contaminated soil sourced from a local refuse dump site situated in Lagos. After 12 weeks, the plant was able to absorb 50.5, 49, 43.3 and 47.4% of the Cd, Cu, Pb and Zn, respectively. In the current work, both plants demonstrated highly efficient Zn removal rates, with water lettuce achieving 80.1% removal and water hyacinth reaching 88% removal within a remarkably short period of just 5 days.

Table 2. Phytoremediation efficiency of various plants for various heavy metals.

Plants Used	Targeted Heavy Metals	Phytoremediation Duration	Removal Efficiency	Reference
Hibiscus cannabinus L. (Kenaf) and *Linum usitatissimum* L. (Flax)	Cr, Co, Cd and Mn	8 weeks	Cr—34% Co—36.5% Mn—17.1% Cd—12% Mn—45.2% Cr—21.2% Co—17% Cd—9.4%	[52]
Clidemia sericea D. Don (Michelang)	Cd, Hg and Pb	12 weeks	Cd—49.2% Hg—18.4% Pb—32.3%	[55]
Lolium perenne L. (Ryegrass)	Cd, Pb and Zn	30 days	Cd—47.8% Pb—37.2% Zn—42.5%	[53]
Typha orientalis (Bulrush)	Cd, Cu, Ni and Pb	60 days	Cd—80% Cu—60% Ni—68.9% Pb— 8.6%	[56]
Cannabis sativa (Indian hemp)	Cu, Fe, Mn, Ni, Pb and Zn	60 days	Cu—75.9% Fe—88.6% Mn—70.8% Ni—78.7% Pb—83.9% Zn—39%	[57]
Amaranthus spinosus (Spiny pigweed)	Cd, Cu, Pb and Zn	12 weeks	Cd—50.5% Cu—49% Pb—43.3% Zn—47.4%	[54]
Pistia stratiotes (water lettuce) and *Eichhornia crassipes* (water hyacinth)	Zn	5 days	Zn—80.1% Zn—88%	Current work

4. Conclusions

This study demonstrated that both water lettuce and water hyacinth are capable of removing Zn contents via a phytoremediation process. SEM images revealed small granular aggregates on the surface of the plants as a result of the formation of stable metal-chelating complexes after phytoremediation. Meanwhile, EDX results proved the existence of Zn metal within the plants after the phytoremediation process, where the roots of both plants were observed to be saturated with Zn. On the other hand, FTIR spectra showed the shifting in bands for both plants after the phytoremediation process, which could be attributed to the interaction between the cationic Zn^{2+} and the hydroxyl group for the metal oxygen binding. The presence of several characteristic functional groups such as O-H, C-H and C=O bonds in both plants suggested the potential binding interaction between the functional groups and Zn metal ions, which could occur while keeping the chemical composition of both plants unchanged. In this context, the water hyacinth has a higher overall Zn content than the water lettuce, possibly due to its larger metal tolerance and phytoremediation capability. These characterization results could prove useful by shedding some light on the understanding of the underlaying complex mechanisms of the interactions between hyperaccumulator plants and heavy metals. By studying the changes in the morphological characteristics of both plants before and after phytoremediation, the phytoremediation performance can be enhanced via modification of the plants, such as through the transgenic method. Until now, to the best of our knowledge, transgenic plants with improved metal tolerance and phytoremediation efficiency have been based on the enhancement of the genes that are involved in the biosynthesis pathway of metal-binding proteins as well as the conversion of toxic ion into a less toxic state [58].

After conducting experimental studies on the phytoremediation of Zn by using both plants under different parameters, the optimum conditions for the highest removal efficiency (80.1% for water lettuce, 88% for water hyacinth) were achieved at a phytoremediation duration of 5 days, Zn concentration of 10 ppm and natural solution pH of 6, as well as salinity concentrations of 6 and 9 ppm for water lettuce and water hyacinth, respectively. Hence, both plants were proven to be strong contenders for phytoremediation of Zn in wastewater. Additionally, this information can be useful for enhancing the phytoremediation performance of plants, especially when selecting suitable hyperaccumulator plants for wastewater remediation applications. However, proper disposal and handling of plant biomasses after phytoremediation also require paying equal attention to fully maximizing the potential of heavy metal phytoremediation without compromising the environmental safety in the aftermath. While some studies have focused on post-phytoremediation plant biomasses, more efforts are still necessary to identify their potential applications. For example, phytomining, which involves recycling precious metals absorbed by hyperaccumulator plants, could offer insights into unlocking the largely untapped capabilities of hyperaccumulator plants and subsequently increasing their economic and scientific value.

Author Contributions: Data curation, methodology, writing—original draft, H.W.T.; resources, conceptualization, Y.L.P.; visualization, S.L.; supervision, W.C.C.; resources, C.W.L.; writing—review and editing, A.Z.A. All authors have read and agreed to the published version of the manuscript.

Funding: This research was supported in part by the Ministry of Education (MOE), Malaysia, which provided the Fundamental Research Grant Scheme (FRGS/1/2022/TK05/UTAR/02/34), the Kurita Asia Research Grant (21Pmy076) provided by Kurita Water and Environment Foundation and the Universiti Tunku Abdul Rahman (UTAR) Research Fund (IPSR/RMC/UTARRF/2020-C2/P01).

Institutional Review Board Statement: Not applicable.

Informed Consent Statement: Not applicable.

Data Availability Statement: Not applicable.

Conflicts of Interest: The authors declare no conflict of interest.

References

1. Wang, Z.; Luo, P.; Zha, X.; Xu, C.; Kang, S.; Zhou, M.; Nover, D.; Wang, Y. Overview Assessment of Risk Evaluation and Treatment Technologies for Heavy Metal Pollution of Water and Soil. *J. Clean. Prod.* **2022**, *379*, 134043. [CrossRef]
2. EPA Potential Well Water Contaminants and Their Impacts. Available online: https://www.epa.gov/privatewells/potential-well-water-contaminants-and-their-impacts#:~:text=Heavymetalscancontami-nate%20private,damage%2Canemia%2Candcancer (accessed on 18 June 2023).
3. Raj, A.R.A.; Mylsamy, P.; Sivasankar, V.; Kumar, B.S.; Omine, K.; Sunitha, T.G. Heavy Metal Pollution of River Water and Eco-Friendly Remediation Using Potent Microalgal Species. *Water Sci. Eng.* **2023**. [CrossRef]
4. Jyoti, D.; Sinha, R.; Faggio, C. Advances in Biological Methods for the Sequestration of Heavy Metals from Water Bodies: A Review. *Environ. Toxicol. Pharmacol.* **2022**, *94*, 103927. [CrossRef]
5. Barakat, M.A. New Trends in Removing Heavy Metals from Industrial Wastewater. *Arab. J. Chem.* **2019**, *4*, 361–377. [CrossRef]
6. Abdel-Raouf, M.S.; Abdul-Raheim, A.R.M. Removal of Heavy Metals from Industrial Waste Water by Biomass-Based Materials: A Review. *J. Pollut. Eff. Control* **2016**, *5*, 15–19. [CrossRef]
7. Yang, L.; Wang, J.; Yang, Y.; Li, S.; Wang, T.; Oleksak, P.; Chrienova, Z.; Wu, Q.; Nepovimova, E.; Zhang, X.; et al. Phytoremediation of Heavy Metal Pollution: Hotspots and Future Prospects. *Ecotoxicol. Environ. Saf.* **2022**, *234*, 113403. [CrossRef] [PubMed]
8. Omar, S.M.; Belal, M.; Safiur, R.; Moshiur, R.; Shafiuddin, A.; Ahmed, R.; Monjurul, H.; Xiao, J.B. Phytoremediation of Toxic Metals: A Sustainable Green Solution for Clean Environment. *Appl. Sci.* **2021**, *11*, 3–48.
9. Yan, A.; Wang, Y.; Tan, S.N.; Mohd Yusof, M.L.; Ghosh, S.; Chen, Z. Phytoremediation: A Promising Approach for Revegetation of Heavy Metal-Polluted Land. *Front. Plant Sci.* **2020**, *11*, 359. [CrossRef] [PubMed]
10. Kumar, V.; Singh, J.; Kumar, P. Heavy Metal Uptake by Water Lettuce (Pistia Stratiotes l.) from Paper Mill Effluent (Pme): Experimental and Prediction Modeling Studies. *Environ. Sci. Pollut. Res.* **2019**, *26*, 14400–14413. [CrossRef]
11. Sundarakumar, M.; Sankar, S.; Pandey, D.; Pandey, B.; George, A.; Karki, B.; Dadeech, P. An Approach to Improve the Water Quality on Industrial Effluent by Phytoremediation with Water Hyacinth (Eichhornia Crassipes). *Bioeng. Bioinf.* **2021**, *1*, 1–8. [CrossRef]
12. Zheng, J.; Liu, H.; Feng, H.; Li, W. Competitive Sorption of Heavy Metals by Water Hyacinth Roots. *Environ. Pollut.* **2016**, *219*, 837–845. [CrossRef]
13. Lamine, S.; Saunders, I. Retraction:Phytoremediation of Heavy-Metals-Contaminated Soils: A Short-Term Trial Involving Two Willow Species from Gloucester WillowBank in the UK. *Minerals* **2023**, *13*, 2023. [CrossRef]
14. Usman, K.; Al-Ghouti, M.A.; Abu-Dieyeh, M.H. The Assessment of Cadmium, Chromium, Copper, and Nickel Tolerance and Bioaccumulation by Shrub Plant Tetraena Qataranse. *Sci. Rep.* **2019**, *9*, 5658. [CrossRef]
15. Heredia-Guerrero, J.A.; Benítez, J.J.; Domínguez, E.; Bayer, I.S.; Cingolani, R.; Athanassiou, A.; Heredia, A. Infrared and Raman Spectroscopic Features of Plant Cuticles: A Review. *Front. Plant Sci.* **2014**, *5*, 305. [CrossRef]
16. Saralegui, A.B.; Willson, V.; Piol, M.N.; Boeykens, S.P. Macrophyte Biomass Productivity for Heavy Metal Adsorption. *J. Environ. Manage.* **2021**, *289*, 112398. [CrossRef]
17. EPA Secondary Drinking Water Standards: Guidance for Nuisance Chemicals. Available online: https://www.epa.gov/sdwa/secondary-drinking-water-standards-guidance-nuisance-chemicals (accessed on 1 September 2023).
18. DOE Environmental Quality (Industrial Effluent) Regulations 2009. Available online: https://www.doe.gov.my/wp-content/uploads/2021/08/Environmental_Quality_Industrial_Effluent_Regulations_2009_-_P.U.A_434-2009.pdf (accessed on 1 September 2023).
19. Ministry of Health Malaysia National Water Quality Standards for Malaysia. Available online: https://www.doe.gov.my/wp-content/uploads/2021/11/Standard-Kualiti-Air-Kebangsaan.pdf (accessed on 1 September 2023).
20. Romero-Guzmán, E.T.; Reyes-Gutiérrez, L.R.; Marín-Allende, M.J.; González-Acevedo, Z.I.; Olguín-Gutiérrez, M.T. Physicochemical Properties of Non-Living Water Hyacinth (*Eichhornia Crassipes*) and Lesser Duckweed (*Lemna Minor*) and Their Influence on the As(V) Adsorption Processes. *Chem. Ecol.* **2019**, *29*, 459–475. [CrossRef]
21. Zhou, J.M.; Jiang, Z.C.; Qin, X.Q.; Zhang, L.K.; Huang, Q.B.; Xu, G.L.; Dionysiou, D.D. Efficiency of Pb, Zn, Cd, and Mn Removal from Karst Water by Eichhornia Crassipes. *Int. J. Environ. Res. Public Health* **2020**, *17*, 5329. [CrossRef]
22. Rodrigues, A.C.D.; do Amaral Sobrinho, N.M.B.; dos Santos, F.S.; dos Santos, A.M.; Pereira, A.C.C.; Lima, E.S.A. Biosorption of Toxic Metals by Water Lettuce (Pistia Stratiotes) Biomass. *Water. Air. Soil Pollut.* **2017**, *228*, 156. [CrossRef]
23. Chieng, H.I.; Priyantha, N.; Linda, B.L. Effective Adsorption of Toxic Brilliant Green from Aqueous Solution Using Peat of Brunei Darussalam: Isotherm, Thermodynamics, Kinetics and Regeneration Studies. *RSC Adv.* **2015**, *1*, 34603–34615. [CrossRef]
24. Kaitlyn, E. The 6 Essential Nutrients For Healthy Plants. Available online: https://www.holganix.com/blog/the-6-essential-nutrients-for-healthy-plants (accessed on 28 August 2023).
25. Sukarni, S.; Zakaria, Y.; Sumarli, S.; Wulandari, R.; Ayu Permanasari, A.; Suhermanto, M. Physical and Chemical Properties of Water Hyacinth (Eichhornia Crassipes) as a Sustainable Biofuel Feedstock. *IOP Conf. Ser. Mater. Sci. Eng.* **2019**, *515*, 12070. [CrossRef]
26. Nayanathara, O.S.; Bindu, A. Effectiveness of Water Hyacinth and Water Lettuce for the Treatment of Greywater—A Review. *Int. J. Innov. Res. Sci. Eng.* **2019**, *3*, 349–355.

27. Rodrigues, A.C.D.; de Rocha, M.V.C.; Lima, E.S.A.; de Pinho, C.F.; dos Santos, A.M.; dos Santos, F.S.; Sobrinho, N.M.B.A. do Potential of Water Lettuce (*Pistia Stratiotes* L.) for Phytoremediation: Physiological Responses and Kinetics of Zinc Uptake. *Int. J. Phytoremediation* **2020**, *22*, 1019–1027. [CrossRef]
28. Gupta, P.; Roy, S.; Mahindrakar, A. Treatment of Water Using Water Hyacinth, Water Lettuce and Vetiver Grass—A Review. *Resour. Environ.* **2018**, *2*, 202–215. [CrossRef]
29. Mohamed, O.I.; Soeprobowati, T.R.; Sudarno, I. The Ability of Aquatic Weed Water Hyacinth and Water Lettuce To Reduce Heavy Metal Manganese (Mn), Zinc (Zn) and Iron (Fe). *J. Pure Appl. Chem.* **2015**, *1*, 101–113.
30. Jayasri, M.; Suthindhiran, K. Effect of Zinc and Lead on the Physiological and Biochemical Properties of Aquatic Plant Lemna Minor: Its Potential Role in Phytoremediation. *Appl. Water Sci.* **2017**, *7*, 1247–1253. [CrossRef]
31. Shingadgaon, S..; Chavan, B. Zinc Uptake Potential in Water Lettuce (*Pistia Stratiotes*, L.). *Int. J. Sci. Res.* **2018**, *7*, 1497–1504. [CrossRef]
32. Dutta, S.; Mitra, M.; Agarwal, P.; Mahapatra, K.; De, S.; Sett, U.; Roy, S. Oxidative and Genotoxic Damages in Plants in Response to Heavy Metal Stress and Maintenance of Genome Stability. *Plant Signal. Behav.* **2018**, *13*, e1460048. [CrossRef]
33. DalCorso, G.; Fasani, E.; Manara, A.; Visioli, G.; Furini, A. Heavy Metal Pollutions: State of the Art and Innovation in Phytoremediation. *Int. J. Mol. Sci.* **2019**, *20*, 3412. [CrossRef] [PubMed]
34. Mishra, V.K.; Tripathi, B.D. Accumulation of Chromium and Zinc from Aqueous Solutions Using Water Hyacinth (Eichhornia Crassipes). *J. Hazard. Mater.* **2019**, *164*, 1059–1063. [CrossRef]
35. Ministry of Environment and Water Malaysia Sewage and Industrial Effluent Discharge Standards. Available online: http://www.water-treatment.com.cn/resources/discharge-standards/malaysia.htm (accessed on 4 August 2021).
36. Nasir, M.; Nur, M.; Pandiangan, D.; Mambu, S.M.; Fauziah, S.; Raya, I.; Fudholi, A.; Irfandi, R. Phytoremediation Study of Water Hyacinth (E Ichhornia Crassipes) on Zinc Metal Ion (Zn2+). *Int. J. Des. Nat. Ecodynamics* **2022**, *17*, 417–422. [CrossRef]
37. Swarnalatha, K.; Radhakrishnan, B. Studies on Removal of Zinc and Chromium from Aqueous Solutions Using Water Hyacinth. *Pollution* **2015**, *1*, 193–202.
38. Emilia, G.; Andrada, M.; Petru, I. Removal of Zinc Ions as Zinc Chloride Complexes from Strongly Acidic Aqueous Solutions by Ionic Exchange. *Cent. Eur. J. Chem.* **2014**, *12*, 821–828. [CrossRef]
39. Sadeghzadeh, B. A Review of Zinc Nutrition and Plant Breeding. *J. Soil Sci. Plant Nutr.* **2019**, *13*, 905–927. [CrossRef]
40. Sena, P.H.; Martin, C.; Albant, S.; Edmond, A.; Martin, P.A. Influence of PH on Water Hyacinth Ponds Treating and Recycling Wastewater. *J. Water Resour. Prot.* **2022**, *14*, 86–99.
41. Wang, X.; Sun, R.; Tian, Y.; Guo, K.; Sun, H.; Liu, X.; Chu, H.; Liua, B. Long-Term Phytoremediation of Coastal Saline Soil Reveals Plant Species-Specific Patterns of Microbial Community Recruitment. *mSystems* **2020**, *5*, 10-1128. [CrossRef]
42. Liu, J.; Fu, C.; Li, G.; Nauman Khan, M.; Wu, H. Ros Homeostasis and Plant Salt Tolerance: Plant Nanobiotechnology Updates. *Sustainability* **2021**, *13*, 3552. [CrossRef]
43. Hongqiao, L.; Suyama, A.; Mitani-ueno, N.; Hell, R. A Low Level of NaCl Stimulates Plant Growth by Improving Carbon and Sulfur Assimilation in Arabidopsis Thaliana. *Plants* **2021**, *10*, 2138. [CrossRef]
44. Litalien, A.A.S.; Rutter, A.; Zeeb, B.A.; Litalien, A.A.S.; Rutter, A.; The, B.A.Z. The Impact of Soil Chloride Concentration and Salt Type on the Excretions of Four Recretohalophytes with Different Excretion Mechanisms. *Int. J. Phytoremediation* **2020**, *22*, 1122–1128. [CrossRef]
45. Filipović, L.; Romić, M.; Romić, D.; Filipović, V.; Ondrašek, G. Organic Matter and Salinity Modify Cadmium Soil (Phyto)Availability. *Ecotoxicol. Environ. Saf.* **2018**, *147*, 824–831. [CrossRef]
46. Tandon, K.; John, M.; Heuss-Aßbichler, S.; Schaller, V. Influence of Salinity and Pb on the Precipitation of Zn in a Model System. *Minerals* **2018**, *8*, 43. [CrossRef]
47. Woraharn, S.; Meeinkuirt, W.; Phusantisampan, T.; Chayapan, P. Rhizofiltration of Cadmium and Zinc in Hydroponic Systems Rhizofiltration of Cadmium and Zinc in Hydroponic Systems. *Water Air Soil Pollut.* **2021**, *232*, 204. [CrossRef]
48. Yanitch, A.; Kadri, H.; Frenette-dussault, C.; Joly, S.; Pitre, F.E.; Labrecque, M. A Four-Year Phytoremediation Trial to Decontaminate Soil Polluted by Wood Preservatives: Phytoextraction of Arsenic, Chromium, Copper, Dioxins and Furans. *Int. J. Phytoremediation* **2020**, *22*, 1505–1514. [CrossRef]
49. Suman, J.; Uhlik, O.; Viktorova, J.; Macek, T. Phytoextraction of Heavy Metals: A Promising Tool for Clean-up of Polluted Environment. *Front. Plant Sci.* **2018**, *9*, 1476. [CrossRef]
50. Sharma, J.; Kumar, N.; Singh, N.; Santal, A. Phytoremediation Technologies and Their Mechanism for Removal of Heavy Metal from Contaminated Soil: An Approach for a Sustainable Environment. *Front. Plant Sci.* **2023**, *14*, 1076876. [CrossRef]
51. Nascimento, C.; Biondi, C.M.; da Silva, F.B.V.; Lima, L.H.V. Using Plants to Remediate or Manage Metal-Polluted Soils: An Overview on the Current State of Phytotechnologies. *Agronomy* **2021**, *43*. [CrossRef]
52. Shehata, S.M.; Badawy, R.K.; Aboulsoud, Y.I.E. Phytoremediation of Some Heavy Metals in Contaminated Soil. *Bull. Natl. Res. Cent.* **2019**, *43*, 1–15. [CrossRef]
53. Zhang, Y.; Gong, J.; Cao, W.; Qin, M.; Song, B. Influence of Biochar and Fulvic Acid on the Ryegrass-Based Phytoremediation of Sediments Contaminated with Multiple Heavy Metals. *J. Environ. Chem. Eng.* **2023**, *11*, 109446. [CrossRef]
54. Njoku, K.L.; Nwani, S.O. Phytoremediation of Heavy Metals Contaminated Soil Samples Obtained from Mechanic Workshop and Dumpsite Using Amaranthus Spinosus. *Sci. African* **2022**, *17*, e01278. [CrossRef]

55. Durante-Yánez, E.V.; Martínez-Macea, M.A.; Enamorado-Montes, G.; Caballero, E.C.; Marrugo-Negrete, J. Phytoremediation of Soils Contaminated with Heavy Metals from Gold Mining Activities Using Clidemia Sericea D. Don. *Plants* **2022**, *11*, 597. [CrossRef]
56. Zhang, Y.; Liu, G.; Gao, S.; Zhang, Z.; Huang, L. Effect of Humic Acid on Phytoremediation of Heavy Metal Contaminated Sediment. *J. Hazard. Mater. Adv.* **2023**, *9*, 100235. [CrossRef]
57. Singh, K.; Tripathi, S.; Chandra, R. Bacterial Assisted Phytoremediation of Heavy Metals and Organic Pollutants by Cannabis Sativa as Accumulator Plants Growing on Distillery Sludge for Ecorestoration of Polluted Site. *J. Environ. Manage.* **2023**, *332*, 117294. [CrossRef] [PubMed]
58. Li, Z.; Tian, Y.; Wang, B.; Peng, R.; Xu, J.; Fu, X.; Han, H.; Wang, L.; Zhang, W.; Deng, Y.; et al. Enhanced Phytoremediation of Selenium Using Genetically Engineered Rice Plants. *J. Plant Physiol.* **2022**, *271*, 153665. [CrossRef] [PubMed]

sustainability

MDPI

Article

Investigation and Optimization of Co-Combustion Efficiency of Food Waste Biochar and Coal

Yoonah Jeong [1,†], Jae-Sung Kim [2,†], Ye-Eun Lee [1], Dong-Chul Shin [1,3], Kwang-Ho Ahn [1], Jinhong Jung [1], Kyeong-Ho Kim [2], Min-Jong Ku [2], Seung-Mo Kim [4], Chung-Hwan Jeon [2,4,*] and I-Tae Kim [1,*]

1 Department of Environmental Research, Korea Institute of Civil Engineering and Building Technology, Goyang-daero 283, Ilsanseo-gu, Goyang-si 10223, Republic of Korea; yoonahjeong@kict.re.kr (Y.J.); yeeunlee@kict.re.kr (Y.-E.L.); dcshin@daejin.ac.kr (D.-C.S.); khahn@kict.re.kr (K.-H.A.); jinhong98@kict.re.kr (J.J.)

2 School of Mechanical Engineering, Pusan National University, 2, Busandaehak-ro 63 beon-gil, Geumjeong-gu, Busan 46241, Republic of Korea; e1223kjs@pusan.ac.kr (J.-S.K.); kkh751@pusan.ac.kr (K.-H.K.); tbvjalswhd@pusan.ac.kr (M.-J.K.)

3 Department of Smart Construction and Environmental Engineering, Daejin University, 1007 Hoguk-ro, Pocheon-si 11159, Republic of Korea

4 Pusan Clean Energy Research Institute, Pusan National University, 2, Busandaehak-ro 63 beon-gil, Geumjeong-gu, Busan 46241, Republic of Korea; kimsm@pusan.ac.kr

* Correspondence: chjeon@pusan.ac.kr (C.-H.J.); itkim@kict.re.kr (I.-T.K.)

† These authors contributed equally to this work.

Abstract: Among the alternative recycling methods for food waste, its utilization as a renewable biomass resource has demonstrated great potential. This study presents empirical findings pertaining to the cofiring of solid biomass fuel and coal for power generation. Various co-combustion ratios involving food waste biochar (FWB) and coal (100:0, 85:15, 90:10, 95:5, and 0:100) were tested to optimize combustion efficiency, monitor the emissions of NO_X, CO, and unburned carbon (UBC), assess ash deposition tendencies, and evaluate grindability. Two types of FWB and sewage sludge were selected as biomass fuels. The results demonstrated that co-combustion involving FWB reduced NO_X and UBC emissions compared to coal combustion alone. In particular, the 10% FWB_A blend exhibited the best combustion efficiency. Notably, FWB demonstrated lower tendencies for ash deposition. The ash fusion characteristics were monitored via thermomechanical analysis (TMA), and the corresponding shrinkage levels were measured. Furthermore, FWB exhibited superior grindability compared to both coal and sewage sludge, reducing power consumption during fuel preparation. This study suggests that FWB is a valuable co-combustion resource in coal-fired power plants, thereby facilitating the efficient recycling of food waste while concurrently advancing clean energy generation. Nevertheless, further research is required to validate its practical applicability and promote its use as a renewable resource.

Keywords: biomass; food waste biochar; co-combustion (co-pyrolysis); thermomechanical analysis; drop tube furnace; slagging; fouling; grindability

Citation: Jeong, Y.; Kim, J.-S.; Lee, Y.-E.; Shin, D.-C.; Ahn, K.-H.; Jung, J.; Kim, K.-H.; Ku, M.-J.; Kim, S.-M.; Jeon, C.-H.; et al. Investigation and Optimization of Co-Combustion Efficiency of Food Waste Biochar and Coal. *Sustainability* **2023**, *15*, 14596. https://doi.org/10.3390/su151914596

Academic Editors: Steven Lim, Shuit Siew Hoong, Santi Chuetor and Pang Yean Ling

Received: 7 September 2023
Revised: 2 October 2023
Accepted: 6 October 2023
Published: 8 October 2023

1. Introduction

The combustion of biomass fuels for power generation has garnered considerable attention because of the carbon-neutral nature of biomass. The benefits of cofiring biomass, including reduced greenhouse gas emissions, low cost, broad availability, and simple application in plant facilities, have promoted its implementation in power generation [1,2]. Although various types of biomass fuels, such as sewage sludge, have been employed [3,4], this study focused on food waste as a renewable biomass resource. The recycling rate of food waste reached 96% in the Republic of Korea in 2019 [5]. However, because of limitations in the recycling approaches, including feedstock, fertilization, and bio-gasification, the development of alternative recycling methods targeting food waste is urgently needed. As an alternative recycling method, food waste is converted into food waste biochar (FWB) via pyrolysis and demineralization. FWB, a solid product of pyrolyzed biomass, has shown potential as a fuel owing to its high calorific value, high recycling rate in Korea, and low chlorine and heavy metal contents [6,7].

The combustion of biomass may cause several issues in facility operations because of the various combustion behaviors of different fuels and difficulties in their preparation. In particular, ash fusion characteristics during the cofiring of biomass and coal have been reported [8]. High inorganic contents such as alkali and alkali earth metals in biomass ash lead to slagging, fouling, and corrosion [9–11]. The operating temperature of a biomass-fired boiler decreases because of the attached biomass ash on its surface. Such inorganic ash compositions in biomass eventually lower the combustion efficiency and cause maintenance issues. Accordingly, when biomass and coal are cofired, ash fusion tests of the biomass by thermomechanical analysis (TMA) must be conducted to predict key technical problems, such as slagging tendencies [12,13]. For example, the shrinkage of biomass fuels with increasing temperatures can be examined using TMA.

Another undesirable characteristic of biomass is its poor grindability [14,15]. A grinding procedure is required for the preparation of fuel to maintain the homogeneous size of both the biomass and coal before combustion. Fuel particle size is a key factor in efficient combustion. In general, grindability is a critical problem in biomass cofiring because of the high grindability index resulting from the fibrous nature of the biomass. Biomass with poor grindability undergoes an additional fragmentation process, which requires more time and cost.

This study set out to assess the feasibility of cofiring FWB and coal. To substantiate the sustainability and environmental friendliness of cofiring FWB, this study investigates the co-combustion of FWB and coal while monitoring the combustibility, ash deposition characteristics, ash fusion tendencies, and FWB grindability. Various mixing ratios between the FWB and coal were tested to determine the optimal operating conditions and maximize combustion efficiency. Considering that most biomass is cofired at approximately 3–5%, mixing ratios of 0%, 5%, 10%, 15%, and 100% FWB were used in this study. Sewage sludge, another biomass fuel that is currently used in coal-fired power plants, was selected to compare its overall co-combustion characteristics with those of FWB. A drop tube furnace (DTF) was used to investigate NO_X, CO, CO_2, and unburned carbon (UBC) emissions from cofiring. The ash deposition and fusion characteristics were also experimentally studied. Because ash fusion characteristics are difficult to estimate from a single TMA, TMA was conducted for all mixing ratios to observe changes in ash fusion trends with increasing cofiring FWB ratio. To the best of our knowledge, this represents the first endeavor to undertake an exhaustive analysis of the cofiring of FWB and to optimize co-combustion ratios between FWB and coal. The authors believe that the results of this study will encourage the use of FWB as a cofiring biomass fuel in coal-fired power plants.

2. Materials and Methods

2.1. Preparation of Solid Fuel

This study used two types of FWBs and sewage sludge as biomass. Food waste was converted into FWB_A and FWB_B via pyrolysis and demineralization. The food waste procured from the Gimpo Urban Management Corporation (Gimpo-si, Republic of Korea) was shredded and dehydrated. FWB_A was produced through the pyrolysis of dried food waste at 500 °C for 20 min. Pyrolyzed FWB was then demineralized using acidified water (i.e., 3% citric acid) to reduce the chlorine content of the FWB, which was referred to as FWB_A. This demineralization step was imperative due to the potential emission of dioxin and dioxin-like substances associated with the high chlorine content in FWB [16]. The detailed preparation process has been described in a previous study [17]. FWB_B was produced following the same protocol, albeit employing a slightly different demineralization method involving deionized water. Before the experiment, FWB_A and

FWB_B were sieved through a 2 mm mesh. Subsequent to the preparation procedures, the yield of FWB ranged between 43.2% and 53.8%.

The sludge and coal were provided by Korea Midland Power. The preparation of all samples involved an initial milling process followed by sieving to achieve a particle size distribution within the range of 75 μm to 125 μm. A sieve shaker (AS 200, Retsch GmbH, Haan, Germany) was employed for this purpose. Subsequent to the size reduction, the samples were subjected to air drying based on the air-dried moisture content. The experimental settings used in this study for biomass and coal are listed in Table 1.

Table 1. Experimental ratios of the coal and biomass fuels.

	Combustion		Co-Combustion				
Coal	100%	0%	95%	90%	85%	85%	85%
FWB_A	0%	100%	5%	10%	15%	0%	0%
FWB_B	0%	0%	0%	0%	0%	15%	0%
Sewage sludge	0%	0%	0%	0%	0%	0%	15%

FWB: food waste biochar.

2.2. Characterization of Biomass and Coal

The fuel properties of biomass and coal were investigated via proximate analyses, heating values, and elemental analyses. A thermogravimetric analyzer (TGA 701, LECO Co., St. Joseph, MI, USA) was used for the proximate analysis based on the American Society for Testing and Materials (ASTM) guidelines: ASTM D3173-11 for analytical moisture, ASTM D3175-11 for volatile matter, ASTM D3174-12 for ash, and ASTM D3172-13 for fixed carbon [18–21]. The net calorific value was measured using a bomb calorimeter (5E-C5508, CKIC, Changsha, China) according to ASTM D5865 [22]. The ash composition was determined using X-ray Fluorescence (XRF) following the German standard (DIN 51729-10) [23].

2.3. Combustion Test Using DTF

In this experiment, a DTF was used to investigate the combustion emission characteristics of NO_X, CO, CO_2, and UBC. A DTF has the advantage of maintaining a uniform temperature and allowing for experiments over a wide range of temperatures. A schematic representation of the experimental apparatus used in this study is shown in Figure 1 [24]. The experimental unit consists of a feeding section for injecting a mixture of pulverized coal and biomass, a reaction section where the combustion reaction occurs, and a collection section for capturing the combustion gases. The feeding part consists of injecting fuel and reaction gas, which are divided into carrier gas and primary reaction gas streams. The flow rate of the main reaction gas, which was directly injected into the reactor, was controlled using a mass flow controller. The fuel and reactive gases meet, and chemical reactions occur in the reaction section. A SiC heater is used to maintain a constant temperature, which can reach up to 1500 °C.

To minimize experimental errors, each experimental condition underwent three iterations. A real-time gas analyzer was used to measure the composition of the exhaust gas at the rear of the DTF once per second. UBC was measured using the ash tracer method [25]. This approach was necessitated by the limited amount of char or ash available for direct comparison with the amount of fuel. The method operates on the assumption of constancy in the ash content of the fuel before and after combustion to determine the UBC of the fuel by comparing its ash content before and after combustion using a proportional formula. The formula used is as follows:

$$\text{UBC }(\%) = [1 - \frac{A_0(100 - A_{char})}{A_{char}(100 - A_0)}] \times 100 \quad (1)$$

where A_0 is the ash content of the raw coal and A_{char} is the ash content of the char captured after the reaction.

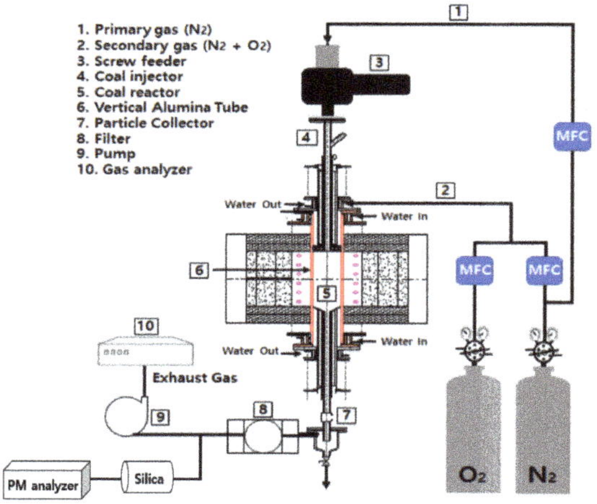

Figure 1. Schematic representation of the experimental drop tube furnace setup.

2.4. Ash Deposition Test Using DTF

An ash deposition experiment was performed using a probe placed at the bottom of the DTF. Ten grams of biomass were supplied at a feeding rate of 0.2 g/min at 1300 °C. Ash deposition was measured using capture efficiency (CE) and energy-based growth rate (GRE) as indicators [26,27]. The CE was obtained by dividing the deposited mass by the mass of ash particles flowing along the surface area of the probe during the experiment (Equation (2)).

$$CE = \frac{m_{Dep}}{m_{Fuel}\left(\frac{A^P}{100}\right)\left(\frac{A_{coupon}}{A_{reactor}}\right)} \times 100\ [\%kg_{Dep}/kg_{Ash}] \quad (2)$$

where m_{Dep} and m_{Fuel} represent the mass (kg) of ash deposited on the coupon surface and fuel supplied during the experiment, respectively. A^P, A_{coupon}, and $A_{reactor}$ are the ash content (wet basis) in the fuel, the projected area of the coupon, and the cross-sectional area of the DTF, respectively, where A denotes the amount of ash in the fuel. While the CE helps in understanding the tendency of the adherent ash fraction, it may inadvertently neglect the influence of the mineral components within the solid fuel and the correlation between the solid fuel and its calorific value. To surmount this limitation, the GRE was calculated as follows (Equation (3)):

$$GRE = \frac{m_{Dep}}{LHV \cdot m_{Fuel}}\ [\%g_{Dep}/MJ] \quad (3)$$

The GRE is an indicator of the amount of ash deposition relative to the total energy supplied to the solid fuel (mass unit: g). *LHV* represents the low heating value (MJ). The characteristics of ash from solid fuels and the ash deposition rate based on energy can be estimated using GRE.

2.5. Ash Fusion Behavior Test Using TMA

Melting experiments were conducted using TMA to analyze the melting tendencies of ash particles in biomass fuel [28]. TMA was designed and fabricated at the Pusan Clean Energy Institute at Pusan National University. The measurement method involved placing a graphite crucible containing the sample on an alumina holder and positioning an alumina rod on top of the sample. The crucible was then heated under a nitrogen atmosphere up to 1600 °C, whereas the shrinkage rate of the sample was measured using a displacement gauge. The shrinkage rate refers to the value obtained by dividing the shrinkage at each temperature by the total shrinkage displacement. The experimental results are denoted as TX%, where X% signifies the temperature at which the shrinkage rate reaches X%. The schematic diagram of the TMA apparatus is presented in Figure 2.

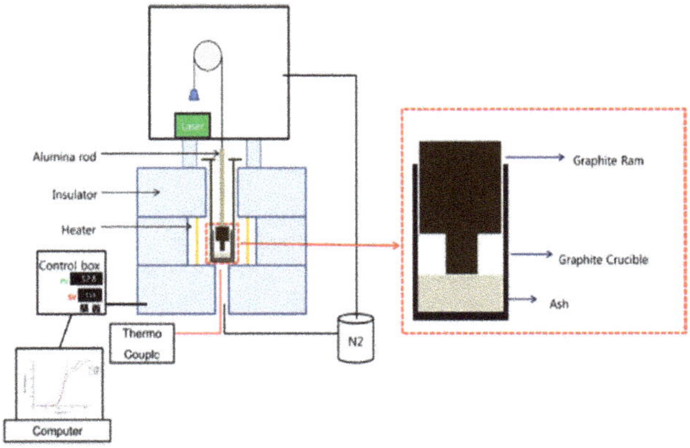

Figure 2. Schematic representation of TMA.

2.6. Evaluation of Grindability

The grindability of the solid fuel was measured using the Hardgrove Grindability Index (HGI), following the guidelines outlined in the Japanese Industrial Standard M8801 [28]. Over 300 g of the solid fuel underwent grinding within the particle size range of 0.6 mm to 1.18 mm. Pulverized solid fuels (50 g) were placed in a grinding bowl under a pressure of 284 N. The grinding bowl was operated at both 20 rpm and 60 rpm. Solid fuels sieved through a 75 μm sieve were recorded to calculate the HGI, which was further modified into the Thermally Treated Biomass Grindability Index (TTBGI) [29,30]. In this study, the grindability of biomass samples was measured via the TTBGI method and leveraging HGI (5E-HA60 × 50, Changsha Kaiyuan Instruments Co., Changsha, China) equipment.

3. Results and Discussion

3.1. Characterization of Biomass and Coal

The proximate and ultimate analyses of the solid fuels are presented in Table 2. FWB_A exhibited the highest fixed carbon content, with a subsequent decrease observed in the following order: coal > FWB_B > sewage sludge. The low fixed carbon content in sewage sludge can be attributed, in part, to its predominant volatile matter, which surpasses that of other solid fuels by approximately twofold. From the ultimate analysis, it becomes evident that FWB_A, FWB_B, and the sludge contain notably elevated N levels (>7%) compared with coal. Furthermore, both FWB variants displayed lower sulfur (S) contents than those of coal and sewage sludge. Intriguingly, FWB_A and coal exhibited similar calorific values, whereas the lowest calorific value observed in the sludge can be attributed to its high oxygen and low carbon contents.

Table 2. Proximate and ultimate analyses of solid fuels.

	Coal	FWB_A	FWB_B	Sewage Sludge
Proximate analysis				
Moisture (%)	2.24	3.49	3.56	5.07
Volatile matter (%)	25.59	30.92	30.93	61.05
Fixed C (%)	51.90	54.84	46.45	9.58
Ash (%)	20.26	10.75	19.07	24.30
Ultimate analysis				
C (%)	78.73	73.43	69.97	48.18
H (%)	5.16	4.35	4.22	7.21
N (%)	4.26	7.27	7.28	7.83
S (%)	0.27	0.04	0.11	0.84
O (%)	11.57	14.91	18.41	35.93
Net calorific value (kcal/kg)	6244	6167	5286	4084

3.2. Gaseous Emissions from the Combustion of Solid Fuels

The gaseous emissions, including NO_X, CO, CO_2, and UBC, from the combustion of solid fuels are shown in Figure 3. As depicted in the figure, all co-combustion cases involving FWB_A and FWB_B exhibited reduced NO_X emissions compared to the combustion of coal or FWB_A alone. The NO_X concentrations ranged from 96.38 ppm (5% FWB_A) to 136.08 ppm (100% coal). Generally, the incorporation of biomass in cofiring significantly mitigated both NO_X and SO_X emissions [31,32]. The combustion of 100% FWB_A yielded the highest NO_X emission, likely attributed to the good combustibility of FWB_A due to its elevated N content and low UBC value. The NO_X emission exhibited a linear increase with increasing FWB_A cofiring ratio. Given the consistency in furnace configuration throughout the testing, discrepancies in NO_X levels likely arise from differences in volatile matter and fuel N contents.

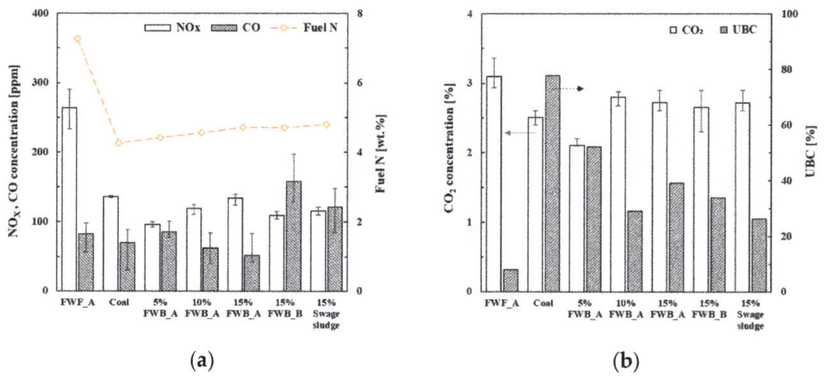

Figure 3. (**a**) Concentrations of NO_X (blank bar), CO (gray bar), and fuel N (yellow dot line). (**b**) CO_2 concentrations (blank bar) and unburned carbon (UBC, gray bar).

CO emissions can be used as an indicator of combustion efficiency [33,34]. Although CO emission levels depend on various factors, such as fuel heterogeneity, size, and distribution, homogeneous fuel generally leads to improved combustion efficiency [35]. For FWB_A, an increased mixing rate corresponded to a reduction in CO concentration. Specifically, CO concentrations for 5% FWB_A, 10% FWB_A, and 15% FWB_A were recorded as 85.23, 62.67, and 51.47 ppm, respectively. Notably, when juxtaposed with the combustion of coal or FWB_A alone, cofiring ratios of 10% FWB_A and 15% FWB_A demonstrated improved combustion efficiencies, as evidenced by their CO concentrations of 70.19 ppm and 83.06 ppm, respectively. Interestingly, the co-combustion of 15% FWB_B and 15% sludge exhibited the highest CO emissions, underscoring the superior combustion efficiency of FWB_A compared to FWB_B and sludge.

The CO_2 emissions (2.10% to 3.09%) remained consistent across all solid fuels. The UBC levels were calculated by measuring the carbon content within the ash after combustion (Equation (1)). In particular, the UBC concentration for 10% FWB_A was lower than that of the other co-combustion ratios. Elevated UBC and CO emissions constitute the primary factors contributing to reduced combustion efficiency [36]. Consequently, the gaseous emission trends suggest that co-combustion with 10% FWB_A was the most effective in achieving minimal NO_X, CO, and UBC emissions.

3.3. Ash Deposition Rate

Reactor fouling and corrosion represent typical challenges inherent to biomass combustion [11,36,37]. Ash deposited on the surface of the reactor functions as an insulating layer, which reduces the heat transfer efficiency. Accordingly, a high ash deposition rate may result in heightened fuel consumption and a reduction in thermal efficiency. Furthermore, excessive ash deposition affects the operation and maintenance of combustion systems. The accumulated ash may cause flame instability and disturb the flow of gases and the combustion process, necessitating more frequent maintenance intervals to uphold the combustion system's dependable operation. Therefore, a DTF deposition experiment was performed to compare the ash deposition for each biomass sample. Figure 4 shows the deposition characteristics of each biomass sample after combustion, obtained from the results of the DTF deposition experiment.

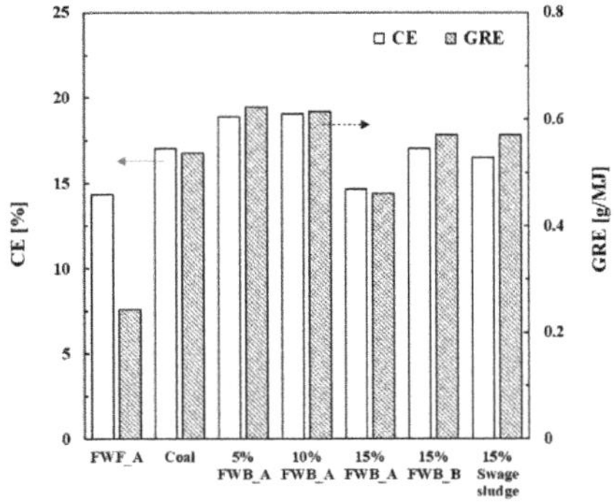

Figure 4. Comparison of capture efficiency (CE, blank bar) and energy-based growth rate (GRE, gray bar) for biomass fuels and coal.

In the context of combustion of coal alone, the CE for coal approached approximately 17%, whereas the CE for 100% FWB_A was 14%. It is noteworthy that the FWBs exhibited relatively lower ash deposition tendencies, as assessed through both the GRE and CE indices. The CE and GRE values for FWB, as determined in this study, are comparably lower than those reported for other biomass [17]. Low ash deposition on the surface of the reactor indicates enhanced heat transfer and thermal efficiency in the combustion process. Additionally, the operational challenges arising from ash deposition constitute a noteworthy consideration in cofiring practices. The co-combustion of biomass and coal can give rise to excessive ash deposition, necessitating heightened maintenance procedures and downtime during combustion. In this regard, FWB presents advantages over other wood and herbaceous biomass.

In the case of co-combustion scenarios involving 15% FWB_A, 15% FWB_B, and 15% sewage sludge, ash CE values were 14.64%, 17.03%, and 16.50%, respectively. Additionally, the ash deposition profiles for 5% FWB_A and 10% FWB_A exhibited similar CE values of 18.86% and 19.05%, respectively. Previous studies have shown that biomass combustion usually yields higher ash deposition rates than those yielded via coal combustion [13]. Our findings in this study align with this pattern, showcasing elevated CE and GRE values for biomass combustion. The smallest accumulated ash mass was observed in the case of FWB_A (100%, 0.062 g). When evaluated in terms of CE and GRE indices, FWB_A at 15% and 100% displayed reduced ash deposition tendencies. It is essential to underline the substantial disparity between GRE and CE in FWB_A combustion, potentially attributable to the distinct definitions of GRE and CE. CE was derived solely from the accumulated ash content without considering the calorific value of the solid fuel, whereas GRE was defined as the deposited ash per calorific value. For instance, complete coal combustion resulted in approximately 0.54 g/MJ, whereas FWB_A yielded 0.24 g/MJ. Co-combustion of 15% FWB_A and coal generated 0.46 g/MJ, while 15% FWB_B and sludge produced 0.57 g/MJ (Figure 4). Furthermore, ash depositions were recorded at 0.62 g/MJ, 0.61 g/MJ, and 0.46 g/MJ for 5% FWB_A, 10% FWB_A, and 15% FWB_A, respectively. These results show that low ash deposition occurred when employing a 15% FWB_A cofiring ratio. Therefore, the combustion of FWB_A appears to be advantageous in terms of ash deposition mitigation.

3.4. TMA Characteristics

The fusion or melting behavior of biomass at elevated temperatures during combustion exerts a direct influence on combustion efficiency. During biomass combustion, inorganic components such as ash and volatile matter undergo thermal transformations. When the ash fusion temperature is low, the ash particles tend to melt, forming sticky deposits on the surface of the combustion reactor. This slag-like residue can disrupt airflow and hinder the overall combustion efficiency. Furthermore, local oxygen-deficient zones can be formed by these slaggy ash deposits, contributing to incomplete combustion of solid fuel and increased emissions of pollutants such as CO, SO_2, particulate matter, and volatile organic compounds [38,39]. Thus, understanding and controlling the ash fusion phenomenon hold paramount importance in reducing pollutant emissions and optimizing combustion conditions. To this end, we conducted an ash fusion experiment using TMA to assess the ash fusion tendencies of biomass fuels.

As shown in Figure 5, both coal and FWBs exhibit secondary to tertiary fusion points rather than a singular, distinctive fusion point. This indicates the heterogeneity and nonuniformity of ash derived from biomass combustion. Coal displayed a primary fusion range between 800 °C and 1100 °C, followed by a rapid secondary fusion between 1400 °C and 1500 °C, as evident in the TMA results (Figure 5). Specific temperature points T25, T50, T75, and T90 for coal were recorded at 1095 °C, 1454 °C, 1494 °C, and 1512 °C, respectively. Conversely, FWB_A exhibited a more intricate fusion behavior than that of general biomass, manifesting at three different fusion rates. Its primary fusion was between 800 °C and 900 °C, followed by a gradual secondary fusion between 1000 °C and 1400 °C, and a tertiary fusion between 1400 °C and 1500 °C. For FWB_A, the temperature values for T25, T50, T75, and T90 were measured at 926 °C, 1258 °C, 1458 °C, and 1504 °C, respectively, depicting fusion characteristics relatively akin to those of coal. FWB_B exhibited a solitary, distinct fusion point occurring between 1400 °C and 1500 °C. The corresponding temperature values for T25, T50, T75, and T90 were 1431 °C, 1484 °C, 1516 °C, and 1532 °C, respectively. In the case of FWB_B, the fusion of the entire ash component occurred at a relatively high temperature, and single fusion occurred at a temperature higher than that of coal. Both FWBs exhibited elevated fusibility levels akin to those of coal, suggesting compatibility in ash fusion temperatures for FWB and coal and implies that cofiring of these fuels can be more efficient and environmentally friendly. Sewage sludge has a lower fusion point than that of coal and FWB, consistent with the general characteristics of biomass fuels [40,41]. The fusion points of sewage sludge were determined to be 910 °C, 951 °C, 977 °C, and 993 °C, which all fall below 1000 °C. In a previous study, the high metal content in sewage sludge was identified as the main factor responsible for the low ash fusion temperature [40].

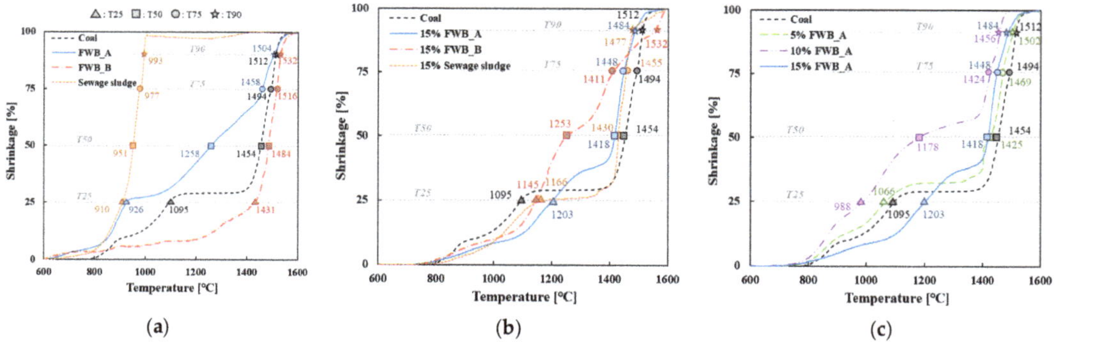

Figure 5. Thermomechanical analysis results of FWB_A (blue line), FWB_B (orange dot line), sewage sludge (yellow dot line), and coal (black dot line). (**a**) Shrinkage (%) of biomass fuels and coal as a function of temperature. (**b**) Shrinkage of co-combustion of biomass fuels and coal. (**c**) Shrinkage of co-combustion of FWB_A and coal with varying ratios.

The fusion tendency of the actual ash was not represented by the TMA results of the single ash components or the combined trends extracted from each TMA result. Furthermore, the fusion temperature of biomass ash depends on its chemical composition [42]. Solid fuels with heterogeneous ash components exhibit inconsistent fusion characteristics during co-combustion, rendering predictions of ash fusion points inherently challenging.

The fusion characteristics during co-combustion were examined by comparing the TMA results of 15% cofiring with each biomass (Figure 5b). The primary fusion of 15% FWB_A was observed between 800 °C and 1300 °C, followed by a distinctive secondary fusion between 1400 °C and 1500 °C (T25 at 1203 °C, T50 at 1418 °C, T75 at 1448 °C, and T90 at 1484 °C). Notably, FWB_A exhibited a higher first fusion temperature (T25) than that of 100% coal, albeit its T75 and T90 were lower than those of coal. A single gradual fusion was observed in 15% FWB_B. Notably, FWB_B showed the lowest fusibility up to T75 during a single combustion and the highest fusion temperature at T90 when cofired at 15% with coal. This trend was attributed to the inherently high fusion temperature of FWB_B during single combustion. A distinct single fusion was observed for 15% sewage sludge between 1400 °C and 1500 °C, similar to the tendency of coal, and in the case of 15% sewage sludge, T25 was at 1166 °C, T50 at 1430 °C, T75 at 1455 °C, and T90 at 1477 °C. Overall, coal played a dominant role in the fusion of FWB_A and sewage sludge, leading to an increase in fusibility.

Finally, a TMA fusion analysis was conducted on the co-combustions of 5% FWB_A, 10% FWB_A, and 15% FWB_A. The co-combustion of 5% FWB_A showed a similar tendency to that of coal but with a slightly lower fusion point. In the case of 5% FWB_A, primary fusion occurred between 800 °C and 1100 °C, followed by a rapid secondary fusion between 1400 °C and 1500 °C (T25 at 1066 °C, T50 at 1425 °C, T75 at 1469 °C, and T90 at 1502 °C). The co-combustion of 10% FWB_A had the lowest fusion point, and this is presumed to be due to the eutectic phenomenon. Primary fusion was observed between 800 °C and 1400 °C, and a drastic secondary fusion was observed between 1400 °C and 1500 °C. The fusion points were relatively low, which is unfavorable for co-combustion. High alkaline oxides such as CaO and MgO in biomass lower the ash melting tendencies, whereas SiO_2 and Al_2O_3 have a negative effect on the ash fluidity of the coal [43]. High CaO content in FWB, compared to coal, may decrease fusion temperature (Table 3). Fusibility did not decrease with an increase in the co-combustion ratio. Furthermore, the fusibility at blending rates of 5% and 15% did not differ significantly from that of the single combustion of coal. In summary, at single fusion, sewage sludge exhibited the lowest fusion point and poor melting properties, whereas relatively superior fusion characteristics were shown for FWB_A and FWB_B. Understanding the ash fusion characteristics of biomass contributes to the establishment of appropriate measures to prevent slagging issues and enhances the combustion efficiency of biomass. To minimize ash deposition and optimize combustion efficiency, the cofiring ratio or blending of solid fuel can be adjusted. These efforts may promote environmentally friendly biomass combustion.

Table 3. XRF analysis results of ash components of solid fuels.

	Coal	FWB_A	FWB_B	Sewage Sludge
SiO_2	86.02	4.36	6.56	25.00
Al_2O_3	9.25	0.00	0.00	16.53
Fe_2O_3	2.23	0.97	1.63	12.21
CaO	0.00	61.18	53.17	9.38
MgO	0.00	2.81	4.04	0.85
Na_2O	0.00	0.00	0.00	0.00
K_2O	0.00	2.30	0.78	2.68
SO_3	0.00	2.68	3.51	0.01
P_2O_5	0.00	20.74	21.43	26.37
Etc.	2.50	5.0	8.9	7.0

3.5. Grindability Evaluation

The grindability of biomass, which is its ability to be pulverized into smaller particles, directly influences the efficiency and effectiveness of biomass combustion. Biomass needs to be ground to improve its combustion efficiency and heat transfer rates. In addition, solid fuels with high grindability have advantages in handling and storage because fuel transportation and storage are more convenient and experience fewer bridging or blockage problems.

The grindability of the biomass was evaluated based on the TTBGI [29]. As shown in Figure 6, the TTBGI values of FWB_A and FWB_B were higher than those of coal and sewage sludge. Coal with TTBGI 36% was classified as very hard, sewage sludge with TTBGI 43% was classified as hard, and FWB_A and FWB_B with TTBGI values were 80% and 75% each, so they were classified as medium-hard. In general, solid biomass fuels have low grindability because of their high moisture content, irregular particle shape, and biomass size. The heterogeneous composition of the biomass, including its inorganic ash content, also results in low grindability. In particular, alkali metals in ash may cause wear and damage to grinding equipment. However, in this study, high grindability using FWB was achieved, likely due to the pyrolysis process. Such preprocessing reduces the moisture and inorganic ash contents of biomass fuel. The higher grindability of the solid fuel suggests that it is an efficient cofiring fuel that requires less power consumption during fuel preparation.

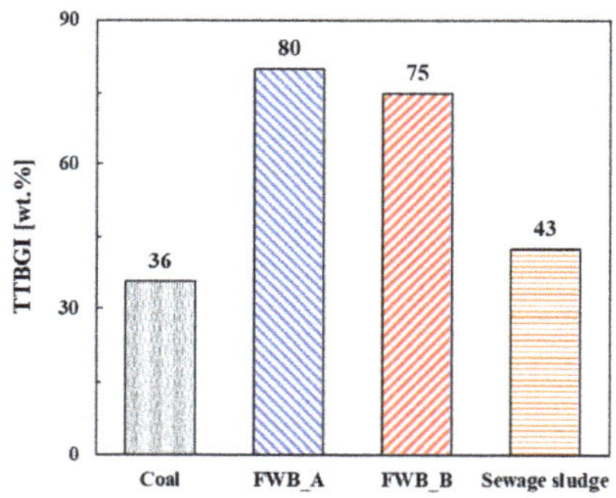

Figure 6. Grindability evaluation of biomass fuels and coal using the Thermally Treated Biomass Grindability Index.

4. Conclusions

FWB was prepared and evaluated as a cofiring biomass fuel in this study. Combustibility, ash deposition characteristics, ash fusion tendencies, and grindability of FWB during co-combustion were investigated. The following conclusions were drawn from this study:

I. FWBs showed similar calorific values with coal, approximately 2000 kcal/kg higher than that of sewage sludge fuel.

II. The gaseous emissions of NO_X and UBC from the combustion of FWB indicated environmental benefits, especially at a 10% mixing ratio.

III. FWB_A exhibited melting behavior at low temperatures, indicating a potential negative impact on ash fouling and slagging. Similar fusion characteristics of FWB_A to coal make it a more suitable cofiring resource. Lower fusion points for sewage sludge imply potential operational issues.

IV. FWBs demonstrated higher grindability compared to traditional biomass fuels due to the pyrolysis process. This high grindability is beneficial for fuel preparation, transportation, and storage, reducing overall operational costs.

The findings of this study suggest that FWB_A is a valuable resource for co-combustion in coal-fired power plants. Because the current study was limited to lab-scale experiments using DTF, the findings cannot be generalized to the actual application of FWB in coal-fired power plants. Therefore, further studies are required to validate the feasibility of using FWB in coal-fired power plants. Continued efforts in both research and energy policy are required to make the FWB a more efficient renewable resource.

Author Contributions: Conceptualization, I.-T.K. and C.-H.J.; methodology, D.-C.S. and S.-M.K.; validation, Y.J., J.-S.K. and Y.-E.L.; formal analysis, K.-H.K. and M.-J.K.; investigation, K.-H.A. and J.J.; writing—original draft preparation, Y.J.; writing—review and editing, Y.J. and J.-S.K.; visualization, J.-S.K.; supervision, I.-T.K. and C.-H.J.; project administration, I.-T.K. All authors have read and agreed to the published version of the manuscript.

Funding: This research was funded by the Korea Institute of Civil Engineering and Building Technology (KICT) (grant number 20230094-001).

Informed Consent Statement: Not applicable.

Data Availability Statement: Not applicable.

Conflicts of Interest: The authors declare no conflict of interest.

References

1. Goerndt, M.E.; Aguilar, F.X.; Skog, K. Drivers of Biomass Co-Firing in U.S. Coal-Fired Power Plants. *Biomass Bioenergy* **2013**, *58*, 158–167. [CrossRef]
2. Savolainen, K. Co-Firing of Biomass in Coal-Fired Utility Boilers. *Appl. Energy* **2003**, *74*, 369–381. [CrossRef]
3. Unchaisri, T.; Fukuda, S. Investigation of Ash Formation and Deposit Characteristics in CFB Co-Combustion of Coal with Various Biomass Fuels. *J. Energy Inst.* **2022**, *105*, 42–52. [CrossRef]
4. García, R.; Pizarro, C.; Lavín, A.G.; Bueno, J.L. Characterization of Spanish Biomass Wastes for Energy Use. *Bioresour. Technol.* **2012**, *103*, 249–258. [CrossRef] [PubMed]
5. Padeyanda, Y.; Jang, Y.-C.; Ko, Y.; Yi, S. Evaluation of Environmental Impacts of Food Waste Management by Material Flow Analysis (MFA) and Life Cycle Assessment (LCA). *J. Mater. Cycles Waste Manag.* **2016**, *18*, 493–508. [CrossRef]
6. Lee, Y.-E.; Shin, D.-C.; Jeong, Y.; Kim, I.-T.; Yoo, Y.-S. Effects of Pyrolysis Temperature and Retention Time on Fuel Characteristics of Food Waste Feedstuff and Compost for Co-Firing in Coal Power Plants. *Energies* **2019**, *12*, 4538. [CrossRef]
7. Pahla, G.; Ntuli, F.; Muzenda, E. Torrefaction of Landfill Food Waste for Possible Application in Biomass Co-Firing. *Waste Manag.* **2018**, *71*, 512–520. [CrossRef] [PubMed]
8. Wei, X.; Lopez, C.; von Puttkamer, T.; Schnell, U.; Unterberger, S.; Hein, K.R.G. Assessment of Chlorine−Alkali−Mineral Interactions during Co-Combustion of Coal and Straw. *Energy Fuels* **2002**, *16*, 1095–1108. [CrossRef]
9. Niu, Y.; Tan, H.; Hui, S. Ash-Related Issues during Biomass Combustion: Alkali-Induced Slagging, Silicate Melt-Induced Slagging (Ash Fusion), Agglomeration, Corrosion, Ash Utilization, and Related Countermeasures. *Prog. Energy Combust. Sci.* **2016**, *52*, 1–61. [CrossRef]
10. Yao, X.; Zhao, Z.; Li, J.; Zhang, B.; Zhou, H.; Xu, K. Experimental Investigation of Physicochemical and Slagging Characteristics of Inorganic Constituents in Ash Residues from Gasification of Different Herbaceous Biomass. *Energy* **2020**, *198*, 117367. [CrossRef]
11. Nutalapati, D.; Gupta, R.; Moghtaderi, B.; Wall, T.F. Assessing Slagging and Fouling during Biomass Combustion: A Thermodynamic Approach Allowing for Alkali/Ash Reactions. *Fuel Process. Technol.* **2007**, *88*, 1044–1052. [CrossRef]
12. Ma, T.; Fan, C.; Hao, L.; Li, S.; Song, W.; Lin, W. Fusion Characterization of Biomass Ash. *Thermochim. Acta* **2016**, *638*, 1–9. [CrossRef]
13. Tortosa Masiá, A.A.; Buhre, B.J.P.; Gupta, R.P.; Wall, T.F. Characterising Ash of Biomass and Waste. *Fuel Process. Technol.* **2007**, *88*, 1071–1081. [CrossRef]
14. Cai, J.; He, Y.; Yu, X.; Banks, S.W.; Yang, Y.; Zhang, X.; Yu, Y.; Liu, R.; Bridgwater, A.V. Review of Physicochemical Properties and Analytical Characterization of Lignocellulosic Biomass. *Renew. Sustain. Energy Rev.* **2017**, *76*, 309–322. [CrossRef]
15. Bridgeman, T.G.; Jones, J.M.; Shield, I.; Williams, P.T. Torrefaction of Reed Canary Grass, Wheat Straw and Willow to Enhance Solid Fuel Qualities and Combustion Properties. *Fuel* **2008**, *87*, 844–856. [CrossRef]
16. Weber, R. Relevance of BFRs and Thermal Conditions on the Formation Pathways of Brominated and Brominated–Chlorinated Dibenzodioxins and Dibenzofurans. *Environ. Int.* **2003**, *29*, 699–710. [CrossRef] [PubMed]
17. Lee, D.-G.; Ku, M.-J.; Kim, K.-H.; Kim, J.-S.; Kim, S.-M.; Jeon, C.-H. Experimental Investigation of the Ash Deposition Characteristics of Biomass Pretreated by Ash Removal during Co-Combustion with Sub-Bituminous Coal. *Energies* **2021**, *14*, 7391. [CrossRef]
18. *ASTM Standard D3173-11*; Standard Test Method for Moisture in the Analysis Sample of Coal and Coke. ASTM International: West Conshohocken, PA, USA, 2011.
19. *ASTM Standard D3175-11*; Standard Test Method for Volatile Matter in the Analysis Sample of Coal and Coke. ASTM International: West Conshohocken, PA, USA, 2011.
20. *ASTM D3174-12*; Standard Test Method for Ash in the Analysis Sample of Coal and Coke from Coal. ASTM International: West Conshohocken, PA, USA, 2011.

21. *ASTM D3172-13(2021)e1;* Standard Practice for Proximate Analysis of Coal and Coke. ASTM International: West Conshohocken, PA, USA, 2021.
22. *ASTM D5865;* Standard Test Method for Gross Calorific Value of Coal and Coke. ASTM International: West Conshohocken, PA, USA, 2011.
23. *DIN 51729-10;* Testing of Solid Fuels-Determination of Chemical Composition of Fuel Ash-Part 10: X-ray Fluorescence Analysis. Deutsches Institut fur Normung: Berlin, Germany, 2011.
24. Wu, Z.; Wang, S.; Zhao, J.; Chen, L.; Meng, H. Product distribution during co-pyrolysis of bituminous coal and lignocellulosic biomass major components in a drop-tube furnace. *Energy Fuels* **2015**, *29*, 4168–4180. [CrossRef]
25. Chen, W.-H.; Du, S.-W.; Tsai, C.-H.; Wang, Z.-Y. Torrefied Biomasses in a Drop Tube Furnace to Evaluate Their Utility in Blast Furnaces. *Bioresour. Technol.* **2012**, *111*, 433–438. [CrossRef]
26. Barroso, J.; Ballester, J.; Ferrer, L.M.; Jiménez, S. Study of Coal Ash Deposition in an Entrained Flow Reactor: Influence of Coal Type, Blend Composition and Operating Conditions. *Fuel Process. Technol.* **2006**, *87*, 737–752. [CrossRef]
27. Blanchard, R. Measurements and Modeling of Coal Ash Deposition in an Entrained-Flow Reactor. Master's Thesis, Brigham Young University, Provo, UT, USA, 2008.
28. *JIS M8801;* Coal–Testing Methods. Japanese Standards Association (JSA): Tokyo, Japan, 2008.
29. Yu, S.; Park, J.; Kim, M.; Kim, H.; Ryu, C.; Lee, Y.; Yang, W.; Jeong, Y.-G. Improving Energy Density and Grindability of Wood Pellets by Dry Torrefaction. *Energy Fuels* **2019**, *33*, 8632–8639. [CrossRef]
30. Huang, X.; Ng, K.W.; Giroux, L. Grindability of biocarbon and coal blends in rolling mill. *Int. J. Coal Prep. Util.* **2022**, *42*, 1651–1663. [CrossRef]
31. Ndibe, C.; Grathwohl, S.; Paneru, M.; Maier, J.; Scheffknecht, G. Emissions Reduction and Deposits Characteristics during Cofiring of High Shares of Torrefied Biomass in a 500 KW Pulverized Coal Furnace. *Fuel* **2015**, *156*, 177–189. [CrossRef]
32. Loeffler, D.; Anderson, N. Emissions Tradeoffs Associated with Cofiring Forest Biomass with Coal: A Case Study in Colorado, USA. *Appl. Energy* **2014**, *113*, 67–77. [CrossRef]
33. Kouprianov, V.I.; Permchart, W. Emissions from a Conical FBC Fired with a Biomass Fuel. *Appl. Energy* **2003**, *74*, 383–392. [CrossRef]
34. Permchart, W.; Kouprianov, V.I. Emission Performance and Combustion Efficiency of a Conical Fluidized-Bed Combustor Firing Various Biomass Fuels. *Bioresour. Technol.* **2004**, *92*, 83–91. [CrossRef] [PubMed]
35. Sun, P.; Hui, S.; Gao, Z.; Zhou, Q.; Tan, H.; Zhao, Q.; Xu, T. Experimental Investigation on the Combustion and Heat Transfer Characteristics of Wide Size Biomass Co-Firing in 0.2 MW Circulating Fluidized Bed. *Appl. Therm. Eng.* **2013**, *52*, 284–292. [CrossRef]
36. Demirbas, A. Potential Applications of Renewable Energy Sources, Biomass Combustion Problems in Boiler Power Systems and Combustion Related Environmental Issues. *Prog. Energy Combust. Sci.* **2005**, *31*, 171–192. [CrossRef]
37. Pronobis, M. The Influence of Biomass Co-Combustion on Boiler Fouling and Efficiency. *Fuel* **2006**, *85*, 474–480. [CrossRef]
38. Wang, L.; Tang, C.; Zhu, T.; Fang, F.; Ning, X.; Che, D. Experimental Investigation on Combustion and NO*x* Formation Characteristics of Low-Ash-Melting-Point Coal in Cyclone Furnace. *ACS Omega* **2022**, *7*, 26537–26548. [CrossRef]
39. Kleinhans, U.; Wieland, C.; Frandsen, F.J.; Spliethoff, H. Ash Formation and Deposition in Coal and Biomass Fired Combustion Systems: Progress and Challenges in the Field of Ash Particle Sticking and Rebound Behavior. *Prog. Energy Combust. Sci.* **2018**, *68*, 65–168. [CrossRef]
40. Zhou, A.; Wang, X.; Magdziarz, A.; Yu, S.; Deng, S.; Bai, J.; Zhang, Q.; Tan, H. Ash Fusion and Mineral Evolution during the Co-Firing of Coal and Municipal Sewage Sludge in Power Plants. *Fuel* **2022**, *310*, 122416. [CrossRef]
41. Namkung, H.; Lee, Y.-J.; Park, J.-H.; Song, G.-S.; Choi, J.W.; Choi, Y.-C.; Park, S.-J.; Kim, J.-G. Blending Effect of Sewage Sludge and Woody Biomass into Coal on Combustion and Ash Agglomeration Behavior. *Fuel* **2018**, *225*, 266–276. [CrossRef]
42. Nunes, L.J.R.; Matias, J.C.O.; Catalão, J.P.S. Biomass Combustion Systems: A Review on the Physical and Chemical Properties of the Ashes. *Renew. Sustain. Energy Rev.* **2016**, *53*, 235–242. [CrossRef]
43. Wang, H.; Cheng, L.; Pu, J.; Zhao, J. Melting Characteristics of Coal Ash and Properties of Fly Ash to Understand the Slag Formation in the Shell Gasifier. *ACS Omega* **2021**, *6*, 16066–16075. [CrossRef]

sustainability

MDPI

Article

Techno-Economic Feasibility Study for Organic and Plastic Waste Pyrolysis Pilot Plant in Malaysia

Mooktzeng Lim [1] and Ee Sann Tan [2,*]

[1] NexantECA (MY) Sdn Bhd, Mid Valley City, Kuala Lumpur 59200, Malaysia; mlim@nexanteca.com
[2] Institute of Sustainable Energy (ISE), Universiti Tenaga Nasional, Jalan Ikram-UNITEN, Kajang 43000, Selangor, Malaysia
* Correspondence: eesann@uniten.edu.my

Abstract: Organic and plastic waste (OPW) is diverted from landfills in order to lower carbon emissions. Nevertheless, modern pyrolysis techniques are frequently utilized in laboratories (using feedstocks that weigh less than 1 kg), which employ costly pure nitrogen gas (N_2). This study developed a fast pyrolysis system to produce pyrolysis oil or liquid (PyOL) from OPW using flue gas as the pyrolysis agent. The added benefits included the efficient value-added chemical extractions and the non-thermal plasma reactor upgraded PyOL. OPW was also pyrolyzed at a pilot scale using flue gas fast pyrolysis in this study. In addition to lowering operational expenses associated with pure N_2, flue gas reduced the lifecycle carbon emissions to create PyOL. The results indicated that considerable material agglomeration occurred during the OPW pyrolysis with an organic-to-plastic-waste (O/P) ratio of 30/70. Furthermore, the liquid yields were 5.2% and 5.5% when O/P was 100/0 (305 °C) and 99.5/0.5 (354 °C), respectively. The liquid yields also increased when polymers (polypropylene) were added, enhancing the aromatics. Two cases were employed to study their techno-economic feasibility: PyOL-based production and chemical-extraction plants. The mitigated CO_2 from the redirected OPW and flue gas produced the highest revenue in terms of carbon credits. Moreover, the carbon price (from RM 100 to 150 per ton of CO_2) was the most important factor impacting the economic viability in both cases. Plant capacities higher than 10,000 kg/h were economically viable for the PyOL-based plants, whereas capacities greater than 1000 kg/h were financially feasible for chemical-extraction plants. Overall, the study found that the pyrolysis of OPW in flue gas is a viable waste-to-energy technology. The low liquid yield is offset by the carbon credits that can be earned, making the process economically feasible.

Keywords: pyrolysis; lifecycle analysis; carbon emissions; techno-economic; feasibility

Citation: Lim, M.; Tan, E.S. Techno-Economic Feasibility Study for Organic and Plastic Waste Pyrolysis Pilot Plant in Malaysia. *Sustainability* **2023**, *15*, 14280. https://doi.org/10.3390/su151914280

Academic Editors: Steven Lim, Shuit Siew Hoong, Santi Chuetor and Pang Yean Ling

Received: 4 July 2023
Revised: 18 August 2023
Accepted: 21 August 2023
Published: 27 September 2023

1. Introduction

The rapid urbanization and development of Malaysia have led to the production of 42 million tons of municipal solid waste (MSW) annually. Landfills receive about 74% of MSW, which contains 40 to 60% of food waste [1]. Approximately 1.8 million tons of plastic wastes are generated each year, where large plastic waste amounts are treated or converted using processes that harm the environment. In 2019, Malaysian factories processed plastic waste without the necessary permits or environmental controls. These plastic wastes were sometimes left unutilized as factories were closed [2].

Although Malaysian economic growth is improving, MSW mismanagement is worsening and 35% of Malaysians in rural areas are still living in poverty [3]. Those who fall under urban poor category could be even higher, depending on the poverty estimation methodology [4,5]. Malaysia generates an estimated 30,000 tons of MSW daily, which could increase to 43,000 tons by 2025. Most of Malaysia's MSW in landfills is a major environmental concern, as greenhouse gas (methane) is 25 times more potent than carbon dioxide. Both organic and plastic wastes (OPW) are included in MSW, composed of carbon-based materials utilized as feedstock to create petroleum products (liquid or pyrolysis oil).

Pyrolysis converts feedstock to liquid oil at a temperature of 300 to 600 °C in an oxygen-deficient (or low-oxygen) environment [6,7]. This process has the potential to substitute and minimizes the amount of crude oil derived from fossil sources while recovering and reducing OPW and diverting it from landfills. Pyrolysis oil consists of a complex mixture of chemical compounds, including oxygen-containing organic compounds derived from OPW. Meanwhile, biomass-based pyrolysis oil comprises cellulose, hemicellulose, and lignin components, which produce organic acids, alcohols, aldehydes, esters, phenols, ketones, and dehydrated carbohydrates [8]. Formic acid, furfural, phenol, and guaiacol are among the important chemicals that are recovered from the produced pyrolysis oil or liquid (PyOL), and their shelf prices are approximately RM 70 to 96, 56 to 303, 125 to 437 RM/L [8–14]. Table 1 lists the application of each chemical extracted from pyrolysis oil and its function.

Table 1. Extractable chemicals from pyrolysis oil and their function.

Extractables	Function	Bulk/Cost (USD/ton)	Estimated Shelf Cost (RM/Liter) [4]
Formic acid	• Preservative and antibacterial agent in livestock feed. • To process organic latex (sap) into raw rubber.	44 [1]	70–96
Furfural	• Solvent for the extraction of aromatics. • As a transportation fuel, jet fuel, gasoline additive, soil enhancer, and organic fertilizer.	900–2000 [2]	56–303
Phenol	• Additive for products to promote hardening, used in paints and varnishes, plastics, etc. • Type of pesticide used to destroy or inhibit the growth of disease-causing mechanisms.	840 [3]	125–437

Generally, the produced PyOL from agricultural wastes or lignocellulosic-based materials is not directly utilized as fuels due to several factors, including high oxygen and low energy contents [15,16]. A study by Sommani et al. reported the liquid fuel synthesis by cracking using vegetable oil (UVO/WCO) mixed with polypropylene (PP) waste, whose activated carbon was applied as a catalyst to increase the quality of the PyOL [17]. The tests were conducted in a batch reactor with a constant hydrogen pressure of 0.1 MPa. Thus, liquid fuels, such as naphthalene and gas oils, were successfully synthesized with activated carbon in the study. A UVO:PP waste ratio of 30:70 obtained the highest total liquid hydrocarbon yield of 80 wt% with selected conditions. These conditions included an activated carbon to raw material ratio of 2.5 wt%, a reaction temperature of 430 °C, and a reaction time of 30 min [17]. These conventional acid–alkali catalytic reactions also required a lengthy reaction period and numerous separation procedures [18]. Hence, the PyOL quality was improved using a novel plasma reactor to replace the traditional approach [15]. Atoms and molecules were also ionized in non-thermal plasma (NTP), producing radicals and excited species and altering the chemical bonds in the exposed matter [19,20].

This study aimed to develop a waste-to-energy fast pyrolysis system for converting OPW into PyOL, which was further applied as biofuels or value-added chemicals. Following previous studies, most pyrolysis laboratory tests were conducted using pure nitrogen (N_2) or N_2 mixed with a little oxygen (O_2), feedstocks weighing less than 1 kg, and low heating rates of 20 °C min^{-1} [21–23]. Nevertheless, these small-scale laboratory tests did not indicate the scalability and economic viability of the pyrolysis process. Therefore, oxidative pyrolysis was performed in this study using a combination of air and flue gas for an oxygen-deficient environment. The PyOL properties were also enhanced in this study using an NTP reactor. Finally, a techno-economic feasibility (TEF) analysis was conducted to ascertain the economic viability of a commercial OPW pyrolysis plant for various processing capacities.

2. Methodology

This study aimed to evaluate the TEF (PyOL-based organic and plastic wastes) of a pilot plant in Malaysia. As oxidative pyrolysis used less energy to reach the pyrolysis temperature, using oxidative pyrolysis rather than pure N_2 pyrolysis produced significant flue gas savings. Conversely, the drawback of this method was lower liquid yield. Meanwhile, the pilot-scale pyrolysis system provided the information for this study has several limitations [24]. These limitations included underestimating the raw material or energy costs and overestimating the product or service demands, which caused inaccurate profitability estimates. Changing policy and market conditions also impacted the assumptions made.

2.1. Oxidative Pyrolysis of OPW

Instead of using pure N_2 for pyrolysis, oxidative pyrolysis was applied, where non-oxidative thermal degradation and heterogeneous oxidation occurred simultaneously [24]. This process was performed using air with flue gas for OPW pyrolysis. Exothermic reactions require oxygen from the air, generating heat for the endothermic pyrolysis reactions. Unlike laboratory studies, the heat for the pyrolysis process is caused by electrical heaters that are not economically viable.

Flue gas has a temperature of between 150 and 200 °C, which is generated by power plants or combustion processes. This gas is an untapped heat source for oxidative pyrolysis, which lowers operational costs and pure N2 consumption [25,26]. A local supplier set the price of N2 at USD 36 per kg (December 2021) or RM 6 per m^3 (January 2022) [27]. Approximately 79% of flue gas is N_2, 14% is CO_2, and 7% is O_2. Compared to N_2-based pyrolysis, oxidative pyrolysis is anticipated to produce a higher gas output [28]. For example, the waste-to-energy system utilizing pyrolysis of food waste assisted with CO_2 has increased CO generation. When CO_2 is used during the pyrolysis of plastics, the formation of polyaromatic hydrocarbons (PAH) and acidic chemicals are reduced compared to N_2-based pyrolysis [29,30]. This outcome is due to CO_2 hindering the aromatization and cyclization reactions while accelerating the thermal cracking of volatiles during pyrolysis. In addition, CO_2-assisted pyrolysis improves thermal efficiency, increases the breakdown and rearrangement of volatile products, and inhibits the reactivity of volatile organic compounds with gas [31,32].

2.2. Lifecycle Assessment (LCA) of the Pyrolysis System and Mitigated CO_2

The CO_2 emission factors for a linear economy were collected from several studies where MSW was in a landfill, while chemicals were manufactured using fossil fuels (conventional routes). Furthermore, the CO_2 emissions per kg were compared to a circular economy-based scenario in which landfill-based MSW produced PyOL (extracted chemicals). For every 2500 tons of MSW, 3117 tons of CO_2 emissions were recorded daily based on the Bukit Beringin Landfill estimates in 2015 [33]. This observation amounted to 1.25 kg of CO_2 per kg of MSW. The "Our World in Data" database contained carbon emission data related to food production [34]. Consequently, shrimp, fish food, chicken meat, and beef produced 27, 14, 10, and 100 kg of CO_2 per kg of MSW, respectively. The CO_2 emission also reported high-value chemical production, ranging between 4 and 5 kg of CO_2 per kg of MSW using conventional and methane-based methods [35].

2.3. TEF Study for a Commercial Pyrolysis Plant

A TEF study assessed the potential of establishing a commercial pyrolysis plant that generated value-added chemicals. This study was evaluated by sending survey forms to potential pyrolysis oil off-takers, involving BASF, PETRONAS Research Sdn. Bhd., Alam Flora, Jengka Advanced Renewable Energy Plant (JAREP), and other chemical companies (see Figure 1). Three of the five comments concerned capital, collection, and operating costs.

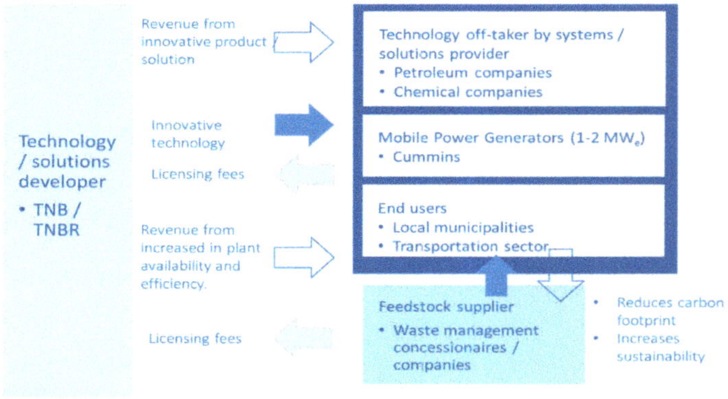

Figure 1. Potential off-takers of technology.

Academics from Curtin University and Manchester University, United Kingdom, conducting TEF studies on industry-based pyrolysis were consulted in this study. Figure 2 portrays the overall process of performing the TEF studies. Capital expense data were sourced from Fivga et al.'s study, which estimated the costs of waste plastic pyrolysis, material balances, energy needs, and utilities using process simulation software (Aspen HYSYS v8.6) [36].

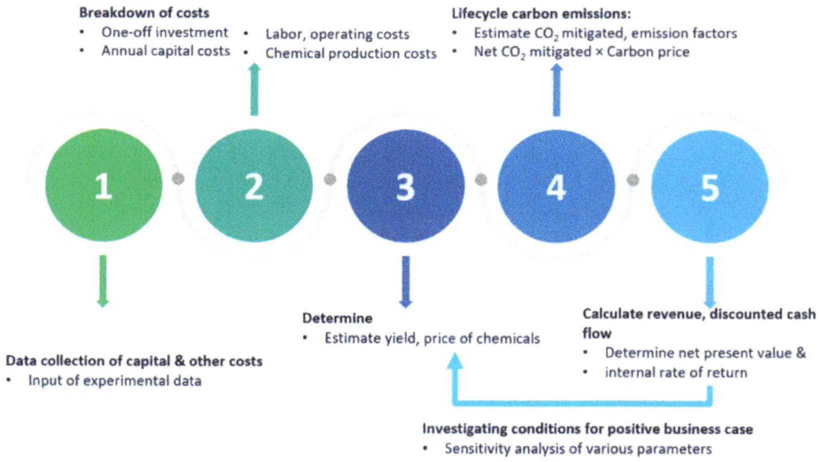

Figure 2. Methodology for techno-economic feasibility studies [36].

Figure 3 illustrates the formic acid, furfural, and phenol extraction methods [10,32,37]. Although the capital expenditures for extracting these compounds were unknown, the average capacity and recent investment values in crude oil-to-chemicals (COTC) plants were utilized instead [38]. The average capital cost was estimated to be RM 2722 per ton of PyOL based on the investment lists and capacities of COTC plants (see Figure 4).

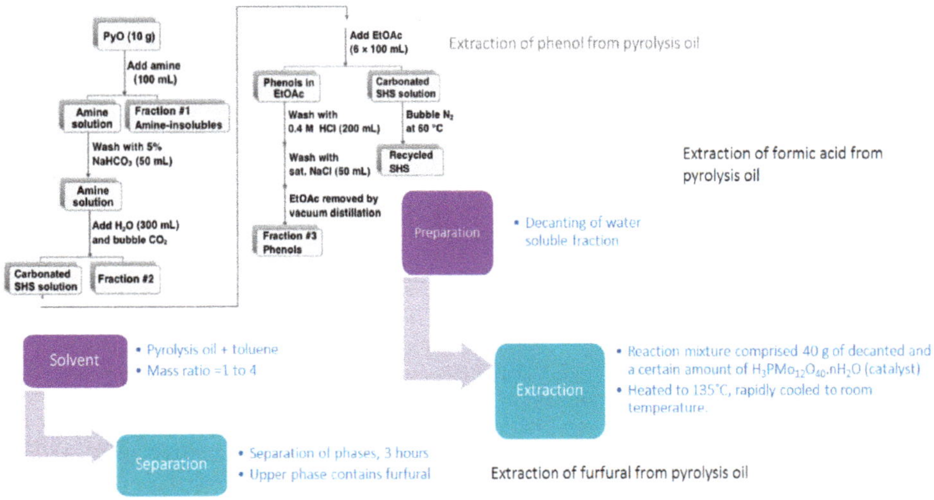

Figure 3. Extraction methods for formic acid, phenol, and furfural.

Project	Refinery Capacity (MMTPA)	Paraxylene Capacity (MMTPA)	Olefin Capacity (MMTPA)	Estimated Chemical Conversion/ barrel of oil (%)	Start Trial Operation
Hengli Petrochemical	20	4.3	1.5	42	Dec 2019
Zhejiang Petroleum and Chemical (ZPC) Phase 1	20	4.0	1.4	45	Q2 2019
Hengyi (Brunei) PMB Refinery-Petrochem	8	1.5	0.5	>40	November 2019
Zhejiang Petroleum and Chemical (ZPC) Phase 2	20	4.8	1.2	50	Estimated 2021
Shenghong Refinery and Integrated Petrochem	16	2.8	1.1	60	H2 2021
Aramco/SABIC JV	20	-	3.0	45	2025

Source: IHS Markit

Figure 4. List of crude oil-to-chemicals (COTC) plant investment and capacities [39].

The TEF analysis was performed for pyrolysis plants handling MSW values of 20, 100, 1000, and 10,000 kg/h in response to Fivga et al.'s study [36]. According to a Malaysian Performance Management and Delivery Unit study, the waste collection cost was approximately RM 150 per ton daily [39]. The waste collection cost was lowered to meet each collector's minimum wage (RM 1500 monthly). Therefore, these costs were reduced from RM 300 ton/day to RM 275, 150, and 100 ton/day for plant capacities of 20, 100, 1000, and 10,000 kg/h, respectively.

The pyrolysis plant data using flue gas as a pyrolysis agent and OPW as feedstock were not readily available. Thus, additional feasibility data were estimated using techno-economic analysis of pyrolysis plants employing woody biomass, rice husks, or plastics as feedstock. These data were obtained from institutions like the National Renewable Energy Laboratory (NREL) [40,41]. The pyrolysis fuel production cost for various plant capacities was then evaluated according to Fivga et al.'s study [36]. Meanwhile, studies by Wright [40], Jahirul [42] and Islam [43] provided the operating breakdown costs for labor, operation, and maintenance (OM) of the proposed pyrolysis plants with an annual operational time of 7008 h/yr. The OM cost involved clearing pipe blockages, repairing conveyors, compressors, and air distribution plates, maintaining pyrolysis agent systems, fixing pipe leaks, and a one-time payment for overhauling major equipment (heating system and large material handling equipment).

The total operational costs increased with plant capacity due to rising energy costs and the demand for consumables, such as liquid petroleum gas (LPG), to heat the reactor. In contrast, the operational costs per kg of OPW decreased when the labor costs did not increase linearly with plant capacity. The operating cost was hypothetically lower than plants that employed pure N2 as the pyrolysis agent. Although the proposed pyrolysis plant utilized flue gas for fluidization and pyrolysis processes, additional pumps and pipe insulations were necessary to raise the pressure while maintaining the temperature.

For plants with capacities of 1000 and 10,000 kg/h, using flue gas at an average temperature of 150 °C for heating the reactor to 350 °C decreased LPG consumption by between RM 40,009 and 400,098 annually. Interestingly, this flue gas-based process also prevented and reduced the CO_2 released into the atmosphere. Depending on the implementation of a domestic carbon trading mechanism and the carbon price (RM 35 to 150 per ton of CO_2), the flue gas fast pyrolysis system provided additional revenue if other carbon-emitting companies purchased carbon credits from the carbon offset of the pyrolysis system [44].

The expected capital, OM expenditures, and revenue for two case studies were considered in the TEF analysis. The first case study demonstrated the PyOL as the sole product, which the PyOL costing approximately RM 0.60 per kg based on a local PyOL manufacturing company survey. Likewise, the pyrolysis technology used in this study yielded a 5% PyOL. A higher yield was also possible (20 to 30%) with low condensation temperatures (10 to −10 °C). Nonetheless, the low condensation temperature was assumed to be unavailable in the commercial plant owing to the high capital investment needed, producing lower PyOL due to the limitation of energy consumption. The second case study reported the chemical extractions of formic acid, furfural, and phenol from PyOLs through processes presented in several studies [10,32,37]. These studies provided the estimated chemical yields with the estimated minimum value prices (obtained from the Chemical Book and Sigma-Aldrich or Merck websites) [45,46]. Hence, the values reported are presented as follows:

- Yield of furfural: 1.0%, at RM 303/kg
- Yield of formic acid: 8.4%, at RM 96/kg
- Yield of phenol: 4.1%, at RM 437/kg

The specific unit pricing (expressed in RM/L) increased with order volume. Thus, the specific unit price for phenol, furfural, and formic acid was averaged over quantities ranging from 1 to 25 kg (or 1 to 25 L). The profitability per year (PF) was calculated as the total annual income minus the total yearly operating, maintenance, and utility costs (PF). Similarly, the actual cash flow per year (CFn) for Year n was calculated by multiplying PF with a present value factor (PVn) and a discount factor r of 8%. Therefore, the actual PF declines over time is as follows:

$$CFn = PF \times PVn \tag{1}$$

Since the construction and commissioning were supposed to take two years and the capital investment (CI) was set at Year n (where n = 0, 1, 2, . . .), Years 1 and 2 demonstrated no cash flow. From Years 3 to 8, the PV value decreased from 0.7938 (PV3) to 0.5403 (PV8).

Subsequently, the net present value (*NPV*) and internal rate of return (IRR) for each scenario are calculated as follows:

$$NPV = \sum_{n=o}^{8} \frac{CF_n}{(1 + IRR)^n} \quad (2)$$

The *NPV* is set to null to determine the IRR.

3. LCA of Carbon Emissions for Pyrolysis and Chemical Production Plant

From the lifecycle assessment (LCA), the maximum possible total CO_2 emissions from the production of food, chemicals, and MSW from landfills is 19.25 kg CO_2 per kg of material, as shown in Figure 5. By utilizing MSW from landfills, the amount of CO_2 that is reduced is equivalent to the amount emitted per kg of MSW (−1.25 kg CO_2/kg MSW). The pyrolysis reactor emits 12% CO_2, with a gas yield of 72% and total flow rate of 22 kg/h (20 kg/h of MSW + 2 kg/h of pyrolysis agent), and the carbon emission from the fluidized bed is estimated at 2.06 kg/h of CO_2 or 0.10 kg CO_2/kg MSW. These data are based on experimental results from the pilot-scale trials.

When 0.012 kg of CO_2/kg MSW was divided by 0.048 kg of chemical/kg MSW, 2.5 kg of CO_2/kg chemical was estimated as the chemical-extracted PyOL emission. By contrast, the LCA did not include the CO_2 quantity of CO_2 emitted per kg chemical. This outcome was observed due to the CO_2 utilization in the pyrolysis process for the industry. If the LCA was based on the weight of the substances, the chemical reduction generated via fossil fuels would significantly reduce CO_2 emissions. Thus, the decline was as much as −4 kg of CO_2/kg chemical (mean value). On average, 0.04 kg of chemicals were predicted to be produced from each kg of pyrolysis oil. Therefore, the CO_2 amount mitigated by chemicals from PyO was insignificant at −0.01 kg of CO_2/kg MSW.

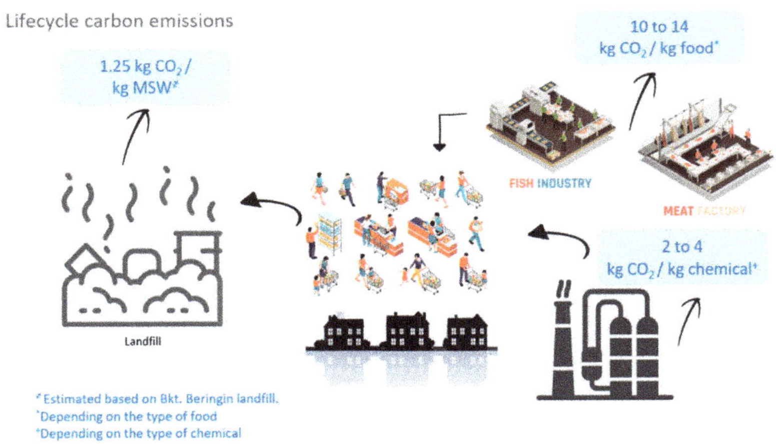

(**a**)

Figure 5. *Cont.*

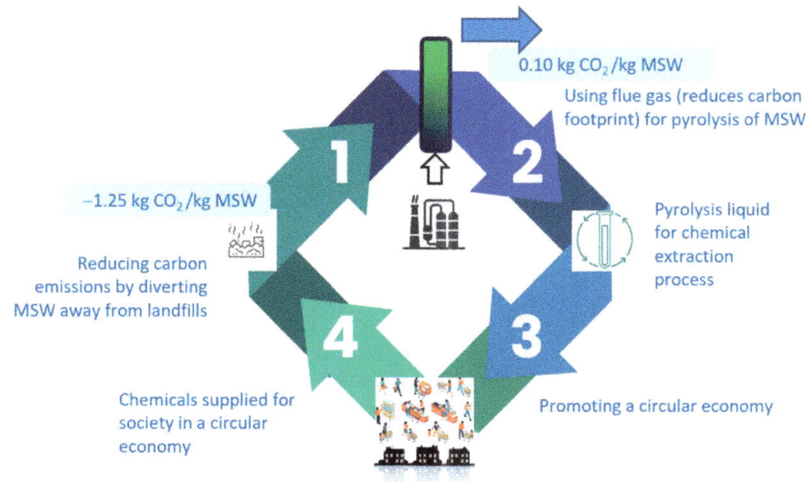

(**b**)

Figure 5. Estimated lifecycle CO_2 emissions for (**a**) a linear economy where municipal solid waste is dumped in landfill ‡—[33] *—[34] +—[35], (**b**) a circular economy where organic waste is pyrolyzed to produce chemicals [29].

According to estimates, the total mitigated CO_2 was −1.00 kg of CO_2/kg MSW, which resulted in lifecycle emissions of 13.79 kg of CO_2 per kg of material (a reduction of 5.46 kg CO_2 per kg of material). For the extraction of formic acid, furfural, and phenol, grid emission factors estimated the carbon emissions from the COTC plant or chemical-extraction procedures [10,32,37]. Using a flue gas fast pyrolysis system could offset indirect CO_2 emissions from the chemical-extraction process (from the supply chain), thus lowering the carbon footprint. Nevertheless, only point emissions were considered in this study.

The MSW removal from landfills generated the highest saved CO_2 amount (see Figure 6). This value was even larger if indirect emissions related to changes in land use and landfill leachates were considered. Figures 5 and 6 present the gross estimations reported in the literature and the comparisons of numerous methods and models, respectively. Thus, an accurate and comprehensive assessment of the LCA was demonstrated, which was vital for further evaluation [47,48].

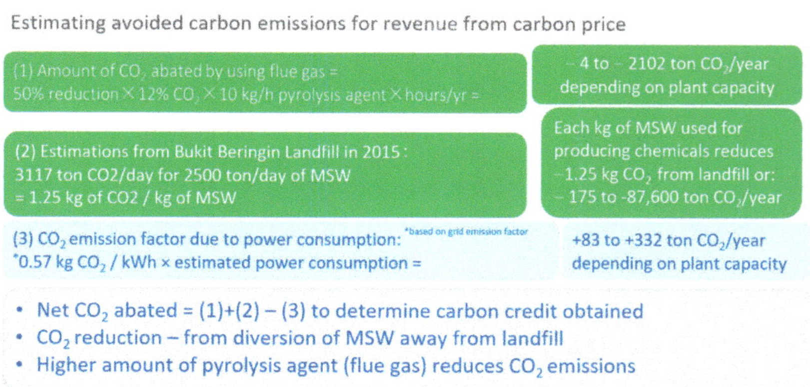

Figure 6. The avoided CO_2 amount using flue gas and MSW for the pyrolysis plant alongside the CO_2 emissions estimated using the grid emission factor [33].

4. TEF Study of Commercial Pyrolysis Oil/Liquid and Chemical-Extraction Plant

Table 2 tabulates the labor cost breakdown for the first PyOL production case study, while Table 3 provides a breakdown of costs for waste collection, operation, maintenance, and utilities. Table 4 lists the PyOL revenue and the averted carbon. Based on data from a local company employing a rotary kiln to produce PyOL at a rate of 15 tons per day, the PyOL cost was RM 0.60 per kg.

Table 2. Summary of the labor cost breakdown for a commercial OPW co-pyrolysis plant.

Organic and Plastic Waste Capacity (kg/h)	20	100	1000	10,000
Capital costs (RM mil)	2.28	2.52	7.20	13.20
Management, production (RM/yr)	100,000.00	100,000.00	100,000.00	100,000.00
Number of engineers	-	-	-	1
Mean salary of engineers (RM/month)	-	-	-	8333.33
Engineers, production (RM/yr)	-	-	-	173,286.21
Engineers, maintenance (RM/yr)	-	-	-	173,286.21
Number of operators and technicians	2	4	6	8
Mean salary operators and technicians (RM/month)	4166.67	4166.67	4166.67	2083.33
Operators and technicians, production (RM/yr)	50,000.00	50,000.00	50,000.00	50,000.00
Operators and technicians, maintenance (RM/yr)	50,000.00	50,000.00	50,000.00	50,000.00
Number of executives	1	1	1	2
Mean salary of executives (RM/month)	2500.00	2500.00	2500.00	2500.00
Executive (RM/yr)	30,000.00	30,000.00	44,362.73	103,971.73
Total Labor Costs	62,000.00	620,000.00	1,073,202.44	2,009,031.75

Table 3. Summary of the waste collection, operational, maintenance, and utility costs concerning plant capacity. Breakdown of labor, collection, OM costs (pyrolysis liquid only).

Organic and Plastic Waste Capacity (kg/h)	20	100	1000	10,000
Waste collection costs (RM/ton/day)	300	275	150	100
Amount of waste (tons/day)	0.24	1.20	12.00	120.00
Number of collectors (staff)	1.00	5.00	25.00	180.00
Mean cost of collectors (RM/month)	1752.00	1606.00	1752.00	1622.22

Table 3. *Cont.*

Total Collection Costs (RM/yr)	21,024.00	96,360.00	525,600.00	3,504,000.00
Clear blockage (RM/yr)	5000.00	7500.00	20,000.00	40,000.00
Repair conveyer (RM/yr)	50,000.00	75,000.00	200,000.00	400,000.00
Maintenance of pyrolyzing agent systems (RM/yr)	25,000.00	37,500.00	100,000.00	200,000.00
Repair compressor (RM/yr)	25,000.00	37,500.00	100,000.00	200,000.00
Repair fluidized bed and auxiliary equipment (RM/yr)	25,000.00	37,500.00	100,000.00	200,000.00
Repair leakages (RM/yr)	20,000.00	30,000.00	80,000.00	160,000.00
Major equipment overhaul (RM/yr)	100,000.00	150,000.00	400,000.00	800,000.00
Total Maintenance Costs (RM/yr)	250,000.00	375,000.00	1,000,000.00	2,000,000.00
Organic and Plastic Waste Capacity (kg/h)	20	100	1000	10,000
Power consumption (RM/yr)	72,829.27	109,243.91	291,317.09	582,634.18
For heating up reactor (RM/yr)	1391.65	6958.23	69,582.30	695,822.96
Total Utilities Costs (RM/yr)	74,220.92	116,202.14	360,899.39	1,278,457.14

Table 4. Summary of the revenue from pyrolysis and carbon credit. Breakdown of revenue from pyrolysis liquid and carbon credit.

Organic and Plastic Waste Capacity (kg/h)	20	100	1000	10,000
Carbon price (RM/ton CO_2)	50 to 150 (proposed by Penang Institute in 2019)			
CO_2 mitigated from landfill (kg CO_2/kg MSW)	1.25 [33]			
CO_2 mitigated from landfill (ton CO_2/yr)	140	701	7008	70,080
Flue gas flow rate (kg/h)	10	50	500	5000
Average CO_2 content in flue gas (%)	12 (based on power plant values)			
CO_2 mitigated from flue gas (ton CO_2/yr)	8	42	420	4205
Power consumption, including OPW prep. (kW)	21	31	42 *	83
Grid emission factor (kg CO_2/kWh)	0.57 [33]			
Grid emission per year (ton CO_2/yr)	83	125	166	332
Net CO_2 mitigated (ton CO_2/yr)	101	794	9014	91,473

Table 4. *Cont.*

Yield of pyrolysis oil (%)	5 (based on exp. in this project published in [49])
Selling price of pyrolysis oil (RM/kg)	0.60 [based on information from local company, 2022]
Tipping fees for pyrolysis plant (RM/ton/day)	0 to 250 [39]

* similar to local company, 1250 kg/h pyrolysis plant.

Table 5 displays the estimated capital costs for various plant capacities for the second chemical manufacturing case study from PyOL. These values were based on the investment amount for COTC plants published by the IHS market, in which the plant capital expense was RM 2722 per ton [50]. The chemical operational cost, power consumption, yield, and price are also provided in Table 6.

Table 5. Summary of the estimated capital costs for the chemical-extraction plant. With value-added chemicals extracted from pyrolysis oil.

Capacity (kg/h)	20	100	1000	10,000
Capital cost (RM million) *	2.66	4.43	26.28	203.97

* Based on min. shelf price, obtained from Sigma-Aldrich (Petaling Jaya, Malaysia) and Chemical Book.

Table 6. Summary of the yield, power consumption, emission factors, and price of chemicals. With value-added chemicals extracted from pyrolysis oil.

Capacity (kg/h)	20	100	1000	10,000
Total operating costs annually (RM/yr)	993,732	1,504,537	5,238,353	17,847,238
Power consumption, including OPW prep. and chemical extraction (kW)	41	62	166	332
Net CO_2 mitigated (ton CO_2/yr)	18	669	8516	90,476
Yield, liquid (kg/kg waste)	0.05 (based on exp. in this study)			
Yield, formic acid (kg/kg waste)	0.084 [37]			
Yield, furfural (kg/kg waste)	0.010 [29]			
Yield, phenol (kg/kg waste)	0.041 [32]			
Price of formic acid (RM/kg)	96 *			
Price of furfural (RM/kg)	303 *			
Price of phenol (RM/kg)	437 *			

* Based on min. shelf price, obtained from Sigma-Aldrich and Chemical Book.

Depending on the carbon price and the tipping fee, the first case study (PyOL production) indicated positive cash flows for the 10,000 kg/h plant (see Figure 7). The IRR was 41% when the carbon price was RM 150 per ton of CO_2, which was reduced to 0% at RM 50 per ton of CO_2. Subsequently, the IRR was 4% when the carbon price was RM 50 per ton of CO_2, with the daily tipping fee of RM 120 per ton MSW. Intriguingly, the IRR was still 4% when there were no carbon prices alongside a daily tipping fee of RM 120 RM per ton MSW. The cash flow was negative for the remaining plant capacities.

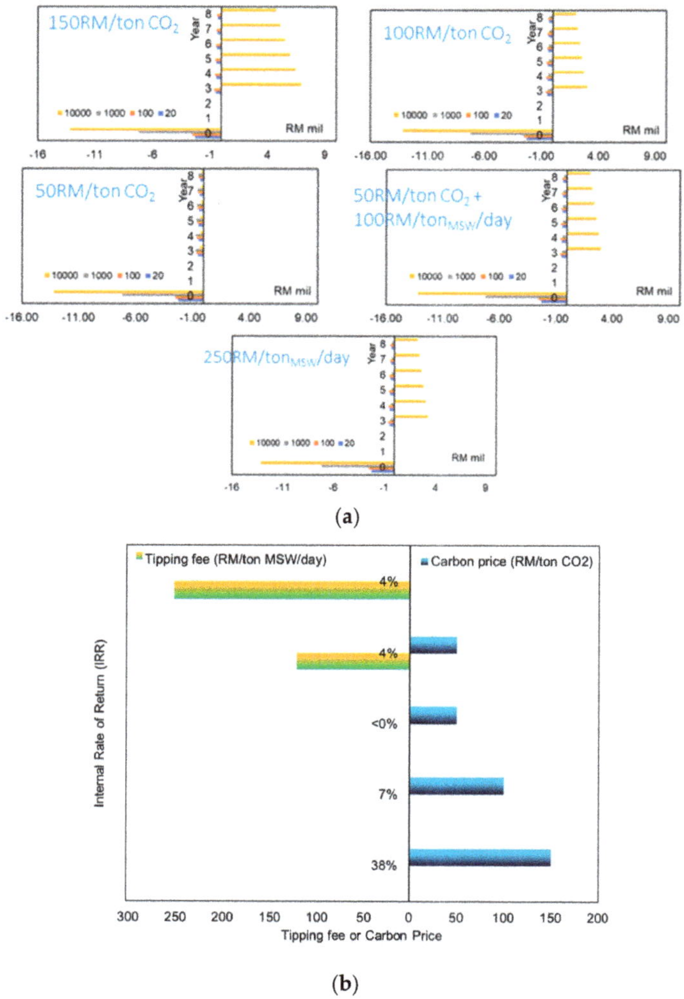

Figure 7. (**a**) Discounted cash flows per year for different plant capacities, carbon price, and tipping fee for pyrolysis liquid only. (**b**) Internal rate of return and different carbon price and tipping fee for pyrolysis liquid only.

Figure 8 shows that for the second case study (chemical extraction from PyOL), there are positive cash flows for the 1000 and 10,000 kg/h plants for carbon prices of RM 35 to 150 per ton CO_2. The internal rate of return is 1 to 7% for the 1000 kg/h system, and 21 to 26% for the 10,000 kg/h plant, for carbon prices of RM 35 to 150 per ton CO_2.

For the second case study (chemical extraction from PyOL) as shown in Figure 9, the carbon prices ranged from RM 35 to 150 per ton of CO_2, and positive cash flows were recorded for the 1000 and 10,000 kg/h plants. Similarly, the IRR were from 1 to 7% and 21 to 26% for the 1000 kg/h and 10,000 kg/h plants, respectively.

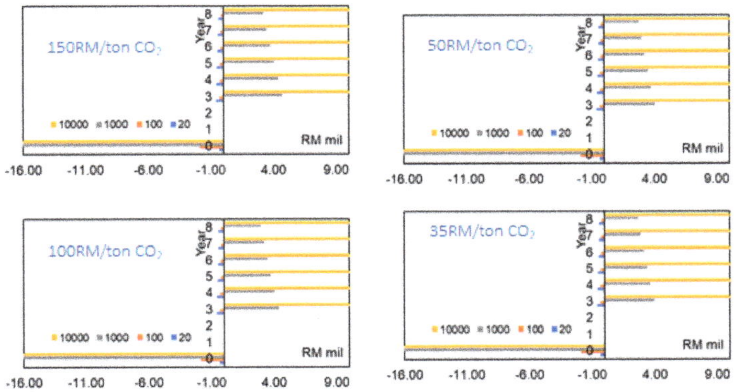

(a)

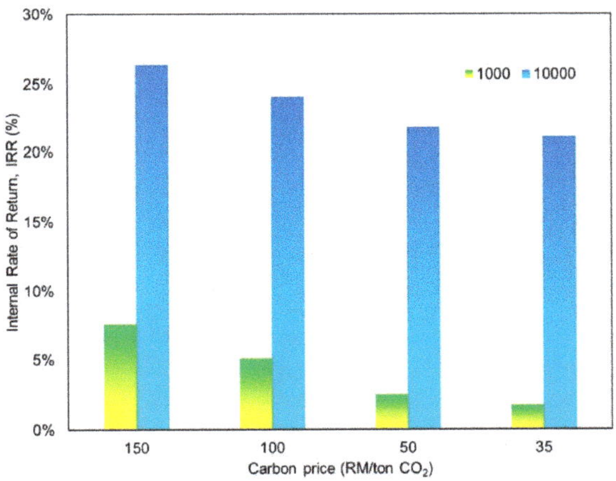

(b)

Figure 8. (**a**) Discounted cash flows per year for different plant capacities, carbon price, and tipping fee for chemical extraction from pyrolysis liquid. (**b**) Internal rate of return and different carbon prices for chemical extraction from pyrolysis liquid.

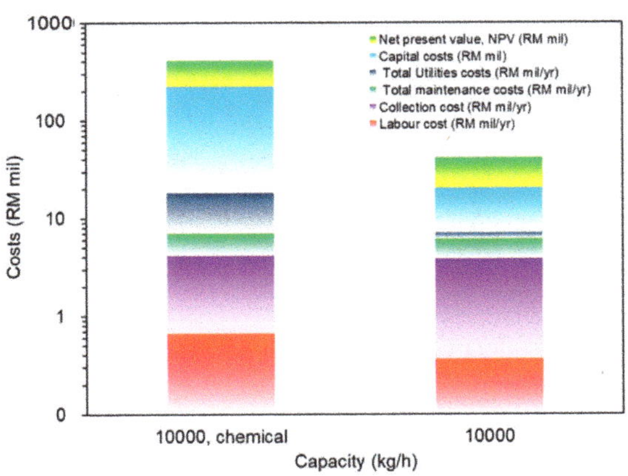

Figure 9. Breakdown of costs for 10,000 kg/h chemical-extraction plant and PyOL-based plant.

Compared to the PyOL-based production plant, the revenue from the chemicals outweighed all other costs of the chemical-extraction plant (excluding the capital costs). Conversely, these figures were indicative and were subject to change based on the supply-and-demand dynamics. The TEF template in this study could be applied to future similar investigations. The assumptions and models employed in this study revealed that Malaysia was increasingly using pyrolysis plants as a more sustainable approach to managing waste-to-energy system while producing electricity. Therefore, pyrolysis facilities were viewed as a key technology in reducing the waste quantity sent to landfills by 30% in 2025.

5. Conclusions

In conclusion, this paper has shown that it is economically feasible to use a waste-to-energy system to produce PyOL from OPW using flue gas as the pyrolysis agent. The flue gas fast pyrolysis system of OPW in a fluidized bed shows that the liquid yield is between 5.2 and 5.5% under the operating conditions of 100/0 (305 °C) and 99.5/0.5 (354 °C), respectively. This low liquid yield shows that it is still economically viable based on the carbon credits obtained by mitigating the carbon emissions from OPW in the landfills and the carbon in the flue gas. There are positive cash flows for the 1000 and 10,000 kg/h plants for carbon prices of RM 35 to 150 per ton CO_2. Meanwhile, the internal rate of return is 1 to 7% for the 1000 kg/h system and 21 to 26% for the 10,000 kg/h plant. This paper has shown that the economic viability of the fast pyrolysis system in developing countries depends on a number of factors, including the cost of the pyrolysis plant, the cost of the feedstock, and the revenue from the sale of pyrolysis products. The development of pyrolysis plants in Malaysia can help to reduce the amount of OPW that ends up in landfills or the environment. It can also create employment opportunities and generate revenue for local communities. This paper has presented key factors that affect the development of pyrolysis plants by analyzing the detailed techno-economic feasibility and serve as a model for other developing countries to address the challenges of waste management and climate change.

Author Contributions: Conceptualization, methodology, investigation, resources, data curation, formal analysis, visualization, project administration, funding acquisition, writing—original draft preparation, M.L.; writing—review and editing, E.S.T. All authors have read and agreed to the published version of the manuscript.

Funding: This work was supported by Tenaga Nasional Berhad under grant number TNBR/SF367/2020 and AAIBE Chair of Renewable Energy at UNITEN with project code 202001KETTHA.

Institutional Review Board Statement: Not applicable.

Informed Consent Statement: Not applicable.

Data Availability Statement: Not applicable.

Acknowledgments: The authors would like to acknowledge the direct and/or indirect assistance and efforts of all related personnel.

Conflicts of Interest: The authors declare no conflict of interest. The funders had no role in the design of the study; in the collection, analyses, or interpretation of data; in the writing of the manuscript; or in the decision to publish the results.

References

1. Bong, C.P.-C.; Goh, R.K.Y.; Lim, J.-S.; Ho, W.S.; Lee, C.-T.; Hashim, H.; Abu Mansor, N.N.; Ho, C.S.; Ramli, A.R.; Takeshi, F. Towards low carbon society in Iskandar Malaysia: Implementation and feasibility of community organic waste composting. *J. Environ. Manag.* **2017**, *203*, 679–687. [CrossRef]
2. Lim, I. Court Fines Two Selangor Factories RM120,000 for Illegal Plastic Recycling. *Malay Mail*, 17 May 2019.
3. Yansaneh, O.Y.; Zein, S.H. Recent Advances on Waste Plastic Thermal Pyrolysis: A Critical Overview. *Processes* **2022**, *10*, 332. [CrossRef]
4. Titone, V.; Gulino, E.F.; La Mantia, F.P. Recycling of Heterogeneous Mixed Waste Polymers through Reactive Mixing. *Polymers* **2023**, *15*, 1367. [CrossRef] [PubMed]

5. Simler, K. An Idea Whose Time Has Come—Raising Malaysia's Poverty Line. *The Star*, 1 September 2019.
6. Sophonrat, N.; Sandström, L.; Zaini, I.N.; Yang, W. Stepwise pyrolysis of mixed plastics and paper for separation of oxygenated and hydrocarbon condensates. *Appl. Energy* **2018**, *229*, 314–325. [CrossRef]
7. Qureshi, K.M.; Lup, A.N.K.; Khan, S.; Abnisa, F.; Daud, W.M.A.W. A technical review on semi-continuous and continuous pyrolysis process of biomass to bio-oil. *J. Anal. Appl. Pyrolysis* **2018**, *131*, 52–75. [CrossRef]
8. Ghorbannezhad, P.; Kool, F.; Rudi, H.; Ceylan, S. Sustainable production of value-added products from fast pyrolysis of palm shell residue in tandem micro-reactor and pilot plant. *Renew. Energy* **2019**, *145*, 663–670. [CrossRef]
9. Brueckner, T.M.; Pickup, P.G.; Hawboldt, K. Improvement of bark pyrolysis oil and value added chemical recovery by pervaporation. *Fuel Process. Technol.* **2019**, *199*, 106292. [CrossRef]
10. Oh, S.-J.; Choi, G.-G.; Kim, J.-S. Fast pyrolysis of corn stover using $ZnCl_2$: Effect of washing treatment on the furfural yield and solvent extraction of furfural. *Energy* **2015**, *88*, 697–702. [CrossRef]
11. Aldrich, S. Catalog of Products. 2020. Available online: https://www.sigmaaldrich.com/catalog/product/aldrich/ (accessed on 17 January 2023).
12. Alibaba. Price of Furfural. 2020. Available online: https://www.alibaba.com/showroom/furfural-price.html (accessed on 2 January 2023).
13. Echemi. Product Information Phenol. 2020. Available online: https://www.echemi.com/productsInformation/pid_Seven3078-phenol.html (accessed on 20 January 2023).
14. BASF. BASF Raises Prices for Formic Acid in Europe and in the Americas. 2016. Available online: https://www.basf.com/global/en/media/news-releases/2016/05/p-16-221.html (accessed on 4 February 2023).
15. Hosseinzadeh, M.B.; Rezazadeh, S.; Rahimpour, H.R.; Taghvaei, H.; Rahimpour, M.R. Upgrading of lignin-derived bio-oil in non-catalytic plasma reactor: Effects of operating parameters on 4-methylanisole conversion. *Chem. Eng. Res. Des.* **2015**, *104*, 296–305. [CrossRef]
16. Xiu, S.; Shahbazi, A. Bio-oil production and upgrading research: A review. *Renew. Sustain. Energy Rev.* **2012**, *16*, 4406–4414. [CrossRef]
17. Sommani, P.; Mankong, N.; Vitidsant, T.; Lothongkum, A.W. Cracking of used vegetabele oil mixed with polypropylene waste in the presence of activated carbon. *ASEAN Eng. J. Part B* **2013**, *4*, 16–24.
18. Buchori, L.; Istadi, I.; Purwanto, P.; Kurniawan, A.; Maulana, T.I. Preliminary Testing of Hybrid Catalytic-Plasma Reactor for Biodiesel Production Using Modified-Carbon Catalyst. *Bull. Chem. React. Eng. Catal.* **2016**, *11*, 59–65. [CrossRef]
19. Lim, M.; Zulkifli, A.Z.S. Investigation of biomass surface modification using non-thermal plasma treatment. *Plasma Sci. Technol.* **2018**, *20*, 115502. [CrossRef]
20. Lim, M.T.; Zulkifli, A.Z.S.; Jayapalan, K.K.; Chin, O. Development of a dimensionless parameter for characterization of dielectric barrier discharge devices with respect to geometrical features. *Plasma Sci. Technol.* **2017**, *19*, 095402. [CrossRef]
21. Al-Asadi, M.; Miskolczi, N.; Eller, Z. Pyrolysis-gasification of wastes plastics for syngas production using metal modified zeolite catalysts under different ratio of nitrogen/oxygen. *J. Clean. Prod.* **2020**, *271*, 122186. [CrossRef]
22. Grycová, B.; Koutník, I.; Pryszcz, A. Pyrolysis process for the treatment of food waste. *Bioresour. Technol.* **2016**, *218*, 1203–1207. [CrossRef]
23. Tang, Y.; Huang, Q.; Sun, K.; Chi, Y.; Yan, J. Co-pyrolysis characteristics and kinetic analysis of organic food waste and plastic. *Bioresour. Technol.* **2018**, *249*, 16–23. [CrossRef]
24. Ding, Y.; Huang, B.; Wu, C.; He, Q.; Lu, K. Kinetic model and parameters study of lignocellulosic biomass oxidative pyrolysis. *Energy* **2019**, *181*, 11–17. [CrossRef]
25. Kim, K.H.; Bai, X.; Rover, M.; Brown, R.C. The effect of low-concentration oxygen in sweep gas during pyrolysis of red oak using a fluidized bed reactor. *Fuel* **2014**, *124*, 49–56. [CrossRef]
26. Ferreira, R.A.d.R.; da Silva Meireles, C.; Assunção, R.M.N.; Barrozo, M.A.S.; Soares, R.R. Optimization of the oxidative fast pyrolysis process of sugarcane straw by TGA and DSC analyses. *Biomass Bioenergy* **2020**, *134*, 105456. [CrossRef]
27. Pharmacompass. API Price Trend Dashboard–Nitrogen. 8 December 2021. Available online: https://www.pharmacompass.com/price/nitrogen (accessed on 8 December 2021).
28. Chen, D. 16—Examples of thermochemical and biological treatment technologies for sustainable waste management in China. In *Substitute Natural Gas from Waste*; Materazzi, M., Foscolo, P.U., Eds.; Academic Press: Cambridge, MA, USA, 2019; pp. 425–454.
29. Oh, J.-I.; Lee, J.; Lee, T.; Ok, Y.S.; Lee, S.-R.; Kwon, E.E. Strategic CO_2 utilization for shifting carbon distribution from pyrolytic oil to syngas in pyrolysis of food waste. *J. CO2 Util.* **2017**, *20*, 150–155. [CrossRef]
30. Kwon, E.E.; Kim, S.; Lee, J. Pyrolysis of waste feedstocks in CO2 for effective energy recovery and waste treatment. *J. CO2 Util.* **2019**, *31*, 173–180. [CrossRef]
31. Chen, Z.; Wu, D.; Chen, L.; Ji, M.; Zhang, J.; Du, Y.; Wu, Z. The fast co-pyrolysis study of PVC and biomass for disposing of solid wastes and resource utilization in N_2 and CO_2. *Process Saf. Environ. Prot.* **2021**, *150*, 489–496. [CrossRef]
32. Fu, D.; Farag, S.; Chaouki, J.; Jessop, P.G. Extraction of phenols from lignin microwave-pyrolysis oil using a switchable hydrophilicity solvent. *Bioresour. Technol.* **2014**, *154*, 101–108. [CrossRef] [PubMed]
33. Tan, S.T.; Ho, W.S.; Hashim, H.; Lee, C.T.; Taib, M.R.; Ho, C.S. Energy, economic and environmental (3E) analysis of waste-to-energy (WTE) strategies for municipal solid waste (MSW) management in Malaysia. *Energy Convers. Manag.* **2015**, *102*, 111–120. [CrossRef]

34. Ritchie, H.; Our World in Data. The Carbon Footprint of Foods: Are Differences Explained by the Impacts of Methane? 2020. Available online: https://ourworldindata.org/carbon-footprint-food-methane (accessed on 10 November 2021).
35. Ren, T.; Patel, M.K. Basic petrochemicals from natural gas, coal and biomass: Energy use and CO_2 emissions. *Resour. Conserv. Recycl.* **2009**, *53*, 513–528. [CrossRef]
36. Fivga, A.; Dimitriou, I. Pyrolysis of plastic waste for production of heavy fuel substitute: A techno-economic assessment. *Energy* **2018**, *149*, 865–874. [CrossRef]
37. Luo, D.; Yin, W.; Han, D.; He, H.; Xia, S. Glycolic acid and formic acid production from pyrolysis oil water-soluble fraction by catalytic oxidation. *Chem. Eng. Sci.* **2021**, *239*, 116644. [CrossRef]
38. IHS Markit. *Crude Oil-to-Chemicals (COTC): A Look Inside Our Technology & Economic Analyses from the 2019–2014 Process and Economics Program (PEP)*; IHS Markit: London, UK, 2019.
39. Performance Management and Delivery Unit Malaysia (PEMANDU), Solid Waste Management Lab 2015. Final Report. 2015; Ministry of Local Government Development, Malaysia. Available online: https://jpspn.kpkt.gov.my/wp-content/uploads/2022/07/slaid_dapatan_makmal.pdf (accessed on 29 December 2022).
40. Wright, M.M.; Daugaard, D.E.; Satrio, J.A.; Brown, R.C. Techno-economic analysis of biomass fast pyrolysis to transportation fuels. *Fuel* **2010**, *89*, S2–S10. [CrossRef]
41. Jaroenkhasemmeesuk, C.; Tippayawong, N. Technical and Economic Analysis of A Biomass Pyrolysis Plant. *Energy Procedia* **2015**, *79*, 950–955. [CrossRef]
42. Jahirul, M.I.; Rasul, M.G.; Chowdhury, A.A.; Ashwath, N. Biofuels Production through Biomass Pyrolysis—A Technological Review. *Energies* **2012**, *5*, 4952–5001. [CrossRef]
43. Islam, M.; Ani, F. Techno-economics of rice husk pyrolysis, conversion with catalytic treatment to produce liquid fuel. *Bioresour. Technol.* **2000**, *73*, 67–75. [CrossRef]
44. The Edge Markets. Domestic Carbon Trading to Be Implemented in Phases from End of Next Year. 2021. Available online: https://www.theedgemarkets.com/article/domestic-carbon-trading-be-implemented-phases-end-next-year-%E2%80%94-minister (accessed on 12 January 2022).
45. Chemical Book. Product Prices from CAS Database List. 2021. Available online: https://www.chemicalbook.com/SupplierPriceList_EN.aspx?cbn=CB4854063&page=3#price (accessed on 25 August 2021).
46. Merck and Sigma-Aldrich. Product Prices for Formic acid, Furfural, and Phenol. 2021. Available online: https://www.sigmaaldrich.com/MY/en/search/phenol?focus=products&page=1&perPage=30&sort=relevance&term=phenol&type=product (accessed on 25 August 2021).
47. Thomas, V.M. The National Academis of Sciences, Engineering and Medicine: Current Methods for Life Cycle Analyses of Low-Carbon Transportation Fuels in the United States. 2021. Available online: https://www.nationalacademies.org/our-work/current-methods-for-life-cycle-analyses-of-low-carbon-transportation-fuels-in-the-united-states (accessed on 10 November 2021).
48. Xikai, M.; Wang, L.; Li, J.; Quan, X.; Wu, T. Comparison of regression models for estimation of carbon emissions during building's lifecycle using designing factors: A case study of residential buildings in Tianjin, China. *Energy Build.* **2019**, *204*, 109519. [CrossRef]
49. Lim, M.Z.; Tan, E.S.; Chai, Y.H.; Chin, B.L.F.; Juwono, F.H.; Hisham, D.H.B.; Anuar, M.A.M. Pilot-scale flue gas pyrolysis system for organic and plastic wastes with improved liquid properties in a non-thermal plasma reactor. *J. Anal. Appl. Pyrolysis* **2023**, *173*, 106062. [CrossRef]
50. Malaysian Investment Development Authority (MIDA). Crude Oil to Chemicals (COTC) Technology: A New Paradigm Shift. Available online: https://www.mida.gov.my/crude-oil-to-chemicals-cotc-technology-a-new-paradigm-shift/ (accessed on 20 July 2023).

sustainability

MDPI

Article

Green Energy Generated in Single-Chamber Microbial Fuel Cells Using Tomato Waste

Segundo Rojas-Flores [1,*], Magaly De La Cruz-Noriega [2], Luis Cabanillas-Chirinos [2], Santiago M. Benites [1], Renny Nazario-Naveda [3], Daniel Delfín-Narciso [4], Moisés Gallozzo-Cardenas [5], Félix Diaz [6], Emzon Murga-Torres [7] and Walter Rojas-Villacorta [8]

1 Vicerrectorado de Investigación, Universidad Autónoma del Perú, Lima 15842, Peru; santiago.benites@autonoma.pe
2 Instituto de Investigación en Ciencias y Tecnología, Universidad Cesar Vallejo, Trujillo 13001, Peru; mdelacruzn@ucv.edu.pe (M.D.L.C.-N.); lcabanillas@ucv.edu.pe (L.C.-C.)
3 Departamento de Ciencias, Universidad Privada del Norte, Trujillo 13007, Peru; renny.nazario@upn.pe
4 Grupo de Investigación en Ciencias Aplicadas y Nuevas Tecnologías, Universidad Privada del Norte, Trujillo 13007, Peru; daniel.delfin@upn.edu.pe
5 Departamento de Ciencias, Universidad Tecnológica del Perú, Trujillo 13011, Peru; c21228@utp.edu.pe
6 Escuela Académica Profesional de Medicina Humana, Universidad Norbert Wiener, Lima 15842, Peru; felix.diaz@uwiener.edu.pe
7 Laboratorio de Investigación Multidisciplinario, Universidad Privada Antenor Orrego (UPAO), Trujillo 13008, Peru; ee_mt_66@hotmail.com
8 Programa de Investigación Formativa, Universidad Cesar Vallejo, Trujillo 13001, Peru; wrojasv33@gmail.com
* Correspondence: segundo.rojas.89@gmail.com

Abstract: This research used tomato waste as a substrate (fuel) in Single Chamber-Microbial Fuel Cells (scMFC) on a small scale. The electrochemical properties were monitored, the functional groups of the substrate were analyzed by Fourier Transform Infrared Spectrophotometry (FTIR) and a microbiological analysis was performed on the electrodes in order to identify the microorganisms responsible for the electrochemical process. The results show voltage peaks and an electrical current of 3.647 ± 0.157 mA and 0.957 ± 0.246 V. A pH of 5.32 ± 0.26 was measured in the substrate with an electrical current conductivity of 148,701 ± 5849 mS/cm and an internal resistance (R_{int}) of 77. 517 ± 8.541 Ω. The maximum power density (PD) displayed was 264.72 ± 3.54 mW/cm^2 at a current density (CD) of 4.388 A/cm^2. On the other hand, the FTIR spectrum showed a more intense decrease in its peaks, with the compound belonging to the phenolic groups being the most affected at 3361 cm^{-1}. The micrographs show the formation of a porous biofilm where molecular identification allowed the identification of two bacteria (*Proteus vulgaris* and *Proteus vulgaris*) and a yeast (*Yarrowia lipolytica*) with 100% identity. The data found show the potential of this waste as a source of fuel for the generation of an electric current in a sustainable and environmentally friendly way, generating in the near future a mechanism for the reuse of waste in a beneficial way for farmers, communities and agro-industrial companies.

Keywords: organic waste; tomato; microbial fuel cells; electric power; biomass; metal electrodes; bioenergy extraction

Citation: Rojas-Flores, S.; De La Cruz-Noriega, M.; Cabanillas-Chirinos, L.; Benites, S.M.; Nazario-Naveda, R.; Delfín-Narciso, D.; Gallozzo-Cardenas, M.; Diaz, F.; Murga-Torres, E.; Rojas-Villacorta, W. Green Energy Generated in Single-Chamber Microbial Fuel Cells Using Tomato Waste. *Sustainability* **2023**, *15*, 10461. https://doi.org/10.3390/su151310461

Academic Editors: Steven Lim, Shuit Siew Hoong, Pang Yean Ling and Santi Chuetor

Received: 13 April 2023
Revised: 8 June 2023
Accepted: 26 June 2023
Published: 3 July 2023

1. Introduction

The uncontrolled increase in the consumption of electrical energy from fossil sources has caused serious damage to the environment, resulting in significant changes in the climate and natural disasters that have affected many countries [1,2]. Humans use this type of energy daily for their activities, becoming a good and primary source worldwide (81%), divided into oil (32%), coal (27.1%), and gas (22%) [3,4]. Each year, more than 130 million tons are processed, and thus far, approximately eight million have already been processed [5]. In addition, recent statistics have shown that the consumption of electrical energy has increased from 13.72 billion KWH in 2000 to 24.57 billion KWH in 2018, which was even higher in the pandemic era [6,7]. On the flip side, the generation of fruits and vegetables has increased exponentially in recent years because human society has begun to consume healthier food, resulting in an economic boost for agro-industrial companies that export these products [8]. However, this increase in production has also generated large amounts of organic waste (agro-industrial waste), which has become a problem for governments because they do not have an organized method of collecting and managing this type of waste [9,10]. It has been reported that municipal waste in 2017 in Europe was 58 million tons, of which 46% was organic waste, which is estimated to increase by 32% by 2030 [11]. Among all the fruits produced by agro-industrial companies, tomato production (*Solanum lycopersicum* L.) is one of the most important; in 2019, its global production was estimated at approximately 37 million tons, and it has been estimated that of every 130 million tons produced, approximately 8 million tons represent the waste generated according to the World Processing Tomato Council (WPTC) [12–14]. This fruit contains natural bioactive compounds and antioxidants and has anti-inflammatory properties, which have beneficial health effects [15].

In the search to solve these two major problems that afflict the world, microbial fuel cell technology is presented as a novel solution because it uses any type of waste as fuel to generate electricity, which is due to the conversion of chemical energy into electricity through the reduction-oxidation process that occurs within cells [16,17]. The MFC design can have two chambers (anodic and cathodic) and a proton exchange membrane that joins them on the inside and an external electrical circuit [18]. A large number of designs and the use of various materials for their construction have been found in the literature, but one of the best-known is the single-chamber microbial fuel cell (scMFCs). The scMFCs is a low-cost cell. with its anode and cathode chambers in the same compartment, and because the anode is in direct contact with the environment (O_2), high power density values are obtained, showing great potential for large-scale scaling [19–21].

In the literature, research has been conducted on the use of different types of organic waste as a substrate for the generation of bioelectricity; for example, Asefi et al. (2019) managed to generate 600 mV voltage spikes in their dual-chamber microbial fuel cells using food waste as fuel and carbon felt as electrodes [22]. Likewise, Mohamed et al. (2020) generated voltage peaks of approximately 585 mV and a maximum power density of 33 mW/m^2 in microbial fuel cells using graphite as electrodes [23]. Yaqoob et al. (2022), using a mixture of rambutan, langsat, and mango residues, managed to generate peaks of 175 mV and a power density (PD) of 0.30 mW/m^2 in their scMFCs with graphite electrodes [24]. In another study, Kondaveeti et al. (2019) used citrus peels in single-chamber microbial fuel cells and generated maximum peak voltages and power densities of 284 mV and 26.41 mW/m^2, respectively, using graphite plates as electrodes [25]. Thus, it has also been observed that the values of electric current can increase with the use of electrodes of a metallic nature due to the intrinsic properties that these have to the passage of electrons for the generation of electricity, the current values have been compared to graphite electrodes, coal and its derivatives with equal amounts of substrate volume, microbial fuel cell design and with equal environmental parameters; managing to generate higher voltage, current and power density peaks [26–28].

This research has as main objective to observe the potential to generate bioelectricity through MFCs at a laboratory scale using tomato residues as a substrate. To achieve this, pH, current, voltage, pH, and electrical conductivity were monitored. Likewise, the values of the power density, current density, and $R_{int.}$ of the microbial fuel cells were measured, as well as the micrographs of the biofilm of the anodic electrode. Finally, the microorganisms present in the anode electrode at the end of the monitoring period were identified, and the FTIR spectra of the initial and final substrates were observed. This research will demonstrate the potential of tomato waste for use as fuel in single-chamber microbial fuel cells, using low-cost materials such as copper and zinc electrodes. Likewise, it will be possible to identify microorganisms that generate electrical energy for their subsequent use as biocatalysts, which will be able to repower microbial fuel cells. The replication of this technology on a large scale could benefit companies dedicated to the sale and purchase of these fruits in the near future.

2. Materials and Methods

2.1. Design of scMFCs

Three single-chamber microbial fuel cells (scMFCs) were manufactured for which copper (Cu) electrodes were used in the anode and zinc (Zn) in the cathode, a 500 mL polymethylmethacrylate tube was used and two 10 cm^2 lids were used. At the ends of the tube, a 5 cm hole was made at one end for the cathode to have contact with the environment (O_2), with an area of 60.12 cm^2. The electrodes were joined by an external resistor joined with a copper wire (0.15 cm diameter) (see Figure 1).

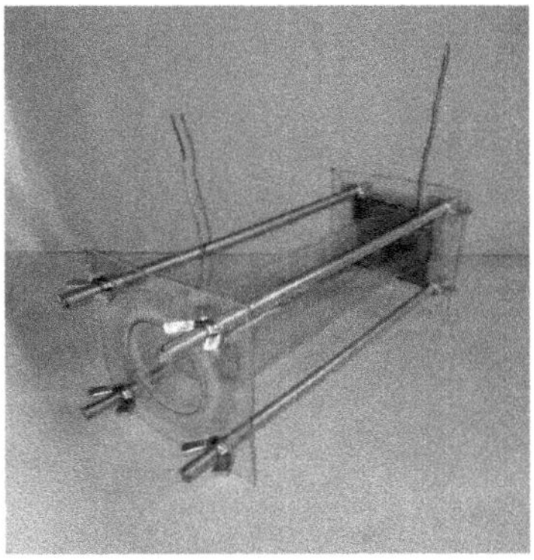

Figure 1. Schematic of the single-chamber microbial fuel cells prototype.

2.2. Collection and Preparation of Tomato Waste

Organic tomato waste was collected from the La Hermelinda plant in Trujillo, Peru, washed several times with distilled water to remove impurities (dust, insects or other impurities) and dried in an oven (Labtron, Camberley, UK, LDO-B10) for 12 h at 25 ± 1 °C. Obtain tomato juice (2 L, 500 mL per scMFC) using an extractor (Maqorito, Lima, Peru, 400 rpm).

2.3. Characterization of the scMFC

The monitoring time for the chemical–electrical parameters was 35 days. The voltage and current were measured using a multimeter (Prasek Premium PR-85) with an external resistance of 1000 Ω. The values of current density (CD) and power density (PD) were calculated according to the method of Segundo et al. (2023), with the same external resistance values [29]. Conductivity and pH were measured using a conductivity meter CD-4301 and a pH meter (Oakton Series 110), the transmittance values were measured by FTIR (Thermo Scientific IS50, Waltham, MA, USA). The electrochemical impedance spectroscopy (IES) measurements were evaluated with a Metrohm Autolab 302N potentiostat/galvanostat system on ensembled MFCs. The impedance spectra were obtained within the frequency range of 10 mHz and 100 kHz with a constant voltage of 10 mV and the Nyquist plot is shown in Figure.

2.4. Isolation and Molecular Identification of Anode Microorganisms

2.4.1. Isolation of Bacteria

From the anode (Copper-Cu plate), a sample was obtained by swabbing with a sterile swab. The sample was transferred to BHI agar, and the sample was extended with a bacteriological loop, making exhaustion streaks over the entire surface of the plate. The culture media were incubated in anaerobic jars at 36 ± 1 °C for 24 to 48 h. The colonies that grew in the culture medium were seeded on Nutrient Agar and MacConkey Agar. From the growth in these cultures, pure cultures were made on slant Nutrient Agar for later identification.

2.4.2. Yeast Isolation

Sampling was carried out in the same way as the previous procedure, the only variation was the culture medium and the incubation time. The medium used both for isolation and for pure culture was Sabouraud Agar with 4% dextrose plus antibiotic (chloramphenicol). While the incubation time was 30 ± 1 °C for 48 to 72 h.

2.4.3. Molecular Identification

The identification was carried out by the Analysis and Research Center of the laboratory "Biodes Laboratorios". To accomplish this, axenic cultures of bacteria and yeasts were sent to extract their genetic material (DNA) using the CTAB method. Amplification was carried out by the PCR technique, and then sequenced in the Macrogen Laboratory (USA). The sequences obtained were analyzed using the bioinformatics software Molecular Evolutionary Genetics Analysis (MEGA). Alignment and comparison with other sequences were carried out using the BLAST program (https://blast.ncbi.nlm.nih.gov/Blast.cgi (accessed on 2 April 2023)), through which percentages of identity were obtained for both bacteria and yeast. Finally, dendrograms of the species identified using MEGA X were constructed.

3. Results and Analysis

Figure 2a shows the voltage values observed during the monitoring of the scMFCs, which showed a voltage increase from day 1 (0.801 ± 0.014 V) to day 12 (0.957 ± 0.246 V) and later a slow decrease until day 35 (0.414 ± 0.297 V). Studies have shown that the highest voltage values in MFCs are directly proportional to the oxidation reactions that occur within these cells, while the decrease in the final monitoring stage is due to the sedimentation of the used substrate [30,31]. In this sense, Aiyer, K. (2020) in their research mentions that fruits with a high sucrose content generate a greater amount of voltage compared to those that do not contain this compound, which is why sucrose can be added to the waste used to observe any increase [32]. Figure 2b shows the values of electric current monitored at the scMFCs, where the values increased from day 1 (2.026 ± 0.008 mA) to day 11 (3.647 ± 0.157 mA) and later decreased until the last day (1.91 ± 0.487 mA). The increase in electric current values is due to the high content of nutrients that contributed to microbial growth and a good biofilm formed on the anode electrode, while the decrease in electric current values is due to the decrease in nutrients and the negative effect of the copper electrode used on microorganisms in the final stage [33–35]. An important factor that negatively affects the electrical values of the MFCs is the parameters of pH, temperature, and electrical conductivity, which in the final stage of monitoring were affected by the sedimentation observed in recent days due to inert organic matter [36].

Figure 3a shows the values of electrical conductivity observed from the substrate used, during the 35 days of monitoring; being able to observe that the values increased up to 148.701 ± 5.849 mS/cm on the eleventh day to later show a decrease until the last day (53.625 ± 3.562 mS/cm) monitored. The results before this investigation gave lower electrical conductivity values because distilled water was used for it; but the values increased using tomato waste as a substrate. The high values of electrical conductivity observed indicate the low internal resistance of the scMFCs [36], which, compared to other works, for example, Yaqoob et al. (2022) showed an electrical conductivity of 81 mS/cm using wastewater as a substrate, managing to generate voltage peaks of 191 mV [37], which is due to the fact that tomato has chemical compounds that help conduct electrons more easily [38]. On the other hand, the decrease in their values is due to the precipitation of tomato waste at the bottom of the scMFCs in the process of generating electrical energy [39,40]. Figure 3b shows the monitoring of the pH values of the scMFCs, being able to observe that the pH increases from the first (slightly acid) to the last (neutral) day, with its optimum operating pH being on day eleven (5.32 ± 0.26). The optimal operating values of each MFC vary, and different types of substrates used with different pH have been found in the literature; one of the main reasons for this phenomenon is that the microorganisms are different and each one grows at specific pH conditions, affecting the system performance [41,42]. For example, Simeon et al. (2020) used sweet potato-shochu waste as a substrate at a pH of approximately 4, managing to generate 0.0002 and 0.08 W/m^3 of power density at pH 4.5 and 4.2, respectively, in their scMFCs [43]. Likewise, Priya A. and Setty Y. (2019) used apple waste at pH 4.1, managing to generate voltage peaks and power density of 0.35 V and 31.58 mW/m^2 in their MFC [40].

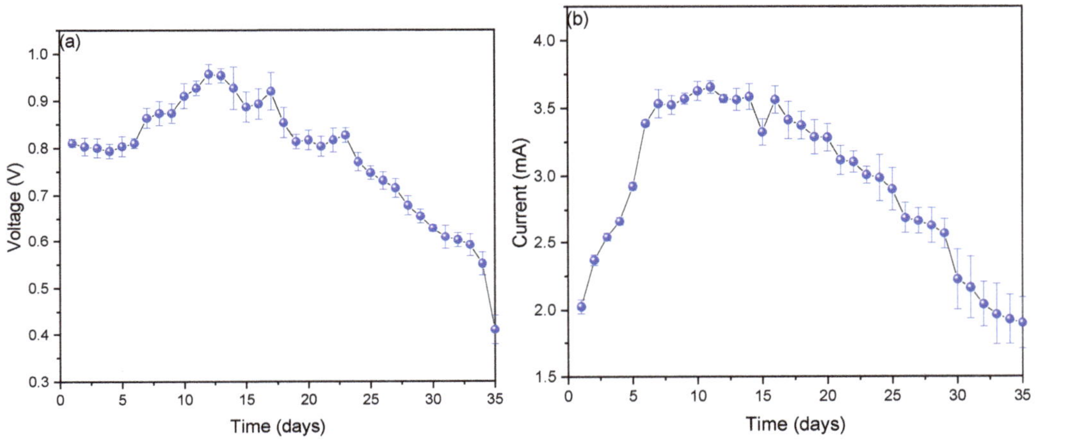

Figure 2. Values of (**a**) voltage and (**b**) electric current obtained during the monitoring of the scMFC.

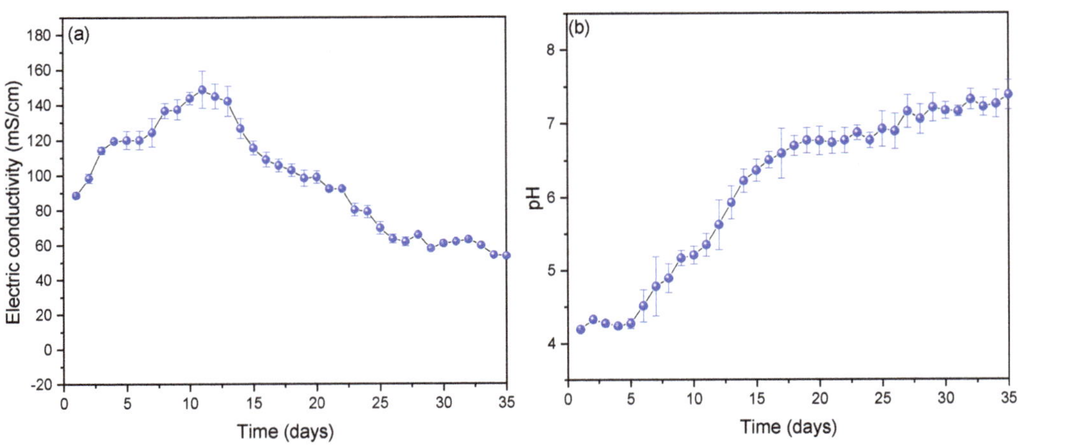

Figure 3. Values of (**a**) electrical conductivity and (**b**) pH obtained from the monitoring of the MFC-SC.

To calculate the internal resistance of the MFC-SC, Ohm's Law (V = IR) was used, where the voltage values were placed on the "y" axis and those of the electric current on the "x" axis, which when performing a linear adjustment, the slope of the line is the internal resistance of the system, see Figure 4a. Based on the work carried out by Christwardana et al. (2020), this method represents an effective tool for the in-depth study of MFCs [16,44], as unlike the electrochemical impedance method (EIS) used in other investigations, internal resistance is obtained more directly. The IES method allows us to find the different components of the internal resistance, such as the resistance of charge transfer, diffusion, and resistance of the substrate, which is obtained with the intersection of the Nyquist curve with the "x" axis measured from the origin. Calculated Internal Resistance (Rint) was measured on the fourteenth day because this was the day in which the highest values of current and voltage were found, giving a value of Rint. of 77.517 ± 8.541 Ω, this value is relatively low compared to other investigations. For example, Antonopoulou et al. (2019) showed an internal resistance of 38 Ω (by electrochemical characterization), managing to generate a peak voltage of 0.488 V using vegetable debris as substrates in their MFC-SC and graphite as electrodes [41]. So too, Daud et al. (2021) used rice waste as a substrate, managing to show an internal resistance of 363.3 Ω (by Electrochemical tests) using graphite rods as electrodes, generating 510 mV on the fourteenth day [42]. One of the highest values found in the literature was the one reported by Simeon et al. (2020), where they were able to generate 725 mV peaks using urine treatment waste as a substrate in their MFC-SC with carbon electrodes, showing an initial and final internal resistance of 269.94 and 1627.89 Ω (by polarization method), respectively, resistance values increase from your control sample to the highest concentration of substrate (368.56 to 676.43 Ω) [43]. Figure 4b shows the power density (PD) values as a function of current density (CD), achieving a maximum power density of 264.72 ± 3.54 mW/cm^2 at a current density of 4.388 A/cm^2 with a peak voltage of 879.56 ± 0.184 V. The high values shown are due to the high inherent conductivity of the electrodes used, since being metallic in nature they facilitate the passage of electrons within the electrical circuit [44]. Other researchers have managed to generate higher values, but with the help of catalysts or biocatalysts, for example, Yaqoob et al. (2022) managed to generate a maximum power density of 41.58 mW/m^2 with an internal resistance of 813.73 Ω using food waste (rice mixed with curry, vegetables, fish, cabbage, bones, and sweet cake pieces) as a substrate in MFC with graphite rod electrodes, compared to his control sample of 14.1 mW/m^2, the increase was notorious [45]. Likewise, Du H. and Shao Z. (2022) managed to generate a maximum power density of 14.1 mW/m^2 at a current density of 320.1 mA/m^2 using potato waste as a substrate in their MFCs using carbon felts as electrodes [46]. Kamperidis et al. (2022) managed to generate a maximum power density of 14.4 mW/m^2 at a current density of approximately 90 mA/m^2 in their MFC-SCs using MnO_2-coated graphites as electrodes and a condensed food substrate, but these values do not show much of a difference from their control sample, where a PD of 11.5 mW/m^2 was observed [47].

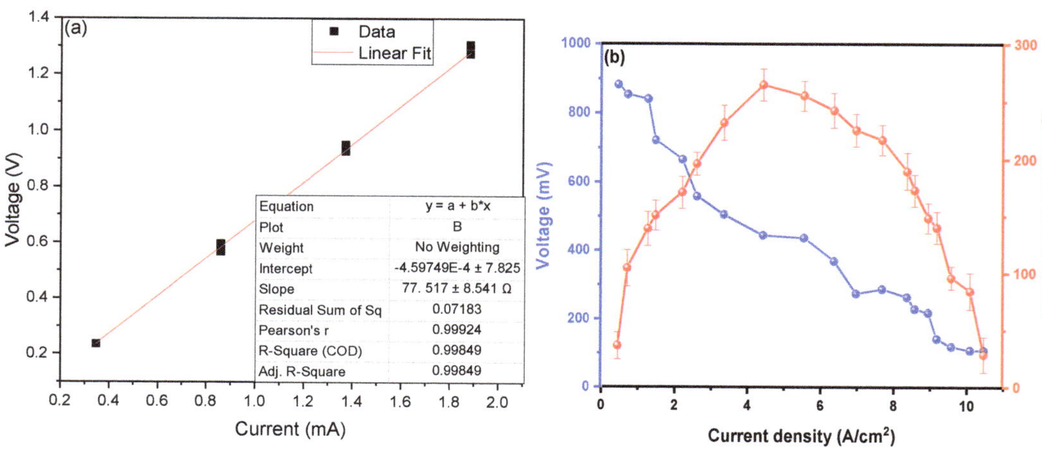

Figure 4. (**a**) $R_{int.}$ and (**b**) PD and voltage versus CD of the scMFC.

Figure 5 shows the FTIR spectra of the initial and final tomato waste, being able to observe the most intense peak at the 3361 cm^{-1} wavelength belonging to the N-H bond, O-H groups, carboxylic acids and phenolic groups, while the peak at 2917 cm^{-1} belongs to the C-H stretch of alkanes, ketones and aldehyde compounds [46]; while the 1712 and 1616 cm^{-1} peaks are associated with amine stretching and C=C [47]. Likewise, the 1404 cm^{-1} peak confirms the presence of the C-H, C=C, C==N, and C-N stretching belonging to the alkane, alkene, primary and secondary amine, respectively; finally, the 1052 cm^{-1} peak belongs to the C-H stretch of the alkane and the amide [48–50]. The decrease in the FTIR spectrum is mainly due to the decrease in the chemical components in the process of generating electrical energy by the microorganisms [51,52], as well as the degradation of the organic material that was observed in the last days of monitoring of the scMFC. Microorganisms consume carbon-rich compounds for their metabolism, this process causes electrons to be released and captured by the anode electrode and flow to the cathode, thus generating electric current [53]. In Figure 6, the micrographs of the initial and fine anodic electrode are observed, being able to observe a smooth surface belonging to the copper electrode; while the image belonging to the final state shows the formation of a porous biofilm due to the adhesion of the microorganisms present in the substrate, which were the electrical energy generators of the electrical device, observe the presence of carbon, oxygen, sodium, magnesium, calcium, copper and zinc; mainly in the EDS of the electrode in its final state. Hou et al. (2022) reported that the porosity in the form of round globules was due to the biomass forming an external layer on top of the biofilm, generating a double biolayer [54]. Hirose et al. (2023) also mentioned that the formation and adhesion of the biofilm to the electrodes optimizes the generation of electricity because the microorganisms generate the greatest number of electrons for the generation of electric current, but the durability of this depends on the environmental conditions given by the researchers [55].

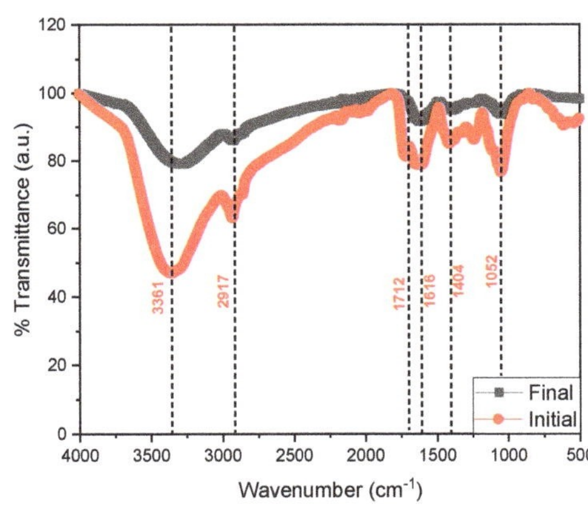

Figure 5. Fourier transform infrared spectrum of tomato waste.

The Blast characterization performed at the anode electrode can be seen in Table 1, where two bacterial isolates (MT-B01 and MT-B02) and one yeast (MT-L01) were obtained from the anodic growth within the MFC. The two bacteria were identified as *Proteus vulagris* (100.00% identity), while the yeast corresponded to *Yarrowia lipolytica* (100.00% identity). Meanwhile, in Figures 7 and 8, the dendrogram of the identified species and the phylogenetic relationships with other species support the Blast characterization. In this sense, the microorganisms identified (Table 1) correspond to the bacterium *P. vulgaris* (100.00% identity) and the *Y. lipolytic* yeast (100.00% identity). The microorganisms were isolated from tomato waste, and their origin may be due to contamination with soil or contaminated water. For example, *P. vulgaris*, is a Gram-negative rod-shaped bacterium commonly present in habitats such as soil and fecal matter [54]. In the same way, *Y. lipolytica* is a yeast belonging to the genus *Candida*, which can be isolated from food products and soil. This species stands out for its ability to degrade lipids and proteins [55].

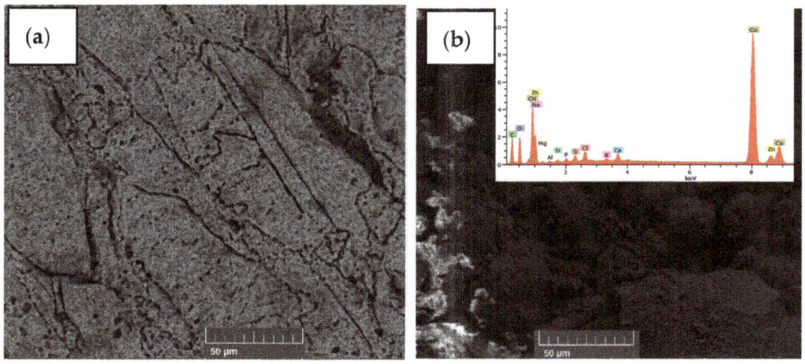

Figure 6. Micrographs (**a**) initial and (**b**) final of the surface belonging to the anode electrode of the scMFC.

Table 1. Blast characterization of the rDNA sequence of microorganisms isolated from the anode plate of the scMFC with tomato waste substrates.

ID Sample	BLAST Characterization	Length of Consensus Sequence (nt)	% Identity	Accession Number	Phylogeny
			Bacteria		
MT-B01	*Proteus vulgaris*	1467	100.00	CP023965.1	*Cellular organisms; Bacteria; Proteobacteria; Gammaproteobacteria; Enterobacterales; Morganellaceae; Proteus*
MT-B02	*Proteus vulgaris*	1466	100.00	CP023965.1	*Cellular organisms; Bacteria; Proteobacteria; Gammaproteobacteria; Enterobacterales; Morganellaceae; Proteus*
			Yeast		
MT-L01	*Yarrowia lipolytica*	368	100.00	MN124085.1	*Cellular organisms; Eukaryota; Opisthokonta; Fungi; Dikarya; Ascomycota; saccharomyceta; Saccharomycotina; Saccharomycetes; Saccharomycetales; Dipodascaceae; Yarrowia*

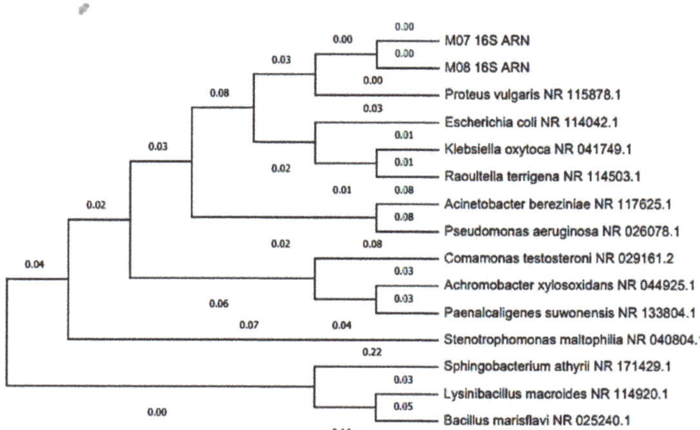

Figure 7. Dendrogram of bacteria isolated from the scMFCs anode plate with tomato waste substrate.

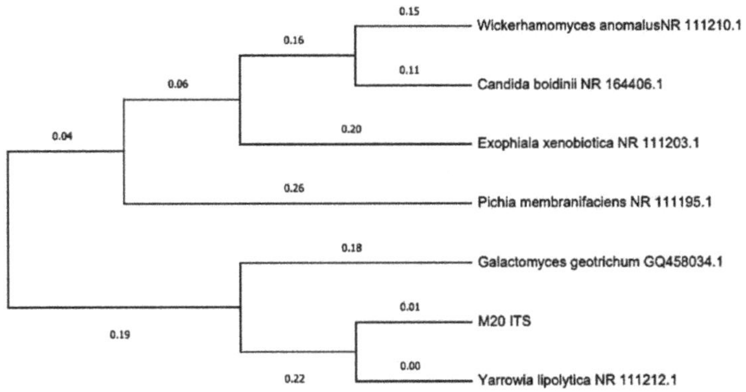

Figure 8. Dendrogram based on the ITS regions of *Yarrowia lipolytica* isolated from the scMFC anode plate with tomato waste substrate.

The identification of both bacteria and yeast indicates that they are part of a microbial community that developed on the MFC anode, and to which electricity generation can be attributed [56,57]. Figure 7 shows that *P. vulgaris* is phylogenetically related to other Enterobacteriaceae. Some studies indicate that some enterobacteria, such as *P. vulgaris*, may be associated with the generation of electricity in MFC with organic waste substrates [58]. On the other hand, it is known that in most studies on microbial communities of MFC anodes, bacteria belonging to the Proteobacteria phylum are frequently isolated [59,60]. In other investigations, *P. vulgaris* has been used in MFCs to generate bioelectricity due to its metabolic reducing power, which through electron mediators can improve the energy efficiency of the MFC [54,61,62].

Regarding the identified yeast, *Y. lipolitica* belongs to the Ascomycota phylum. Other species within this phylum have been studied in MFCs to generate bioelectricity, such as *Candida boidinii*, and *Saccharomyces cerevisiae*, among others [63–65]. However, these yeasts are phylogenetically distant as shown in Figure 8. Studies have shown that using yeast in MFCs has advantages over bacteria, being classified as ideal biocatalysts since most are not pathogenic and grow in a wide range of organic substrates [63,66]. These can transfer electrons to the anode through the use of mediators [63]; however, there is evidence that there may be transfer by direct contact [67].

4. Conclusions

An electric current was successfully generated using tomato residue as a substrate in scMFC at a laboratory scale using zinc and copper electrodes. It was possible to generate voltage and electric current peaks of 0.957 $\pm$ 0.246 V and 3.647 $\pm$ 0.157 mA on days 12 and 11, respectively. The optimum operating pH was 5.32 $\pm$ 0.26 with an electrical conductivity of 148.701 $\pm$ 5.849 mS/cm. Thus, it also showed a low internal resistance of 77.517 $\pm$ 8.541 Ω with a power density of 264.72 $\pm$ 3.54 mW/cm^2 at a current density of 4.388 A/cm^2. On the other hand, the FTIR spectra showed a decrease in their peaks between the initial and final spectrum, being the peak belonging to the phenolic groups (3361 cm^{-1}) the one that presented the greatest decrease, while the SEM micrographs show the formation of a porous biofilm. Finally, the molecular identification of the microorganisms showed the presence of two bacteria (*Proteus vulgaris* and *Proteus vulgaris*) and one yeast (*Yarrowia lipolytica*) with 100% identity on the anode electrode. This research contributes greatly to the sustainability of electric power generation in remote locations, where this type of fruit or its derivatives are harvested so that they can use their own waste as fuel in combination with larger-scale microbial fuel cells, with efficient enough to power a house overnight. Likewise, in the near future, companies will be able to use this technology as a fuel to generate electricity and reduce their expenses, all in an environmentally sustainable environment. For future work, it is recommended to standardize the pH value (5.32) to obtain the maximum potential of this residue, based on the results of this investigation; as well as covering the metallic electrodes with some type of chemical compound that is not harmful to microorganisms and the use of a catalyst or biocatalyst to enhance the generation of electrical energy from this device.

Author Contributions: Conceptualization, S.R.-F., L.C.-C. and M.G.-C.; methodology S.M.B. and W.R.-V.; software, R.N.-N.; validation, W.R.-V. and F.D.; formal analysis, S.R.-F., E.M.-T. and M.D.L.C.-N.; investigation S.R.-F. and W.R.-V.; data curation, M.D.L.C.-N. and M.G.-C.; writing—original draft preparation, D.D.-N.; writing—review and editing, S.R.-F. and E.M.-T.; project administration, S.R.-F. and R.N.-N. All authors have read and agreed to the published version of the manuscript.

Funding: This research received no external funding.

Institutional Review Board Statement: Not applicable.

Informed Consent Statement: Not applicable.

Data Availability Statement: Not applicable.

Conflicts of Interest: The authors declare no conflict of interest.

References

1. Opoku, R.; Obeng, G.Y.; Osei, L.K.; Kizito, J.P. Optimization of industrial energy consumption for sustainability using time-series regression and gradient descent algorithm based on historical electricity consumption data. *Sustain. Anal. Model.* **2022**, *2*, 100004. [CrossRef]
2. Li, X.; Wang, Y.; Ma, G.; Chen, X.; Fan, J.; Yang, B. Prediction of electricity consumption during epidemic period based on improved particle swarm optimization algorithm. *Energy Rep.* **2022**, *8*, 437–446. [CrossRef]
3. Ramos, D.; Faria, P.; Vale, Z.; Mourinho, J.; Correia, R. Industrial Facility Electricity Consumption Forecast Using Artificial Neural Networks and Incremental Learning. *Energies* **2020**, *13*, 4774. [CrossRef]
4. vom Scheidt, F.; Medinová, H.; Ludwig, N.; Richter, B.; Staudt, P.; Weinhardt, C. Data analytics in the electricity sector–A quantitative and qualitative literature review. *Energy AI* **2020**, *1*, 100009. [CrossRef]
5. Rojas-Flores, S.; De La Cruz-Noriega, M.; Nazario-Naveda, R.; Benites, S.M.; Delfín-Narciso, D.; Rojas-Villacorta, W.; Romero, C.V. Bioelectricity through microbial fuel cells using avocado waste. *Energy Rep.* **2022**, *8*, 376–382. [CrossRef]
6. Masebinu, R.; Kambule, N. Electricity consumption data of a middle-income household in Gauteng, South Africa: Pre and Post COVID-19 lockdown (2019–2021). *Data Brief* **2022**, *43*, 108341. [CrossRef]
7. Hadjout, D.; Torres, J.; Troncoso, A.; Sebaa, A.; Martínez-Álvarez, F. Electricity consumption forecasting based on ensemble deep learning with application to the Algerian market. *Energy* **2022**, *243*, 123060. [CrossRef]
8. Chew, K.W.; Chia, S.R.; Yen, H.-W.; Nomanbhay, S.; Ho, Y.-C.; Show, P.L. Transformation of Biomass Waste into Sustainable Organic Fertilizers. *Sustainability* **2019**, *11*, 2266. [CrossRef]
9. Wainaina, S.; Awasthi, M.K.; Sarsaiya, S.; Chen, H.; Singh, E.; Kumar, A.; Ravindran, B.; Awasthi, S.K.; Liu, T.; Duan, Y.; et al. Resource recovery and circular economy from organic solid waste using aerobic and anaerobic digestion technologies. *Bioresour. Technol.* **2020**, *301*, 122778. [CrossRef]
10. Ahuja, I.; Dauksas, E.; Remme, J.F.; Richardsen, R.; Løes, A.-K. Fish and fish waste-based fertilizers in organic farming–With status in Norway: A review. *Waste Manag.* **2020**, *115*, 95–112. [CrossRef]
11. Rocamora, I.; Wagland, S.T.; Villa, R.; Simpson, E.D.; Fernández, O.; Bajón-Fernández, Y. Dry anaerobic digestion of organic waste: A review of operational parameters and their impact on process performance. *Bioresour. Technol.* **2020**, *299*, 122681. [CrossRef] [PubMed]
12. Ronga, D.; Caradonia, F.; Parisi, M.; Bezzi, G.; Parisi, B.; Allesina, G.; Pedrazzi, S.; Francia, E. Using Digestate and Biochar as Fertilizers to Improve Processing Tomato Production Sustainability. *Agronomy* **2020**, *10*, 138. [CrossRef]
13. Trombino, S.; Cassano, R.; Procopio, D.; Di Gioia, M.L.; Barone, E. Valorization of Tomato Waste as a Source of Carotenoids. *Molecules* **2021**, *26*, 5062. [CrossRef] [PubMed]
14. Løvdal, T.; Van Droogenbroeck, B.; Eroglu, E.C.; Kaniszewski, S.; Agati, G.; Verheul, M.; Skipnes, D. Valorization of Tomato Surplus and Waste Fractions: A Case Study Using Norway, Belgium, Poland, and Turkey as Examples. *Foods* **2019**, *8*, 229. [CrossRef]
15. Domínguez, R.; Gullón, P.; Pateiro, M.; Munekata, P.E.S.; Zhang, W.; Lorenzo, J.M. Tomato as Potential Source of Natural Additives for Meat Industry. A Review. *Antioxidants* **2020**, *9*, 73. [CrossRef]
16. Christwardana, M.; Hadiyanto, H.; Motto, S.A.; Sudarno, S.; Haryani, K. Performance evaluation of yeast-assisted microalgal microbial fuel cells on bioremediation of cafeteria wastewater for electricity generation and microalgae biomass production. *Biomass Bioenergy* **2020**, *139*, 105617. [CrossRef]
17. Yaqoob, A.A.; Al-Zaqri, N.; Yaakop, A.S.; Umar, K. Potato waste as an effective source of electron generation and bioremediation of pollutant through benthic microbial fuel cell. *Sustain. Energy Technol. Assess.* **2022**, *53*, 102560. [CrossRef]

18. Obileke, K.; Onyeaka, H.; Meyer, E.L.; Nwokolo, N. Microbial fuel cells, a renewable energy technology for bio-electricity generation: A mini-review. *Electrochem. Commun.* **2021**, *125*, 107003. [CrossRef]
19. Yaqoob, A.A.; Ibrahim, M.N.M.; Rodríguez-Couto, S. Development and modification of materials to build cost-effective anodes for microbial fuel cells (MFCs): An overview. *Biochem. Eng. J.* **2020**, *164*, 107779. [CrossRef]
20. Prathiba, S.; Kumar, P.S.; Vo, D.-V.N. Recent advancements in microbial fuel cells: A review on its electron transfer mechanisms, microbial community, types of substrates and design for bio-electrochemical treatment. *Chemosphere* **2022**, *286*, 131856. [CrossRef]
21. Gul, H.; Raza, W.; Lee, J.; Azam, M.; Ashraf, M.; Kim, K.-H. Progress in microbial fuel cell technology for wastewater treatment and energy harvesting. *Chemosphere* **2021**, *281*, 130828. [CrossRef] [PubMed]
22. Asefi, B.; Li, S.-L.; Moreno, H.A.; Sanchez-Torres, V.; Hu, A.; Li, J.; Yu, C.-P. Characterization of electricity production and microbial community of food waste-fed microbial fuel cells. *Process. Saf. Environ. Prot.* **2019**, *125*, 83–91. [CrossRef]
23. Mohamed, S.N.; Hiraman, P.A.; Muthukumar, K.; Jayabalan, T. Bioelectricity production from kitchen wastewater using microbial fuel cell with photosynthetic algal cathode. *Bioresour. Technol.* **2020**, *295*, 122226. [CrossRef] [PubMed]
24. Yaqoob, A.A.; Guerrero–Barajas, C.; Ibrahim, M.N.M.; Umar, K.; Yaakop, A.S. Local fruit wastes driven benthic microbial fuel cell: A sustainable approach to toxic metal removal and bioelectricity generation. *Environ. Sci. Pollut. Res.* **2022**, *29*, 32913–32928. [CrossRef]
25. Kondaveeti, S.; Mohanakrishna, G.; Kumar, A.; Lai, C.; Lee, J.-K.; Kalia, V.C. Exploitation of Citrus Peel Extract as a Feedstock for Power Generation in Microbial Fuel Cell (MFC). *Indian J. Microbiol.* **2019**, *59*, 476–481. [CrossRef]
26. Segundo, R.-F.; De La Cruz-Noriega, M.; Nazario-Naveda, R.; Benites, S.M.; Delfín-Narciso, D.; Angelats-Silva, L.; Díaz, F. Golden Berry Waste for Electricity Generation. *Fermentation* **2022**, *8*, 256. [CrossRef]
27. Zafar, H.; Peleato, N.; Roberts, D. A comparison of reactor configuration using a fruit waste fed two-stage anaerobic up-flow leachate reactor microbial fuel cell and a single-stage microbial fuel cell. *Bioresour. Technol.* **2023**, *374*, 128778. [CrossRef]
28. Rahman, W.; Yusup, S.; Mohammad, S.A. Screening of fruit waste as substrate for microbial fuel cell (MFC). In *AIP Conference Proceedings*; AIP Publishing LLC: Arau, Malaysia, 2021; Volume 2332, p. 020003.
29. Rojas-Flores, S.; De La Cruz-Noriega, M.; Cabanillas-Chirinos, L.; Nazario-Naveda, R.; Gallozzo-Cardenas, M.; Diaz, F.; Murga-Torres, E. Potential Use of Coriander Waste as Fuel for the Generation of Electric Power. *Sustainability* **2023**, *15*, 896. [CrossRef]
30. Prasidha, W. Electricity Production from Food Waste Leachate (Fruit and Vegetable Waste) using Double Chamber Microbial Fuel Cell: Comparison between Non-aerated and Aerated Configuration. *ROTASI* **2020**, *22*, 162–168.
31. Naveenkumar, M.; Senthilkumar, K. Microbial fuel cell for harvesting bio-energy from tannery effluent using metal mixed biochar electrodes. *Biomass Bioenergy* **2021**, *149*, 106082. [CrossRef]
32. Aiyer, K.S. How does electron transfer occur in microbial fuel cells? *World J. Microbiol. Biotechnol.* **2020**, *36*, 19. [CrossRef] [PubMed]
33. Koo, B.; Lee, S.-M.; Oh, S.-E.; Kim, E.J.; Hwang, Y.; Seo, D.; Kim, J.Y.; Kahng, Y.H.; Lee, Y.W.; Chung, S.-Y.; et al. Addition of reduced graphene oxide to an activated-carbon cathode increases electrical power generation of a microbial fuel cell by enhancing cathodic performance. *Electrochim. Acta* **2019**, *297*, 613–622. [CrossRef]
34. Yaqoob, A.A.; Serrà, A.; Ibrahim, M.N.M.; Yaakop, A.S. Self-assembled oil palm biomass-derived modified graphene oxide anode: An efficient medium for energy transportation and bioremediating Cd (II) via microbial fuel cells. *Arab. J. Chem.* **2021**, *14*, 103121. [CrossRef]
35. Hung, Y.-H.; Liu, T.-Y.; Chen, H.-Y. Renewable Coffee Waste-Derived Porous Carbons as Anode Materials for High-Performance Sustainable Microbial Fuel Cells. *ACS Sustain. Chem. Eng.* **2019**, *7*, 16991–16999. [CrossRef]
36. Guan, C.-Y.; Tseng, Y.-H.; Tsang, D.C.; Hu, A.; Yu, C.-P. Wetland plant microbial fuel cells for remediation of hexavalent chromium contaminated soils and electricity production. *J. Hazard. Mater.* **2019**, *365*, 137–145. [CrossRef]
37. Yaqoob, A.Y.; Hussian, N.A.; Alhejuje, M.M. Application of Heavy metals pollution index on two types of constructed wetland. *Marsh Bull.* **2022**, *17*, 22–29.
38. Sreelekshmy, B.R.; Basheer, R.; Sivaraman, S.; Vasudevan, V.; Elias, L.; Shibli, S.M.A. Sustainable electric power generation from live anaerobic digestion of sugar industry effluents using microbial fuel cells. *J. Mater. Chem. A* **2020**, *8*, 6041–6056. [CrossRef]
39. Iigatani, R.; Ito, T.; Watanabe, F.; Nagamine, M.; Suzuki, Y.; Inoue, K. Electricity generation from sweet potato-shochu waste using microbial fuel cells. *J. Biosci. Bioeng.* **2019**, *128*, 56–63. [CrossRef]
40. Priya, A.D.; Setty, Y.P. Cashew apple juice as substrate for microbial fuel cell. *Fuel* **2019**, *246*, 75–78. [CrossRef]
41. Antonopoulou, G.; Ntaikou, I.; Pastore, C.; di Bitonto, L.; Bebelis, S.; Lyberatos, G. An overall perspective for the energetic valorization of household food waste using microbial fuel cell technology of its extract, coupled with anaerobic digestion of the solid residue. *Appl. Energy* **2019**, *242*, 1064–1073. [CrossRef]
42. Daud, N.N.M.; Ahmad, A.; Yaqoob, A.A.; Ibrahim, M.N.M. Application of rotten rice as a substrate for bacterial species to generate energy and the removal of toxic metals from wastewater through microbial fuel cells. *Environ. Sci. Pollut. Res.* **2021**, *28*, 62816–62827. [CrossRef] [PubMed]
43. Simeon, M.I.; Asoiro, F.U.; Aliyu, M.; Raji, O.A.; Freitag, R. Polarization and power density trends of a soil-based microbial fuel cell treated with human urine. *Int. J. Energy Res.* **2020**, *44*, 5968–5976. [CrossRef]
44. Rojas-Flores, S.; Nazario-Naveda, R.; Benites, S.M.; Gallozzo-Cardenas, M.; Delfín-Narciso, D.; Díaz, F. Use of Pineapple Waste as Fuel in Microbial Fuel Cell for the Generation of Bioelectricity. *Molecules* **2022**, *27*, 7389. [CrossRef] [PubMed]

45. Yaqoob, A.A.; Bin Abu Bakar, M.A.; Kim, H.-C.; Ahmad, A.; Alshammari, M.B.; Yaakop, A.S. Oxidation of food waste as an organic substrate in a single chamber microbial fuel cell to remove the pollutant with energy generation. *Sustain. Energy Technol. Assess.* **2022**, *52*, 102282. [CrossRef]
46. Du, H.; Shao, Z. Synergistic effects between solid potato waste and waste activated sludge for waste-to-power conversion in microbial fuel cells. *Appl. Energy* **2022**, *314*, 118994. [CrossRef]
47. Kamperidis, T.; Pandis, P.K.; Argirusis, C.; Lyberatos, G.; Tremouli, A. Effect of Food Waste Condensate Concentration on the Performance of Microbial Fuel Cells with Different Cathode Assemblies. *Sustainability* **2022**, *14*, 2625. [CrossRef]
48. Segundo, R.-F.; Magaly, D.L.C.-N.; Benites, S.M.; Daniel, D.-N.; Angelats-Silva, L.; Díaz, F.; Luis, C.-C.; Fernanda, S.-P. Increase in Electrical Parameters Using Sucrose in Tomato Waste. *Fermentation* **2022**, *8*, 335. [CrossRef]
49. Rehal, J.K.; Aggarwal, P.; Dhaliwal, I.; Sharma, M.; Kaushik, P. A Tomato Pomace Enriched Gluten-Free Ready-to-Cook Snack's Nutritional Profile, Quality, and Shelf Life Evaluation. *Horticulturae* **2022**, *8*, 403. [CrossRef]
50. Ramya, S.; Chandran, M.; King, I.J.; Jayakumararaj, R.; Loganathan, T.; Pandiarajan, G.; Kaliraj, P.; Pushpalatha, G.G.L.; Abraham, G.; Vijaya, V.; et al. Phytochemical Screening, GCMS and FTIR Profile of Bioactive Compounds in Solanum lycopersicum Wild Fruits collected from Palani Hill Ranges of the Western Ghats. *J. Drug Deliv. Ther.* **2022**, *12*, 56–64. [CrossRef]
51. Zhang, Z.; Yang, Y.; Xi, H.; Yu, Y.; Song, Y.; Wu, C.; Zhou, Y. Evaluation methods of inhibition to microorganisms in biotreatment processes: A review. *Water Cycle* **2023**, *4*, 70–78. [CrossRef]
52. Su, F.; Wang, F.; Zhang, C.; Lu, T.; Zhang, S.; Zhang, R.; Qi, X.; Liu, P. Ameliorating substance accessibility for microorganisms to amplify toluene degradation and power generation of microbial fuel cell by using activated carbon anode. *J. Clean. Prod.* **2022**, *377*, 134481. [CrossRef]
53. Nawaz, A.; Haq, I.U.; Qaisar, K.; Gunes, B.; Raja, S.I.; Mohyuddin, K.; Amin, H. Microbial fuel cells: Insight into simultaneous wastewater treatment and bioelectricity generation. *Process. Saf. Environ. Prot.* **2022**, *161*, 357–373. [CrossRef]
54. Hou, B.; Liu, X.; Zhang, R.; Li, Y.; Liu, P.; Lu, J. Investigation and evaluation of membrane fouling in a microbial fuel cell-membrane bioreactor systems (MFC-MBR). *Sci. Total Environ.* **2022**, *814*, 152569. [CrossRef]
55. Hirose, S.; Inukai, K.; Nguyen, D.T.; Taguchi, K. Use of loofah electrodes coated with rice husk smoked charcoal and Japanese ink in a microbial fuel cell for muddy water treatment. *Energy Rep.* **2023**, *9*, 160–167. [CrossRef]
56. Mbugua, J.K.; Mbui, D.N.; Mwaniki, J.M.; Mwaura, F.B. Electricity Generation by Clostridiumspp and Proteus Vulgaris from Rotten Tomatoes in a Double Chamber Microbial Fuel Cell. *Int. J. Res. Stud. Microbiol. Biotechnol.* **2018**, *4*, 32–33.
57. Heard, G.M.; Fleet, G.H. CANDIDA | Yarrowia (Candida) Lipolytica. In *Encyclopedia of Food Microbiology*; Elsevier: Amsterdam, The Netherlands, 1999; pp. 360–365. ISBN 9780122270703.
58. Logan, B.E.; Regan, J.M. Electricity-Producing Bacterial Communities in Microbial Fuel Cells. *Trends Microbiol.* **2006**, *14*, 512–518. [CrossRef]
59. Gatti, M.N.; Milocco, R.H. A Biofilm Model of Microbial Fuel Cells for Engineering Applications. *Int. J. Energy Environ. Eng.* **2017**, *8*, 303–315. [CrossRef]
60. Rojas-Flores, S.; Cabanillas-Chirinos, L.; Nazario-Naveda, R.; Gallozzo-Cardenas, M.; Diaz, F.; Delfin-Narciso, D.; Rojas-Villacorta, W. Use of Tangerine Waste as Fuel for the Generation of Electric Current. *Sustainability* **2023**, *15*, 3559. [CrossRef]
61. Greenman, J.; Gajda, I.; You, J.; Mendis, B.A.; Obata, O.; Pasternak, G.; Ieropoulos, I. Microbial Fuel Cells and Their Electrified Biofilms. *Biofilm* **2021**, *3*, 100057. [CrossRef] [PubMed]
62. Cao, Y.; Mu, H.; Liu, W.; Zhang, R.; Guo, J.; Xian, M.; Liu, H. Electricigens in the Anode of Microbial Fuel Cells: Pure Cultures versus Mixed Communities. *Microb. Cell Fact.* **2019**, *18*, 39. [CrossRef]
63. Park, D.H.; Zeikus, J.G. Electricity Generation in Microbial Fuel Cells Using Neutral Red as an Electronophore. *Appl. Environ. Microbiol.* **2000**, *66*, 1292–1297. [CrossRef]
64. Kim, N.J.; Choe, Y.J.; Jeong, S.H.; Kim, S.H. Development of Microbial Fuel Cells Using Proteus Vulgaris. *Bull. Korean Chem. Soc.* **2000**, *21*, 44–48. [CrossRef]
65. Verma, M.; Mishra, V. Recent Trends in Upgrading the Performance of Yeast as Electrode Biocatalyst in Microbial Fuel Cells. *Chemosphere* **2021**, *284*, 131383. [CrossRef] [PubMed]
66. Rojas Flores, S.; Pérez-Delgado, O.; Naveda-Renny, N.; Benites, S.M.; De La Cruz-Noriega, M.; Delfin Narciso, D.A. Generation of Bioelectricity Using Molasses as Fuel in Microbial Fuel Cells. *Environ. Res. Eng. Manag.* **2022**, *78*, 19–27. [CrossRef]
67. Gunawardena, A.; Fernando, S.; To, F. Performance of a Yeast-Mediated Biological Fuel Cell. *Int. J. Mol. Sci.* **2008**, *9*, 1893–1907. [CrossRef] [PubMed]

sustainability

MDPI

Article

Investigation of Microwave Irradiation and Ethanol Pre-Treatment toward Bioproducts Fractionation from Oil Palm Empty Fruit Bunch

Ashvinder Singh Gill [1,2], Kam Huei Wong [1,2], Steven Lim [1,2,*], Yean Ling Pang [1,2], Lloyd Ling [1] and Sie Yon Lau [3]

1 Department of Chemical Engineering, Lee Kong Chian Faculty of Engineering and Science, Universiti Tunku Abdul Rahman, Kajang 43000, Selangor, Malaysia; ashvindergill@1utar.my (A.S.G.); wongkh@utar.edu.my (K.H.W.); pangyl@utar.edu.my (Y.L.P.); linglloyd@utar.edu.my (L.L.)
2 Centre for Advanced and Sustainable Materials Research, Lee Kong Chian Faculty of Engineering and Science, Universiti Tunku Abdul Rahman, Kajang 43000, Selangor, Malaysia
3 Department of Chemical and Energy Engineering, Faculty of Engineering and Science, Curtin University Malaysia, CDT250, Miri 98009, Sarawak, Malaysia; johnlsy@curtin.edu.my
* Correspondence: stevenlim@utar.edu.my or stevenlim_my2000@hotmail.com

Abstract: Lignocellulosic biomass (LCB), such as the oil palm empty fruit bunches (OPEFB), has emerged as one of the sustainable alternative renewable bioresources in retrieving valuable bioproducts, such as lignin, cellulose, and hemicellulose. The natural recalcitrance of LCB by the disarray of lignin is overcome through the combinative application of organosolv pre-treatment followed by microwave irradiation, which helps to break down LCB into its respective components. This physicochemical treatment process was conducted to evaluate the effect of ethanol solvent, microwave power, and microwave duration against delignification and the total sugar yield. The highest delignification rate was achieved, and the optimum level of total sugars was obtained, with the smallest amount of lignin left in the OPEFB sample at 0.57% and total sugars at 87.8 mg/L, respectively. This was observed for the OPEFB samples pre-treated with 55 vol% of ethanol subjected to a reaction time of 90 min and a microwave power of 520 W. Microwave irradiation functions were used to increase the temperature of the ethanol organic solvent, which in turn helped to break the protective lignin layer of OPEFB. On the other hand, the surface morphology supported this finding, where OPEFB samples pre-treated with 55 vol% of solvent subjected to similar microwave duration and power were observed to have higher opened and deepened surface structures. Consequently, higher thermal degradation can lead to more lignin being removed in order to expose and extract the total sugars. Therefore, it can be concluded that organosolv pre-treatment in combination with microwave irradiation can serve as a novel integrated method to optimize the total sugar yield synthesized from OPEFB.

Keywords: oil palm empty fruit bunch; organosolv pre-treatment; microwave irradiation; delignification; total sugars

check for updates

Citation: Gill, A.S.; Wong, K.H.; Lim, S.; Pang, Y.L.; Ling, L.; Lau, S.Y. Investigation of Microwave Irradiation and Ethanol Pre-Treatment toward Bioproducts Fractionation from Oil Palm Empty Fruit Bunch. *Sustainability* **2024**, *16*, 1275. https://doi.org/10.3390/su16031275

Academic Editor: Changhyun Roh

Received: 28 April 2023
Revised: 20 December 2023
Accepted: 22 January 2024
Published: 2 February 2024

1. Introduction

According to the Global Status Report (GSR) on energy consumption, approximately 80% of energy consumption is met by non-renewable fossil fuels, such as coal, petroleum, and natural gases, as opposed to merely 18% of the energy that is obtained from renewable energy sources, such as solar, wind, hydropower, and biomass [1]. Despite this fact, there are several limitations that are associated with the dependency on fossil fuels, such as finite reserves being located in politically unstable regions of the world [2], fluctuating and uncertain fossil fuel prices, fossil fuels being non-renewable energy resources, and the emission of greenhouse gases that pollutes the environment. These drawbacks on fossil fuel use limit its applicability in meeting future energy and sustainable product requirements.

Currently, the industry, as well as society, is moving toward the concept of "reuse and recovery" of resources from the primitive conception of "take, make and dispose of" resources [3]. This is to ensure a healthy and environmentally healthy society, at the same time ensuring socio-economic prosperity. Biomass has emerged as one of the sustainable alternative renewable bioresources. As such, due to the renewable nature of biomass, a term for a parallel economy based on bio-based products has been coined by state policy decision-makers, economists, and scientists: "bio-based economy" [4].

Lignocellulosic biomass (LCB) is composed of various polymers, such as polysaccharides (comprising cellulose and hemicellulose) and phenol aldehyde lignin polymer, along with some non-polar and polar substances, in which they are all interwoven together. Hence, it can be deduced that the three main components of lignocellulosic biomass are cellulose, hemicellulose, and lignin [5]. The natural recalcitrance of LCB by the disarray of lignin can be overcome through the application of suitable pre-treatment methods, which help to break down LCB into its respective components. Consequently, the accessibility of hydrolyzing enzymes to the cellulosic and hemicellulosic components can be further improved from the pre-treatment methods of LCB, in which the sugars can be fermented to produce biofuels [6]. On the other hand, the polyphenolic lignin and other hydrolysates that are extracted from the biomass can be converted into value-added chemicals that may eventually be applied as building blocks for economically vital chemicals in the industry [7].

According to Diyanilla et al. (2020), the most popular biomass is the oil palm empty fruit bunch (OPEFB) [8]. This is due to the ease of performing pre-treatment on such biomass as a result of having a comparatively low lignin content ranging from 14.2% to 38.4%. There are various pre-treatment methods that can be employed on OPEFB [8]. These include physical, chemical, physicochemical, and biological pre-treatment methods. A widely applied pre-treatment method for the environmentally friendly and efficient fractionation of bioproducts from biomass, such as OPEFB, is biological pre-treatment, such as enzymatic pre-treatment, due to its ability to maximize the sugar yield. However, studies on different pre-treatment methods have shown that the combination of chemical pre-treatments with an increase in physical conditions, where higher temperatures and pressures are applied, can lead to a better delignification process [8]. Examples of combined pre-treatment methods include steam explosion with alkaline delignification, hot water with alkaline pre-treatment, alkaline pre-treatment with ultrasonication, electron beam irradiation with ionic liquid pre-treatment, and steam-alkali-chemical pre-treatment.

Extensive research has been conducted over recent years on the bioproducts recovery from LCB pre-treated with microwave-assisted irradiation or the organosolv pre-treatment. A study was conducted on the sugar extraction from bamboo wood subjected to microwave pre-treatment at 121 °C, where a 36.9% conversion with 16.8 g/100 g of sugar yield was obtained [9]. Furthermore, a three-component deep eutectic solvent (DES) comprising of lactic acid, glycerol, and choline chloride, subjected under microwave irradiation at 393 K for a duration of 30 min, with a solid-to-liquid ratio of 1:50, was able to achieve a delignification rate of 45 wt% for the pre-treatment of wheat straw [10]. Moreover, the use of a physical pre-treatment such as the microwave irradiation on wheat bran, corn stalks, and miscanthus stalks was able to achieve hemicellulose solubilization of approximately 30%, which in turn led to an enhanced recovery of cellulose in these low-lignin yielding plants [11].

On the other hand, organosolv pre-treatment is regarded as a promising technique for the pre-treatment of biomass through the application of organic solvents, such as ethanol, glycols, acetone, and small amounts of acids, to stimulate the pre-treatment efficiency [11]. Wei et al. (2021) compared the acid- and alkali-catalyzed ethylene glycol organosolv pre-treatment for the production of sugars from bagasse. The results showed the efficiency of the ethylene glycol-hydrochloric acid (HCl) pre-treatment in the removal of lignin (67.1%) and hemicellulose (99.3%) due to the synergistic effect of the ethylene glycol solvent with the HCl acid when compared with the single ethylene glycol pre-treatment [12]. Furthermore, the organosolv pre-treatment of mustard biomass with acetone and acid catalyst at a constant autoclave temperature of 121 °C was able to yield 80% glucose [13]. Additionally, the acid-catalyzed ethanol pre-treatment of poplar biomass was able to release up to 78% of the polysaccharides [14]. A two-stage acid hydrolysis on ethylene glycol pre-treated degraded OPEFB was also conducted on the sugar-based substrate recovery, where stage I 45 wt% of the acid coupled with high cellulose loading of 21.25 *w*/*v*% with 12 wt% of acid at 65 °C for a duration of 30 min was able to release 89.8% of optimum sugar yield [15].

Thus, it can be realized that microwave-assisted pre-treatment methods have not been widely explored and employed for the fractionation of biomass. As a result, this research project was conducted to investigate this novel and facile process intensification to address the issue that such a knowledge gap exists, especially for the re-utilization of biomass waste such as OPEFB to remove the lignin while simultaneously facilitating higher total sugar yield. As such, the main purpose of this research was to investigate the effectiveness of the synergistic relationship between microwave irradiation and organosolv pre-treatment (a physicochemical process) on OPEFB for delignification and total sugar yield, as well as to characterize the physicochemical properties of the bioproducts extracted from the pre-treatment method through various characterization studies. Consequently, the novelty of this research work lies in the focus on the effectiveness of the microwave-assisted organosolv pre-treatment method as opposed to the commonly applied singular approach of either microwave irradiation (physical process) or organosolv pre-treatment (chemical method) on OPEFB samples.

2. Experimental Material and Methods

2.1. Materials

The OPEFB samples were provided and collected from the Research and Development Centre at Universiti Putra Malaysia-Malaysian Technology Development Corporation (UPM-MTDC, Serdang, Malaysia). Chemicals such as phenol solution 89%, glucose standard solution, iodine solution (starch indicator), sodium thiosulphate ($Na_2S_2O_3$) 99%, potassium permanganate ($KMnO_4$) 99%, and potassium iodide (KI) 99% were obtained from Sigma-Aldrich, St. Louis, MO, USA.

2.2. Materials Preparation

The OPEFB was cut into smaller pieces before being dried in the oven at 80 °C overnight (for 24 h). This was done to avoid fungal or microbial growth, as well as to remove moisture from samples prior to grinding and blending. The dried pieces of biomass waste were then ground, blended, and sieved using an appropriate mesh with a size of 150 μm to ensure a uniform size of feedstock material before the pre-experiment feedstock characterization and microwave-assisted organosolv pre-treatment were conducted.

2.3. Microwave-Assisted Organosolv Pre-Treatment

The pre-treatment methods for the conversion of OPEFB biomass into the required bioproducts are regarded as the most critical step due to the complicated structure of the lignocellulosic biomass [16]. The pre-treatment method carried out in this research project is the combination of the organosolv pre-treatment and microwave-assisted irradiation. The organosolv pre-treatment of OPEFB samples using ethanol was added to the samples before being placed into a modified microwave reactor. The solutions were pre-treated using 35%, 55%, and 75% volume in aqueous solution of ethanol concentrations with various microwave powers (65, 130, 260, 390, and 520 Watts) at various reaction times (30, 60, and 90 min). The residue obtained from the pre-treatment process was separated from the solvent via filtration, and the solid residues were dried at 80 °C for 24 h to obtain a constant weight. The pre-treated solid residues were then subjected to lignin analysis via the Kappa number analysis, as well as the subsequent characterization. The schematic design of the research activity and the research methodology flow chart (inclusive of the pre-treatment stage) are displayed in Figure 1 and Figure 2, respectively.

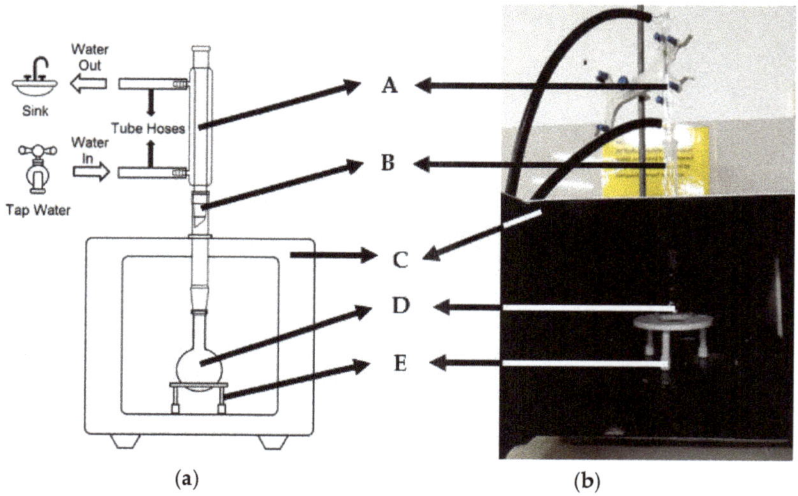

Figure 1. The experimental rig set-up of the microwave reactor: (**a**) schematic diagram of experimental set-up; (**b**) actual experimental rig set-up of microwave reactor: A: condenser tube; B: straight adapter; C: microwave reactor; D: round-bottomed flask; E: PTFE plate.

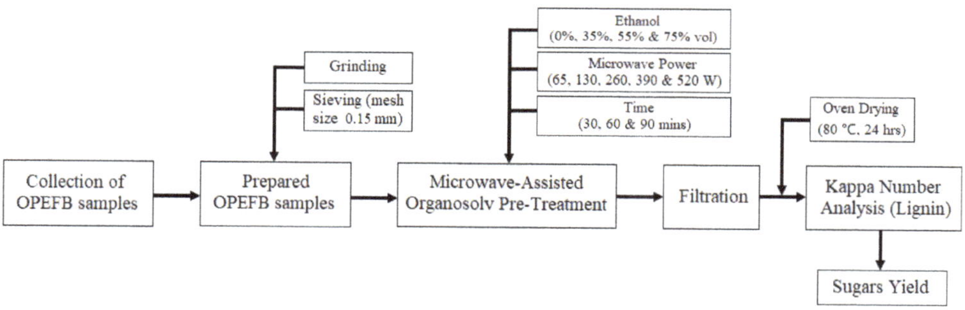

Figure 2. The research methodology flow chart, including the OPEFB samples collection, the pre-treatment processes and conditions, up to lignin and sugars analysis prior to further characterization (such as SEM and FTIR).

Each set of experiments was conducted in triplicates, where a mean value was obtained for the characterization of lignin and total sugars analysis. Additionally, the standard deviation was computed for each set of data using the mean value. The calculated standard deviation did not exceed a value of 10, ensuring that the repeatability tests are reliable.

2.4. Lignin Analysis

Kappa number analysis was conducted to analyze the lignin content based on the different conditions employed during the pre-treatment process. A 250 mL solution was prepared (200 mL distilled water with 0.1 g pre-treated biomass waste + 25 mL of 0.1 N potassium permanganate ($KMnO_4$) + 25 mL of 4.0 N sulphuric acid (H_2SO_4)). After obtaining 10 min of reaction from the solutions by gently stirring the solutions contained inside the volumetric flask, potassium iodide (KI) was added to the solution. Blank (standard) was carried out using the same steps without the inclusion of the OPEFB sample.

Titration was then carried out using 0.5 N sodium thiosulphate ($Na_2S_2O_3$) and 1.0 N indicator starch. The lignin content was calculated using the Equations (1)–(3):

$$p = \frac{(b-a) \times N}{0.1}, \tag{1}$$

$$\text{Log}\, k = Log\left(\frac{p}{W}\right) + 0.00093\,(p - 12.5), \tag{2}$$

$$\%\ \text{Lignin} = 0.147 \times k, \tag{3}$$

where a and b are the volume of $Na_2S_2O_3$ for OPEFB and blank titration, respectively (in mL), N is the normality of $Na_2S_2O_3$, p is the volume of $Na_2S_2O_3$ (in mL), W is the weight of the OPEFB sample used, and k is the Kappa number (volume of 0.1 N $KMnO_4$ consumed by 1 g of OPEFB).

2.5. Total Sugars Analysis

The total sugars analysis was carried out using the phenol–sulfuric acid method. A glucose standard solution of 2 mL was prepared in various concentrations (0, 10, 20, 30, and 60 ppm) mixed with 1 mL phenol solution (5% vol concentration) in respective test tubes. After that, 5 mL of concentrated H_2SO_4 was dropped to the surface of the solution vertically, and the mixture was left for 10 min before the measurement of the UV-Vis spectrophotometer (PG Instruments T60) using deionized water as blank at wavelength 480 nm, with this result of the measurement known as the standard curve. Furthermore, the sugar solution was prepared by mixing the OPEFB sample and distillate water at a composition ratio of 1:250 as a sample solution. A 1 mL sample solution was extracted and mixed with 1 mL distilled water in a test tube. 5 mL of H_2SO_4 and 1 mL of phenol solution (5% vol concentration) were added simultaneously to a test tube containing 1 mL sample solution mixed with distilled water prior to their UV-Vis spectrophotometer measurement.

2.6. Sample Characterizations

All the prepared samples (both pre-treated and without) were characterized using a scanning electron microscope (SEM, Hitachi, S-3400N, Tokyo, Japan) to observe their surface morphology. The samples were sputter-coated with gold prior to imaging with SEM, with a magnification power ranging from ×100 up to ×2000.

Furthermore, Fourier transform infrared spectroscopy (FTIR, Nicolet, IS10, Waltham, MA, USA) was used to characterize and determine the functional groups that were attached to the samples. The FTIR spectra were obtained using a potassium bromide (KBr) pellet/disc containing the grounded sample. The sample was then scanned, with the spectra being recorded between 400 cm^{-1} and 4000 cm^{-1} with a resolution of 4 cm^{-1}.

3. Results and Discussion

3.1. SEM Characterization Study

Figure 3 shows the surface morphological changes of the untreated and microwave-treated OPEFB samples in the SEM images through the observation of the external surfaces of the samples. Figure 3a shows that the OPEFB fibers are made up of many oriented microfibrils adhered to each other, with their surface morphology observed as having smooth surfaces with few flakes, as well as no opened structures. Figure 3b,c also display similar morphologies observed on the surface of OPEFB samples without being subjected to microwave power and/or organosolv pre-treatment. Contrary to expectations, a flakier surface was observed on the surface of OPEFB samples when microwaved at 130 W for 90 min, as seen in Figure 3d. Moreover, an opened and deepened structure was observed in Figure 3e when the samples were microwaved at a slightly higher power of 390 W for 90 min. Furthermore, Figure 3f demonstrates a more opened and deepened surface structure along with flakier surfaces when the samples are microwaved at an even higher power of 520 W for the same duration. Cavities were observed for OPEFB samples when subjected to microwave irradiation, revealing opened and deepened surface structures. This is due to the microwave energy that heats up the inner part of OPEFB samples, resulting in the disruption of the plant material's cell wall as the pressure increases with higher power [17].

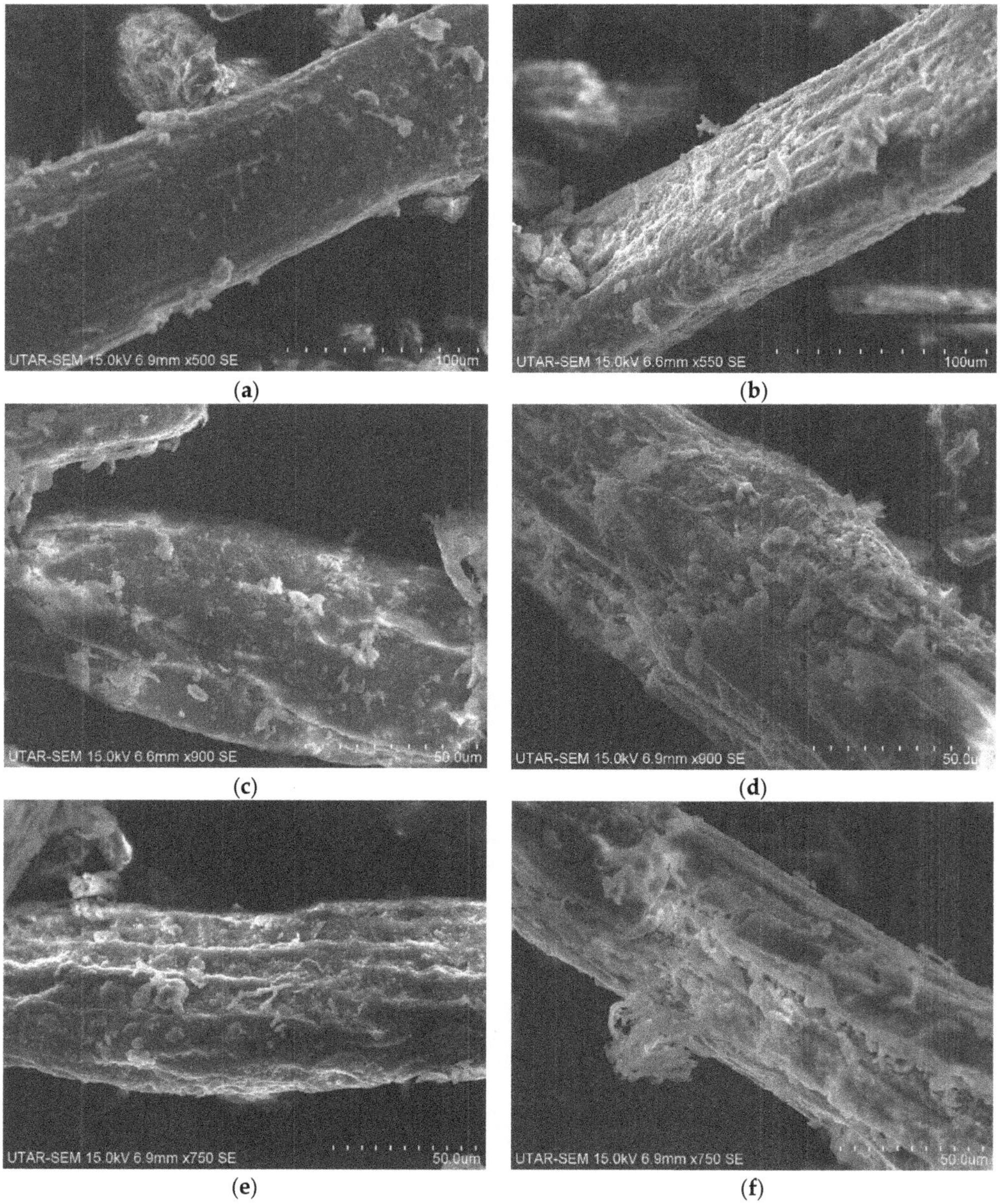

Figure 3. SEM images of OPEFB samples: (**a**) (magnification of ×500), (**c**) (magnification of ×900), (**e**) (magnification of ×750) without being subjected to microwave power and/or organosolv pretreatment; (**b**) at 130 W (magnification of ×550); (**d**) at 390 W (magnification of ×900); (**f**) at 520 W (magnification of ×750) for 90 min.

Microwave-irradiation heating can be explained via the thermal effect and athermal (non-thermal) effect. The lignin and cell wall components are disrupted through an increase in the temperature and pressure applied during the thermal effect, whereas relaxation and polarization of dielectric substances under the influence of electromagnetic fields generate non-thermal effect forces. The disruption of hydrogen bonds is caused by displacement resulting from the realignment of the polar molecules, where the cell wall components, such as the lignin and cellulose crystallinity breakdown, are destroyed, enhancing the hydrolysis process. Hence, it can be said that the dielectric property is the principal factor that affects the heating property of the biomass under microwave irradiation [2].

A study by Xiaokang et al. (2020) compared freeze-drying and microwave-assisted extraction on mushroom samples, where the images obtained from the SEM showed that all extractions led to the deformation of mushroom samples to some extent [17]. In the case of the microwave extraction method, some cavities were observed due to the microwaves that help heat up the inner part of the samples, which led to pressure build-up inside the cells and resulted in cell wall disruption [18]; hence, this demonstrated that the opened surface structure is indicative of the biomass cell walls being disrupted, which was in agreement with the literature.

Figures 4 and 5 show the surface morphological changes of the microwave-assisted organosolv pre-treatment of OPEFB samples in the SEM images through the observation of external surfaces of the samples at the lower and higher microwave power of 65 W and 520 W, respectively. Flaky surfaces were seen on the OPEFB samples in Figure 4a when being pre-treated with lower ethanol concentration (35 vol%) at lower microwave power (65 W), whereas flakier surfaces with more opened and deepened surface structures were observed on the surface of OPEFB samples in Figure 5b when pre-treated with higher ethanol concentration (55 vol%) at higher microwave power (520 W).

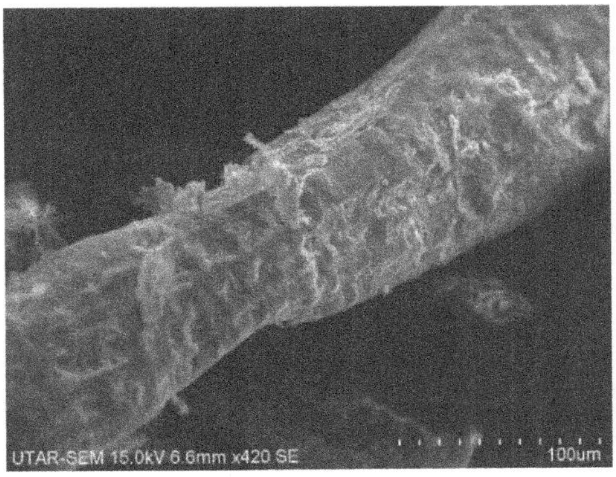

(a)

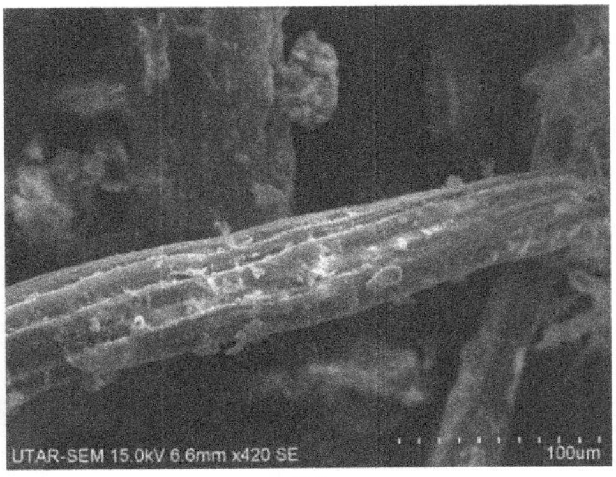

(b)

Figure 4. SEM images of OPEFB samples: (**a**) pre-treated with 35 vol% ethanol; (**b**) pre-treated with 55 vol% ethanol at a microwave power of 65 W for 90 min.

A deepened surface structure indicated that the delignification process successfully took place in order for the total sugar analysis to be conducted for subsequent analysis. A more deepened surface structure indicated that a higher delignification rate was achieved for samples pre-treated with a slightly higher ethanol concentration of 55 vol% at higher microwave powers due to the higher thermal degradation attacking the lignin protective layer. Since OPEFB is a lignocellulosic biomass comprised mainly of cellulose, hemicellulose, and lignin, the functional groups in the ethanol organic solvent break through the lignin layer, allowing for the extraction of cellulose and hemicellulose. On the other hand, microwave power assists in increasing the temperature of the organic solvent to degrade the OPEFB, helping the ethanol solvent to be more active in breaking the lignin protective layer of the biomass. The inclusion of an organic solvent such as ethanol as a pre-treatment process in the microwave extraction of biomass helps shield the samples from the high-penetrating power of the microwaves. Subsequently, this can lead to a higher recovery of the sugar content from the biomass.

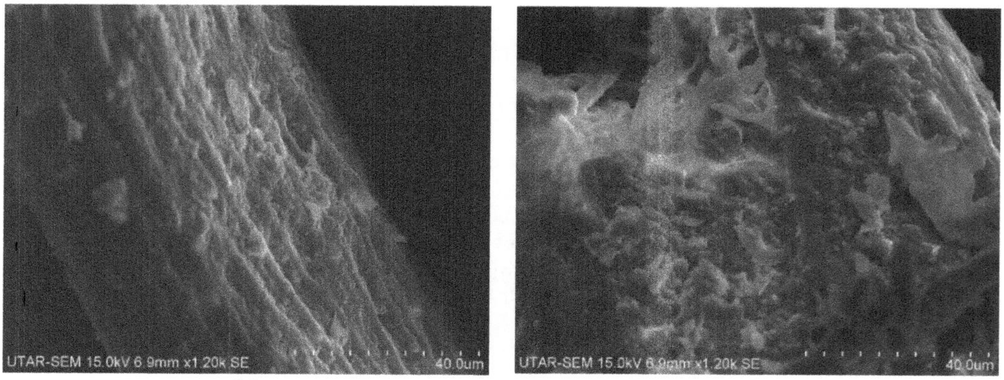

Figure 5. SEM images of OPEFB samples: (**a**) pre-treated with 35 vol% ethanol; (**b**) pre-treated with 55 vol% ethanol at a microwave power of 520 W for 90 min.

A study by Gonçalves et al. (2015) observed that untreated coconut shell samples showed a more highly ordered structure of the coconut fibers, whereas treated coconut shell samples showed modified disoriented sample structures [19]. Another study by De et al. (2014) used microwave irradiation in the presence of glycerol solvent for the pre-treatment of bagasse, where they found that the physicochemical properties of the biomass changed considerably before and after the pre-treatment processes, and a large amount of sugar was released during hydrolysis [20]. Binod et al. (2012) studied the effect of microwave, microwave-acid, microwave-alkali, and microwave-acid-alkali methods on enzymatic saccharification, as well as the removal of lignin from bagasse. Their results pointed out that microwave-assisted extraction methods, as opposed to microwave irradiation alone, can lead to a significant yield of fermentable sugars and lignin removal [21]. This further substantiates the efficiency of a physicochemical method such as the microwave-assisted organosolv pre-treatment in the extraction of valuable bioproducts from OPEFB. The results from the SEM images clearly demonstrated the disruption of the biomass cell wall, hence indicating the successful removal of the lignin layer in order to extract the sugar content.

3.2. FTIR Characterization Study

It is imperative to first understand the structures of the cellulose, hemicellulose, and lignin components of OPEFB before analyzing the FTIR bands. The structures of the three main components of an LCB such as OPEFB, cellulose, hemicellulose, and lignin fibers, are illustrated in Figure 6. The lignocellulosic polymers are linked to each other via specific bonds and contribute to over 70% of the total biomass [2]. Cellulose is a beta β-(1,4)-linked chain of glucose molecules. The resistance of the crystalline cellulose being subjected to degradation is contributed by the hydrogen bonds between the different layers of this particular polysaccharide.

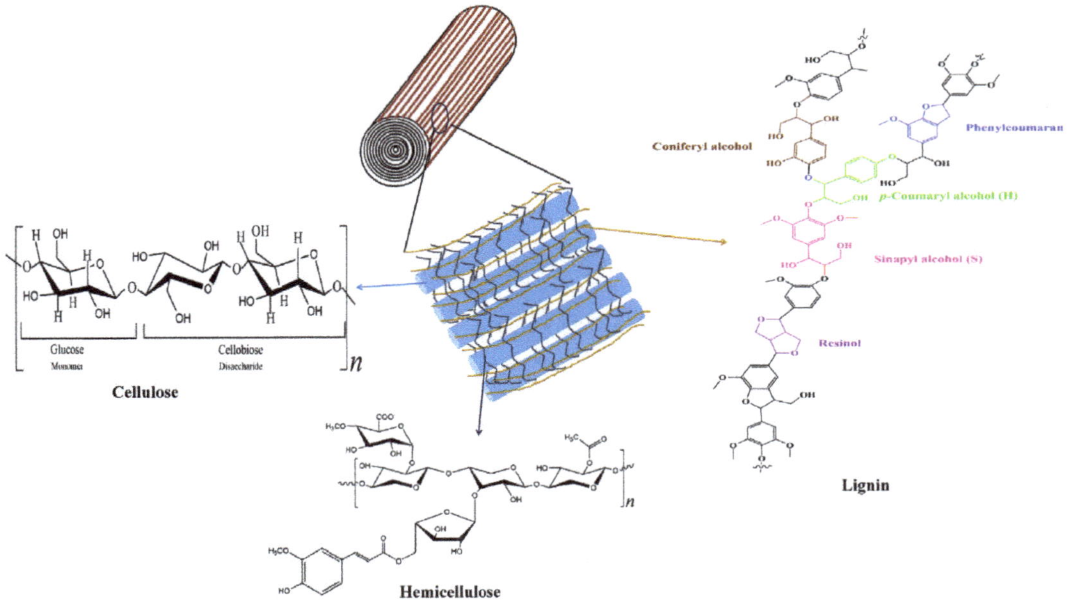

Figure 6. Structure of the main components of LCB (OPEFB): cellulose, hemicellulose, and lignin [2].

On the other hand, hemicellulose is composed of various five- and six-carbon sugars, such as galactose, arabinose, glucose, xylose, and mannose. As for lignin, the three major phenolic compounds that constitute this particular component include *p*-Coumaryl alcohol, Coniferyl alcohol, and Sinapyl alcohol. Lignin is synthesized through the polymerization of these three major phenolic compounds, where their ratio within the polymer varies according to the different parts of the plants, such as the wood tissues and cell wall layers [2]. The linkage between lignin and polysaccharides is regarded as a lignin–polysaccharide complex due to the composite links that connect lignin to the polyoses side groups of arabinose, galactose, and 4-*O*-methylglucoronic acid, as depicted in Figure 7 [22]. Hence, the difficulty in achieving a complete separation of lignin from lignocellulosic materials resulted from the sterically structured component of lignin. Thus, a combination of methods, such as a physicochemical method, is typically employed in order for the separation of lignin from lignocellulosic biomass material to be accomplished [22]. The cellulose, hemicellulose, and lignin components form structures known as microfibrils, which are then structurally organized into macrofibrils that are responsible for mediating the structural ability in the cell wall of plants.

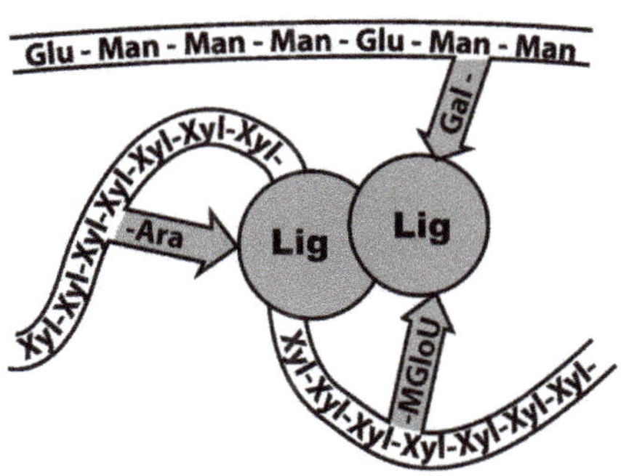

Figure 7. A schematic representation of the linkage between lignin and polysaccharides [22].

The FTIR spectra of untreated OPEFB samples, as well as OPEFB samples pre-treated with 55 vol% of ethanol at the lower and higher microwave power of 65 W and 520 W, respectively, for a duration of 90 min, are presented in Figure 8. The absorption peak at 2850 cm^{-1} represented the acetyl group ($CH_3C{=}O$) of hemicellulose or methyl oxide group (OCH_3) of lignin; hence, this indicated that the ethanol solvent at both the lower and higher microwave power with a longer duration was able to achieve a considerable delignification rate, with a higher intensity of absorbance peak observed for 520 W compared with 65 W. It can also be observed that the absorbance peak at 3330 cm^{-1} was intensified. This indicated that the ethanol solvent at the higher microwave power with a longer duration was able to achieve a significant delignification rate, thus further supporting the correspondence of the absorbance peak at 3330 cm^{-1} to the hydrogen bond of cellulose [23].

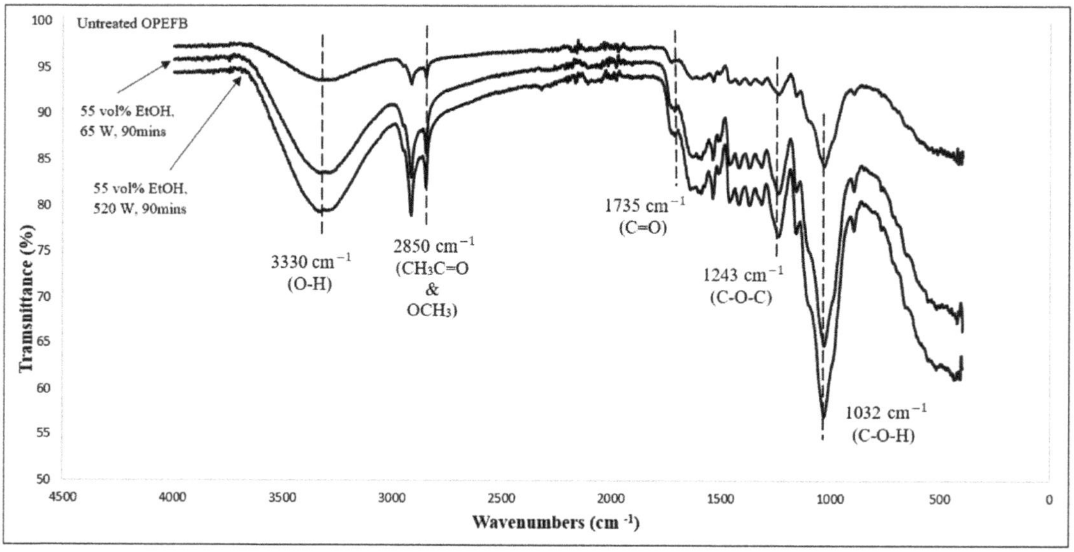

Figure 8. FTIR spectra of untreated OPEFB samples and samples pre-treated with 55 vol% ethanol (EtOH) at microwave powers of 65 W and 520 W for 90 min.

The broadening of the band at 3200 to 3400 cm^{-1} was associated with the O-H stretching of the cellulose hydrogen bonds [24]. Furthermore, the absorption peak at 1243 cm^{-1} indicated the presence of lignin for the OPEFB samples pre-treated with both microwave irradiation and organosolv pre-treatment due to the presence of the aryl alkyl ether (C-O-C) bond [23]. Moreover, the absorbance peak at 1243 cm^{-1} observed for both the fresh and pre-treated OPEFB samples was also attributed to both the C-O-C and OCH_3 groups for lignin [25]. The absorption bands at 1032 cm^{-1} were also present in both untreated and microwave-assisted ethanol pre-treated OPEFB samples, attributed to the cellulose and hemicellulose components of the biomass material [23].

However, a prominent difference was observed at an absorbance peak of 1735 cm^{-1} for both untreated and treated OPEFB samples, where a more intense peak was observed for the untreated samples when compared to the treated samples. This observation corresponds to the C=O stretching vibration in the ester groups of hemicellulose [26], where a higher hemicellulose composition is associated with an untreated biomass sample, with the hemicellulose content still present within the biomass material. For the OPEFB samples that undergo microwave-assisted organosolv pre-treatment with ethanol solvent, the microwave power increases the temperature of the solvent in order to allow for the functional groups in the solvent to break through the lignin layer, allowing for the extraction of cellulose and hemicellulose in OPEFB. However, some of the cellulose and hemicellulose dissolved in the solvent during the microwave irradiation process; hence, the smaller amount of hemicellulose in the treated OPEFB samples was in line with the observation of the lower intensity peak of 1735 cm^{-1}.

3.3. Lignin Analysis

Figure 9 shows the amount of lignin left for OPEFB samples subjected to microwave irradiation without organosolv pre-treatment, where a decreasing trend in the amount of lignin left in the sample was observed when the samples were subjected to higher microwave powers and longer duration. A higher microwave power of 520 W recorded a higher delignification rate with 1.15% of lignin when compared to 4.61% of lignin for a lower microwave power of 65 W, both at a longer duration of 90 min. This was attributed to the working principle of microwave, where higher penetrating powers and energies are associated with higher microwave powers, thus allowing for more disruption to take place in the OPEFB cell walls, subsequently breaking the lignin protective layer and resulting in a higher delignification rate and smaller lignin amount left in the OPEFB sample. Moreover, the longer microwave duration allows for more lignin to be broken down and removed, hence resulting in a higher lignin removal.

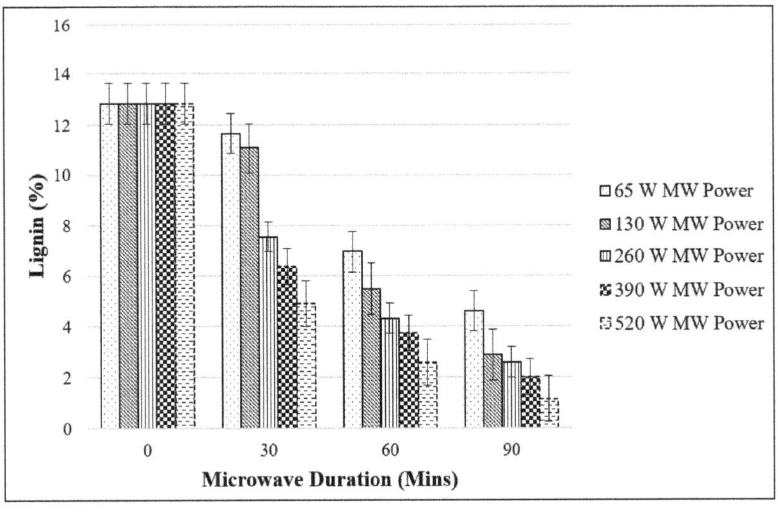

Figure 9. Lignin analysis of OPEFB samples at different microwave (MW) powers without organosolv pre-treatment.

For the OPEFB samples pre-treated with different concentrations of ethanol solvent without being subjected to microwave irradiation, a significant decreasing trend in the amount of lignin left in the sample was observed with an increase in ethanol concentrations from 35 to 55 vol%, as seen in Figure 10. However, a slight increase in the amount of lignin left in the sample was also observed when the ethanol concentration increased from 55 to 75 vol%. The OPEFB samples pre-treated with 35 vol% ethanol recorded a higher delignification rate, with 9.87% of lignin compared with 4.90% of lignin when pre-treated with 55 vol% ethanol; however, the delignification rate slightly increased by 7.52% of lignin when pre-treated with 75 vol% ethanol. This observation can be explained by the effect of the different ethanol concentrations on the Kappa number.

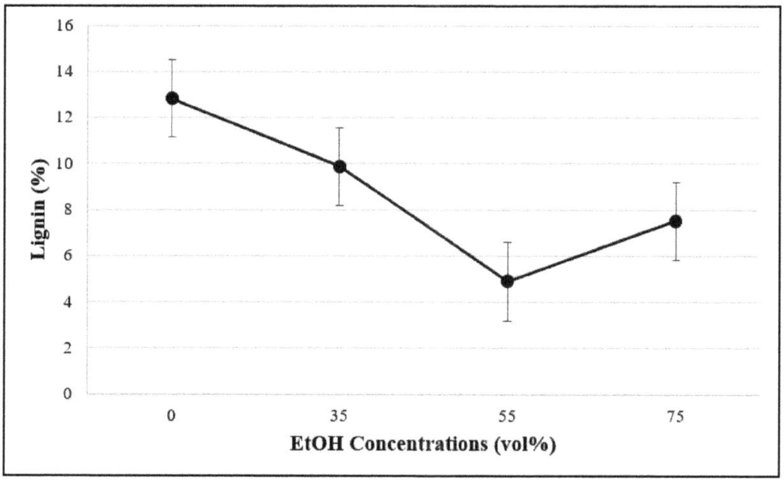

Figure 10. Lignin analysis of OPEFB samples at different ethanol (EtOH) concentrations (35, 55, and 75 vol%) without microwave irradiation.

The Kappa number analysis estimates the number of lignin left in the sample [5]. The amount of lignin left in the sample (%) can then be calculated using the Kappa number, as displayed in Equations (1)–(3). The amount of lignin left in the OPEFB samples decreases significantly after a 35 vol% ethanol solvent is applied. This indicated an increase in the number of degraded lignin by increasing the ethanol concentration. Therefore, this observation was important for future cellulose and hemicellulose content determination. Cellulose and hemicellulose were able to be converted into sugars through acid hydrolysis, allowing us to investigate the effect of organosolv pre-treatment on sugar yield. The study conducted by Nurfahmi et al. (2016) on the effects of organosolv pre-treatment and acid hydrolysis on OPEFB reported a decrease in the Kappa number from 38 to 25.8 when the ethanol concentration increased from 35 to 55 vol%, and the Kappa number increased slightly from 25.8 to 26.4 when the ethanol concentration increased from 55 to 75 vol% [5].

Figure 11 displays the lignin amount left for OPEFB samples subjected to microwave-assisted organosolv pre-treatment. A decreasing trend in the amount of lignin was observed when the ethanol concentration increased from 35 to 55 vol%, with a smaller amount of lignin also being observed for higher microwave powers with a longer pre-treatment period. The OPEFB samples subjected to a higher microwave power of 520 W pre-treated with 55 vol% ethanol recorded a higher delignification rate, with 0.57% of lignin compared with 2.30% of lignin for the lower microwave power of 65 W pre-treated with the same concentration of ethanol, both over a longer duration of 90 min. A similar observation was also noted, with a slight increase in the amount of lignin when the ethanol concentration increased from 55 to 75 vol%.

The trend in observation for the amount of lignin left in OPEFB samples after undergoing the microwave-assisted organosolv pre-treatment process was also attributed to the strong relationship between lignin solubility with microwave power and duration [27]. The high surface area of the OPEFB samples resulted in them being more susceptible to chemical attacks, which enhanced the delignification and degradation of lignin. The mechanism of the lignin depolymerization of organic solvents (such as ethanol) was mainly conducted through the cleavage of aryl ether linkages [28]. The delignification achieved by the ethanol solvent can be explained by the Hildebrand solubility parameter [29]. The highest delignification achieved by ethanol is attributed to its closer Hildebrand solubility parameter (a value of 26.2 MPa^{-1}) when compared to that of lignin (a general value of 22.5 MPa^{-1}) [30]. It was further supported that the smaller differences in the Hildebrand solubility between the lignin and organic solvent used would result in the largest possible solubility [31]. A higher delignification rate (lower lignin amount left in the sample) is achieved through a higher solubility of lignin in the ethanol solvent.

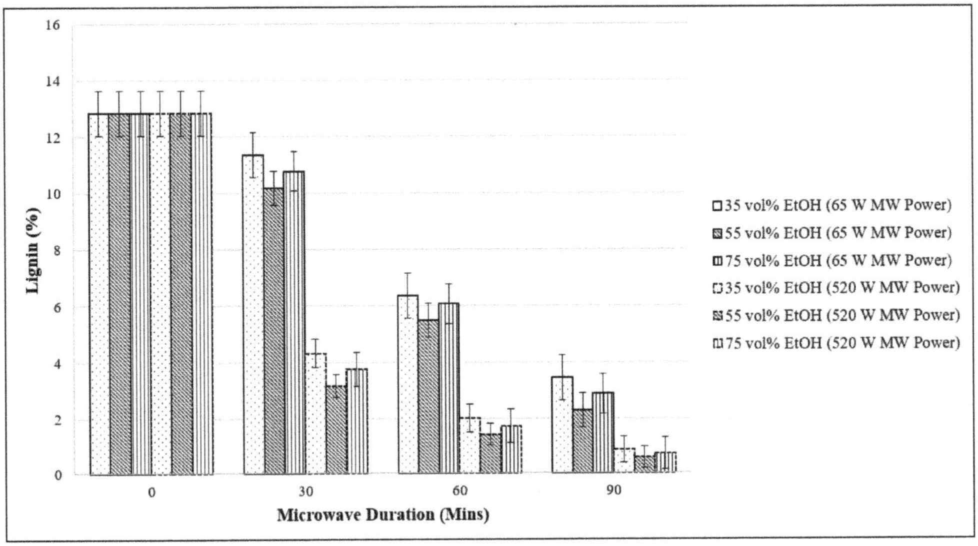

Figure 11. Lignin analysis of OPEFB samples at 65 W and 520 W microwave (MW) powers pre-treated with different ethanol (EtOH) concentrations.

It was also further reported that the lignin solubility increases when the ethanol concentration reaches a maximum of 70 vol% [32]. This explains the slightly higher amount of lignin left in the OPEFB samples subjected to a 75 vol% of ethanol pre-treatment due to a slightly lower solubility of the lignin with higher ethanol concentrations, leading to a slightly lower delignification rate. From the microwave-only OPEFB pre-treatment data recorded in Figure 9, it was observed that the highest delignification rate with the lowest amount of lignin left in the sample (1.15%) was achieved with the OPEFB sample subjected to the highest microwave power of 520 W for the longer duration of 90 min. Contrary to expectations, from the organosolv-only OPEFB pre-treatment data recorded in Figure 10, it was observed that the highest delignification rate with the lowest amount of lignin left in the sample (4.90%) was achieved with the OPEFB sample subjected to the 55 vol% ethanol concentration. In comparison to the physicochemical process intensification of the microwave-assisted organosolv pre-treatment of OPEFB samples, which recorded the highest delignification rate with the lowest amount of lignin left in the sample (0.57%), it was observed that the singular approaches of pre-treatment were less effective in breaking the lignin protective layer, with the microwave-only pre-treatment being able to remove the lignin approximately four times more effectively than the organosolv-only pre-treatment. This observation was attributed to the penetrative ionizing powers of the microwave that facilitated the breakdown of the lignin protective layer. With the inclusion of a combinative approach by first pre-treating the OPEFB samples with ethanol before subjecting them to microwave irradiation, the microwave powers helped increase the temperature of the ethanol solvent, which in turn helped the solvent to be more active in attacking the OPEFB lignin layer, subjecting it to a thermal degradation process that is less severe. Hence, the microwave-assisted organosolv pre-treatment was proven to be more effective in achieving the highest delignification rate, with the lowest amount of lignin left in the OPEFB sample of 0.57%.

3.4. *Total Sugars Analysis*

Figure 12 presents the total sugar yield obtained for OPEFB samples pre-treated with 35, 55, and 75 vol% of ethanol concentrations while subjected to various microwave powers for different durations. It showcased the optimum (highest) yield of total sugars, which was obtained at 87.8 mg/L using 55 vol% of ethanol concentration at a higher microwave power and a longer microwave duration of 520 W and 90 min, respectively. Contrary to expectations, the OPEFB samples pre-treated with 35 vol% of ethanol and subjected to a lower microwave power of 65 W for a shorter treatment duration of 30 min yielded the lowest total sugar content of 30.1 mg/L. This observation was attributed to the correlation between microwave power and temperature [33], along with the mechanism or working principle of microwave-assisted organosolv (ethanol) pre-treatment. The optimum sugar yield was obtained for the highest microwave power of 520 W, which corresponds to approximately 180 °C (see Figure 13) [17]. A temperature higher than 180 °C could lead to the degradation of OPEFB bioactive compounds, such as hemicellulose [5]. Therefore, the maximum microwave power used in this study is within the suggested range. Furthermore, the mechanism of the microwave-assisted organosolv pre-treatment entails the microwave power increasing the temperature of the ethanol solvent, effectively breaking down the lignin protective layer. Consequently, this process leads to a lower amount of lignin left in the OPEFB sample, subsequently necessitating the extraction of the total sugars from within the OPEFB biomass sample. Thus, the optimum sugar yield was attained using the combinative approach of the microwave-assisted organosolv process intensification pre-treated with 55 vol% ethanol and subjected to the maximum microwave power with a longer duration of 520 W and 90 min, respectively.

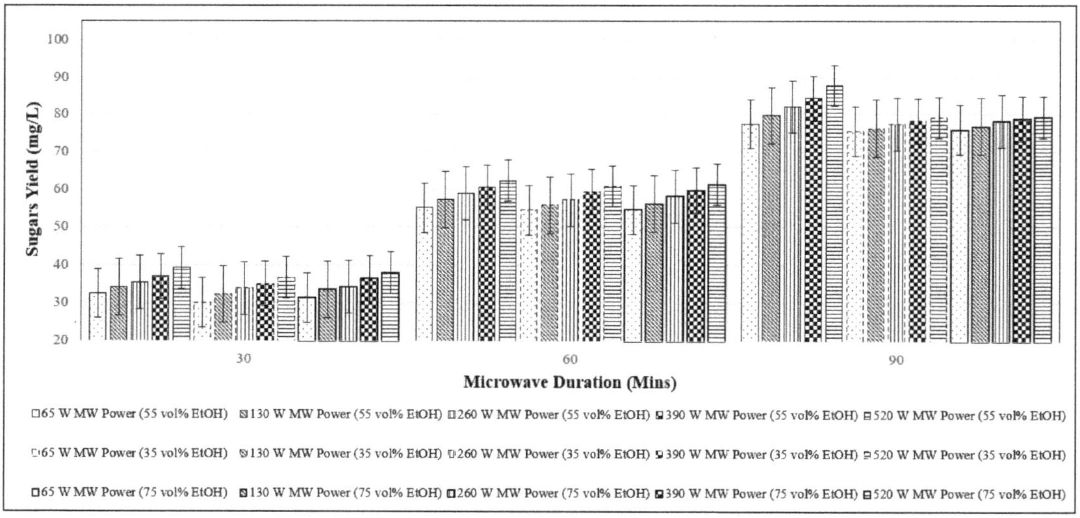

Figure 12. The effect of different microwave powers (MW) on total sugar yield for samples pre-treated with 35, 55, and 75 vol% ethanol solvent at different microwave durations.

The hydroxyl groups of solvents can influence the organosolv pre-treatment process. Ethanol solvent has one hydroxyl group, where the mechanism of the hydroxyl group involves the β-O-4 linkage cleavage of lignin; as such, this degradative pathway resulted in the formation of a dissolved organosolv lignin. This relates back to the solubility justification, where a higher delignification rate was achieved with higher solubility of lignin in a higher concentration of ethanol solvent (55 vol%). However, the solubility of lignin increases when the ethanol concentration reaches a maximum of 70 vol% [32], as previously discussed in Section 3.3. Therefore, the slightly higher amount of lignin left in the OPEFB samples can be explained by the lower solubility of lignin at higher ethanol concentrations, leading to a slightly lower delignification rate. Consequently, a comparison could be made among the samples pre-treated with the lowest concentration (35 vol%), slightly higher concentration (55 vol%), and highest concentration (75 vol%) of the ethanol solvent. Samples pre-treated with 55 vol% of ethanol yielded the highest amount of sugar extracted due to the higher delignification rate achieved, followed by those pre-treated with 75 vol% and 35 vol% of ethanol.

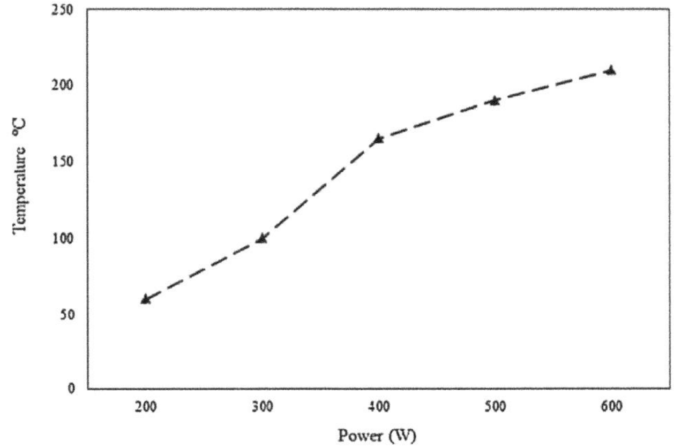

Figure 13. The correlation between microwave power (W) and temperature (°C) [17].

A further process parameter study was also conducted by further increasing the microwave powers (beyond the maximum of 520 W), duration (beyond the maximum of 90 min), and ethanol concentrations (beyond the maximum of 75 vol%) in order to study the effect of these three parameters on the amount of lignin and, consequently, the total sugar yield. With an increase in microwave power to 100% (650 W) for the longer durations of 105 and 120 min, the amount of lignin left in the 85 vol% ethanol pre-treated OPEFB samples were recorded to be 0.67% and 0.65%, respectively; as a result, the total sugar yield was recorded to be 76.7 and 77.9 mg/L, respectively. As such, it was observed that there was a higher amount of lignin left in the OPEFB samples, with an increase in the process parameters, consequently leading to a lower amount of sugar extracted. This observation was attributed to the degradation of hemicellulose above 180 °C since the microwave power of 650 W corresponds to a temperature exceeding the hemicellulose stability range, which is between 180 °C and 340 °C [5,17]. Thus, this temperature range affects the total sugars extracted, leading to a lower amount of sugars yield achieved. Moreover, the crucial physicochemical property that became the subject matter of this research work, leading to the effectiveness in extracting the total amount of sugars from OPEFB samples, is the higher solubility of organosolv lignin in the ethanol solvent used. Organosolv lignin has a tendency to aggregate in most solvents, affecting the process of lignin recovery. In a study on Alcell lignin and its solubility in ethanol-water mixtures, it was demonstrated that the solubility of lignin increased as the concentration of ethanol increased to a maximum of 70 vol% [32]. When the precipitation process of lignin was conducted by diluting solvent with water, decreasing the amount of organic solvent, the solubility of lignin decreased. This led to less lignin being recovered, with more lignin left in the sample. The decrease in lignin solubility was attributed to the increased coagulation degree of lignin in the ethanol solvent [32], with the larger particle size of lignin subsequently leading to a lower surface area reaction [32] for the precipitation process to take place, thereby resulting in a lower lignin recovery; hence, a lower amount of sugar was extracted due to the presence of lignin that was not successfully recovered.

There is a correlation between microwave power and temperature, whereby the temperature increases as microwave power increases, as depicted in Figure 13. Increasing the temperature helps promote the separation of bioactive compounds from the OPEFB, which leads to a higher reaction rate to elevate and promote the separation and cleavage of lignin bonds. However, high extraction temperatures may result in the degradation of some thermolabile bioactive compounds. Nurfahmi et al. (2016) reported that the thermal stability of hemicellulose was between 180 °C and 340 °C [5]. The maximum temperature that was used in this research project was 80% microwave power (520 W), corresponding to approximately 180 °C, as observed in Figure 13 [17]; hence, the maximum microwave power did not exceed the suggested range.

Similarly, there is a correlation between microwave duration and microwave temperatures, as depicted in Figure 14. The OPEFB sample, being pre-treated with 55 vol% ethanol subjected to 80% of microwave power (520 W), was taken as an example of the correlation demonstration. This was due to the temperatures of the samples being recorded instead of the temperature of the microwave oven after each sample was subjected to microwave irradiation, as the microwave oven used had no temperature probes for the measurement of the corresponding microwave powers, resulting in different temperatures being recorded for each sample (with different pre-treatment conditions). It was observed that a cubic equation best represented the relationship between these two parameters, with a higher correlation coefficient (R^2 of 0.9057) being achieved in comparison to a slightly lower R^2 of 0.9015 achieved for a quadratic equation. The cubic correlation indicated that, with a higher microwave duration, higher temperatures were recorded for the OPEFB samples. Naturally, as seen in Figure 13, higher microwave powers corresponded to higher microwave temperatures. Likewise, from the data recorded for all the experimental sets, the temperature of the OPEFB samples increased with a rise in the microwave powers applied. Therefore, this observation, as well as the cubic relationship plotted in Figure 14, justified the correlation between microwave duration and temperatures.

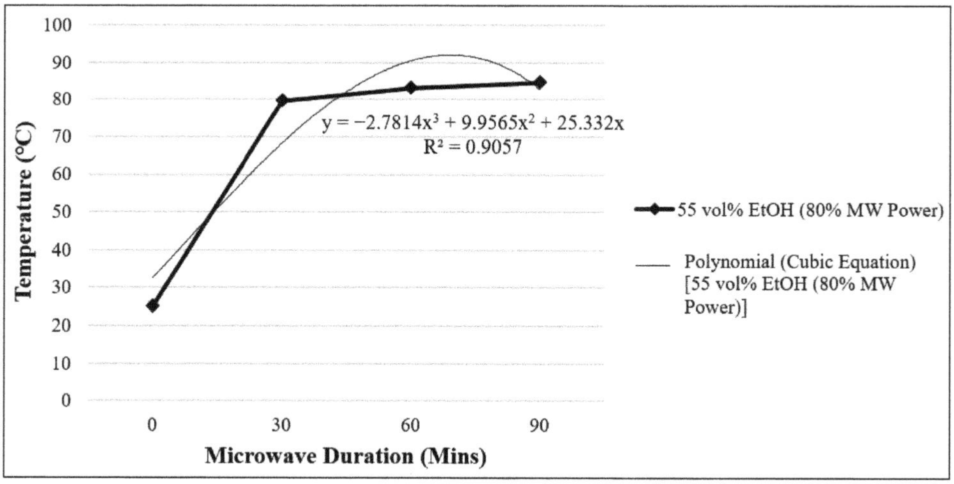

Figure 14. The cubic correlation between microwave duration and microwave temperature (°C) for samples pre-treated with 55 vol% ethanol (EtOH).

Under the working principle of microwave radiation, the occurrence of dipole rotation takes place in dielectric materials containing polar molecules that encompass an electrical dipole moment. Consequently, energy in the form of electromagnetic radiation is converted into heat energy in the materials as a result of the interactions between the dipole and the electromagnetic field [33]. In order to interlink the microwave power and treatment temperature, it is imperative to understand the phenomenon of thermal runaway in a microwave irradiation process. Thermal runaway is defined as the dramatic change in temperature resulting from small changes in the geometrical sizes of the heated material or the applied microwave power [33]. This phenomenon may lead to positive feedback in the material, where warmer areas are able to better accept more energy than the colder areas.

Theoretically, the relationship between temperature and the applied microwave field strength (or microwave power) follows a characteristic "S"-shaped curve, as reported by Brodie (2011) and represented in Figure 15 [33]. As the microwave power increases steadily, the temperature also rises steadily along the stable lower arm of the power curve. In the region of the critical power range, a small increase in the electric field strength may result in the equilibrium temperature shifting from the lower limb to the higher limb of the curve, for which the resulting change in temperature could be rather rapid and substantial. This temperature jump gives rise to the phenomenon of thermal runaway. The organic solvent, along with microwave irradiation (which provides high penetrating microwave power and, thus, higher temperature), was able to extract the lignin and hydrolyses of hemicellulose through the breakage of the lignin wall and the loosening of the crystalline structure of cellulose and hemicellulose, enabling the extraction of total sugars.

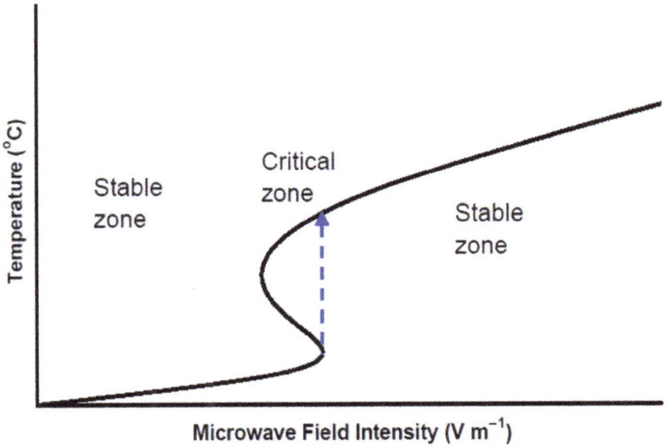

Figure 15. The characteristic "S"-shaped curve correlating the relationship between temperature and the applied microwave field strength or power [25].

According to Kumar et al. (2020), the selection of suitable pre-treatments of the biomass material is one of the most crucial factors in extracting valuable bioproducts for various applications, in which the choice of the pre-treatment relies on the physicochemical nature or properties of the biomass. This statement can be justified by the fact that different biomasses require different methods to overcome the natural recalcitrance to ensure that a higher sugar yield can be achieved [2]. Moreover, Tsubaki and Azuma (2011) compared the conventional steam explosion method to the microwave irradiation process and reported that the removal efficiency of lignin is comparable between the two methods; however, the microwave irradiation recorded a higher sugar yield since this physical pre-treatment method specifically affects the polysaccharide fraction, as it has higher hydrophilicity, and thus, is more sensitive to hydrothermal catalysis when compared to the steam explosion [34].

An experiment conducted by Tang et al. (2017) used an organic amine catalyst for their pre-treatment method of corn stover, where the synergistic effect of n-propylamine and aqueous ethanol reported a sugar yield of 83.2% [35]. Through the research conducted by Pan et al. (2005) on the biorefining of softwoods by making use of ethanol as the solvent for organosolv pulping, it was observed that more than 90% of the sugars were retrieved by using a combination of the organosolv pre-treatment and enzymatic hydrolysis on both softwoods and hardwoods [36]. Furthermore, Ravindran et al. (2018) reported that the use of an organic amine catalyst that is capable of breaking the bond joining the lignin and hemicelluloses, proving it to be an efficient use of an organic solvent for high lignin-containing biomass. The sugar yield was achieved at 29.05 mg/g through the use of ethanol-assisted organosolv pre-treatment of biomass (as spent coffee waste) [37].

Gümüşkaya investigated the chemical content of alkali sulfite pulp at temperatures ranging between 120 °C and 200 °C and reported that the sugar content in the pulp slurries increased with a rise in the treatment temperature [38]. Additionally, the lignin content was reported to decrease with increasing heating treatment temperature. Therefore, it is a good indicator to set the heating temperature above 120 °C during the delignification process to obtain a higher sugar yield [38]. A high-temperature reaction can help promote a higher rate of reaction, thus allowing for the separation of lignin from the biomass in order to extract the sugar content. However, it was also reported that the thermal stability of hemicellulose degrades at temperatures above 340 °C [5]; hence, a temperature higher than 180 °C may decrease the total sugar yield due to hemicellulose degradation [39,40]. It was also reported that a long duration of the biomass heating process by using ethanol as a solvent seemed to dissolve some amount of hemicellulose and, at the same time, led to an increase in the solubility of lignin in pulp slurries [39].

The increase in the total sugar yield as a result of increasing the ethanol concentration was also attributed to the ether chain breaking of lignin and hemicellulose molecules. This occurrence can be justified through a higher conversion of cellulose and hemicellulose into sugars being achieved as the ethanol concentrations increase [5]. However, the total sugars seemed to decrease when 75 vol% of ethanol concentration was applied in the OPEFB pre-treatment. This was due to the effect of heavy ethanol concentration that caused some hemicellulose content to be decomposed at above 60 vol% concentration of the ethanol solvent [41].

4. Conclusions

In summary, it was concluded that OPEFB samples pre-treated with 55 vol% ethanol concentration and subjected to a higher microwave power of 520 W for a longer duration of 90 min led to the highest delignification rate, with 0.57% of lignin remaining in the OPEFB samples after the microwave-assisted organosolv pre-treatment, further corresponding to the optimum yield of total sugars obtained as 87.8 mg/L. The data on surface morphology supports this finding, revealing that OPEFB samples pre-treated with similar pre-treatment conditions have a higher degree of opened and deepened surface structure. These observations were attributed to the higher thermal degradation achieved, where the microwave power functions to an increase in the temperature of the ethanol solvent, which in turn helps facilitate the breakdown of the OPEFB protective lignin layer, leading to more lignin being removed in order to expose and extract the total sugars content.

The potential decomposition mechanism behind these observations on the lignin amount and total sugar yield is the disruption of hydrogen bonds that is caused by the displacement reaction resulting from the realignment of polar molecules, where cell wall components, such as cellulose crystallinity, are destroyed, enhancing the hydrolysis process of extracting the sugar. When LCB such as OPEFB is irradiated through microwave-assisted irradiation, the atoms or molecules within the biomass are thermally vibrated, corresponding to a rise in temperature of the organic material without rapid degradation of the biomass samples, as the presence of the ethanol organic solvent along with the working principle of microwave irradiation ensures a less severe process of thermal degradation, thereby preserving the valuable bioproducts to be extracted. At the same time, the function of the ethanol solvent is to allow for the cleavage of β-O-4 linkages and ester bonds, which are the major mechanisms of lignin cleavage.

To conclude, these results suggest that OPEFB subjected to microwave-assisted ethanol pre-treatment can help to extract the total sugars for various uses in industrial applications, such as its emergence as one of the sustainable alternative bioresources that are renewable in order to reduce the dependency on non-renewable resources, such as fossil fuels. For instance, the impact of this research project could give rise to the application of bioproducts extracted from OPEFB toward bioenergy technologies, such as biofuels (biodiesel), biopower in the generation of heat and electricity, and biochemicals such as bioplastics. However, more research on other organic solvents for OPEFB microwave-assisted pre-treatment is required and necessary to extract a higher yield of total sugars.

Author Contributions: Conceptualization, A.S.G. and S.L.; methodology, A.S.G.; research and investigation, A.S.G.; writing—original draft, A.S.G.; writing—review and editing, K.H.W., S.L., Y.L.P., L.L. and S.Y.L.; supervision, K.H.W., S.L. and Y.L.P.; funding acquisition, K.H.W. and S.L. All authors have read and agreed to the published version of the manuscript.

Funding: This work was supported by the Universiti Tunku Abdul Rahman Research Fund (UTARRF) [IPSR/RMC/UTARRF/2022-C2/S04].

Institutional Review Board Statement: Not applicable.

Informed Consent Statement: Not applicable.

Data Availability Statement: All the data are available inside this paper.

Conflicts of Interest: The authors declare no conflicts of interest.

References

1. REN21. *Renewables 2020 Global Status Report*; REN21 Secretariat: Paris, France, 2020.
2. Kumar, B.; Bhardwaj, N.; Agrawal, K.; Chaturvedi, V.; Verma, P. Current perspective on pretreatment technologies using lignocellulosic biomass: An emerging biorefinery concept. *Fuel Process. Technol.* **2020**, *199*, 106244. [CrossRef]
3. Hassan, S.S.; Williams, G.A.; Jaiswal, A.K. Emerging technologies for the pretreatment of lignocellulosic biomass. *Bioresour. Technol.* **2018**, *262*, 310–318. [CrossRef] [PubMed]
4. Mohan, S.V.; Modestra, J.A.; Amulya, K.; Butti, S.K.; Velvizhi, G. A circular bioeconomy with biobased products from co_2 sequestration. *Trends Biotechnol.* **2016**, *34*, 506–519. [CrossRef]
5. Nurfahmi; Ong, H.C.; Jan, B.M.; Tong, C.W.; Fauzi, H.; Chen, W.-H. Effects of organosolv pretreatment and acid hydrolysis on palm empty fruit bunch (PEFB) as bioethanol feedstock. *Biomass Bioenergy* **2016**, *95*, 78–83. [CrossRef]
6. Sun, S.; Sun, S.; Cao, X.; Sun, R. The role of pretreatment in improving the enzymatic hydrolysis of lignocellulosic materials. *Bioresour. Technol.* **2016**, *199*, 49–58. [CrossRef]
7. Werpy, T.; Petersen, G. *Top Value-Added Chemicals from Biomass: Volume I—Results of Screening for Potential Candidates from Sugars and Synthesis Gas*; U.S. Department of Energy Office of Scientific and Technical Information: Oak Ridge, TN, USA, 2004; pp. 1–76. [CrossRef]
8. Diyanilla, R.; Hamidon, S.; Suryanegara, L.; Hussin, H. Overview of pretreatment methods employed on oil palm biomass in producing value-added products: A Review. *Bioresources* **2020**, *15*, 9935. [CrossRef]
9. Bensah, E.C.; Kádár, Z.; Mensah, M.Y. Alkali and glycerol pretreatment of West African biomass for production of sugars and ethanol. *Bioresour. Technol. Rep.* **2019**, *6*, 123–130. [CrossRef]
10. Schieppati, D.; Patience, N.A.; Galli, F.; Dal, P.; Seck, I.; Patience, G.S.; Fuoco, D.; Banquy, X.; Boffito, D.C. Chemical and Biological Delignification of Biomass: A Review. *Ind. Eng. Chem. Res.* **2023**, *62*, 12757–12794. [CrossRef]
11. Ashokkumar, V.; Venkatkarthick, R.; Jayashree, S.; Chuetor, S.; Dharmaraj, S.; Kumar, G.; Chen, W.-H.; Ngamcharussrivichai, C. Recent advances in lignocellulosic biomass for biofuels and value-added bioproducts—A critical review. *Bioresour. Technol.* **2022**, *344*, 126195. [CrossRef]
12. Wei, W.; Wang, B.; Wang, X.; Ling, R.; Jin, Y. Comparison of acid and alkali catalyzed ethylene glycol organosolv pretreatment for sugar production from bagasse. *Bioresour. Technol.* **2021**, *320*, 124293. [CrossRef]
13. Singh, S.; Sinha, R.; Kundu, S. Role of organosolv pretreatment on enzymatic hydrolysis of mustard biomass for increased sac-charification. *Biomass Convers. Biorefinery* **2022**, *12*, 1657–1668. [CrossRef]
14. Chu, Q.; Tong, W.; Chen, J.; Wu, S.; Jin, Y.; Hu, J.; Song, K. Organosolv pretreatment assisted by carbocation scavenger to miti-gate surface barrier effect of lignin for improving biomass saccharification and utilization. *Biotechnol. Biofuels Bioprod.* **2021**, *14*, 136.
15. Chin, D.W.K.; Lim, S.; Pang, Y.L.; Lim, C.H.; Lee, K.M. Two-staged acid hydrolysis on ethylene glycol pretreated degraded oil palm empty fruit bunch for sugar based substrate recovery. *Bioresour. Technol.* **2019**, *292*, 121967. [CrossRef] [PubMed]
16. Rezania, S.; Oryani, B.; Cho, J.; Talaiekhozani, A.; Sabbagh, F.; Hashemi, B.; Rupani, P.F.; Mohammadi, A.A. Different pretreatment technologies of lignocellulosic biomass for bioethanol production: An overview. *Energy* **2020**, *199*, 117457. [CrossRef]
17. Xiaokang, W.; Lyng, J.G.; Brunton, N.P.; Cody, L.; Jacquier, J.-C.; Harrison, S.M.; Papoutsis, K. Monitoring the effect of different microwave extraction parameters on the recovery of polyphenols from shiitake mushrooms: Comparison with hot-water and organic-solvent extractions. *Biotechnol. Rep.* **2020**, *27*, e00504. [CrossRef] [PubMed]

18. Vinatoru, M.; Mason, T.; Calinescu, I. Ultrasonically assisted extraction (UAE) and microwave assisted extraction (MAE) of functional compounds from plant materials. *TrAC Trends Anal. Chem.* **2017**, *97*, 159–178. [CrossRef]
19. Gonçalves, F.A.; Ruiz, H.A.; dos Santos, E.S.; Teixeira, J.A.; de Macedo, G.R. Bioethanol production from coconuts and cactus pretreated by autohydrolysis. *Ind. Crops Prod.* **2015**, *77*, 1–12. [CrossRef]
20. De Souza Moretti, M.M.; Bocchini-Martins, D.A.; Nunes, C.d.C.C.; Villena, M.A.; Perrone, O.M.; da Silva, R.; Boscolo, M.; Gomes, E. Pre-treatment of sugarcane bagasse with microwaves irradiation and its effects on the structure and on enzymatic hydrolysis. *Appl. Energy* **2014**, *122*, 189–195. [CrossRef]
21. Binod, P.; Satyanagalakshmi, K.; Sindhu, R.; Janu, K.U.; Sukumaran, R.K.; Pandey, A. Short duration microwave assisted pre-treatment enhances the enzymatic saccharification and fermentable sugar yield from sugarcane bagasse. *Renew. Energy* **2012**, *37*, 109–116. [CrossRef]
22. Sarip, H.; Hossain, M.S.; Azemi, M.; Allaf, K. A review of the thermal pretreatment of lignocellulosic biomass towards glucose production: Autohydrolysis with DIC technology. *BioResources* **2016**, *11*, 10625–10653. [CrossRef]
23. Zulkiple, N.; Maskat, M.Y.; Hassan, O. Pretreatment of oil palm empty fruit fiber (OPEFB) with aqueous ammonia for high production of sugar. *Procedia Chem.* **2016**, *18*, 155–161. [CrossRef]
24. Nomanbhay, S.M.; Hussain, R.; Palanisamy, K. Microwave-assisted alkaline pretreatment and microwave assisted enzymatic saccharification of oil palm empty fruit bunch fiber for enhanced fermentable sugar yield. *J. Sustain. Bioenergy Syst.* **2013**, *3*, 7–17. [CrossRef]
25. Banerjee, D.; Mukherjee, S.; Pal, S.; Khowala, S. Enhanced saccharification efficiency of lignocellulosic biomass of mustard stalk and straw by salt pretreatment. *Ind. Crops Prod.* **2016**, *80*, 42–49. [CrossRef]
26. Li, X.; Wei, Y.; Xu, J.; Xu, N.; He, Y. Quantitative visualization of lignocellulose components in transverse sections of moso bamboo based on FTIR macro- and micro-spectroscopy coupled with chemometrics. *Biotechnol. Biofuels* **2018**, *11*, 263. [CrossRef] [PubMed]
27. Mondylaksita, K.; Ferreira, J.A.; Millati, R.; Budhijanto, W.; Niklasson, C.; Taherzadeh, M.J. Recovery of high purity lignin and digestible cellulose from oil palm empty fruit bunch using low acid-catalyzed organosolv pretreatment. *Agronomy* **2020**, *10*, 674. [CrossRef]
28. Zhou, X.; Ding, D.; You, T.; Zhang, X.; Takabe, K.; Xu, F. Synergetic dissolution of branched xylan and lignin opens the way for enzymatic hydrolysis of poplar cell wall. *J. Agric. Food Chem.* **2018**, *66*, 3449–3456. [CrossRef] [PubMed]
29. Cheng, F.; Ouyang, T.; Sun, J.; Jiang, T.; Luo, J. Using solubility parameter analysis to understand delignification of poplar and rice straw with catalyzed organosolv fractionation processes. *BioResources* **2019**, *14*, 486–499. [CrossRef]
30. Fink, J.K. *Reactive Polymers: Fundamentals and Applications: A Concise Guide to Industrial Polymers*, 3rd ed.; Ebnesajjad, S., Ed.; A volume in Plastics Design Library; William Andrew: Norwich, NY, USA, 2018.
31. Zhang, K.; Pei, Z.; Wang, D. Organic solvent pretreatment of lignocellulosic biomass for biofuels and biochemicals: A review. *Bioresour. Technol.* **2016**, *199*, 21–33. [CrossRef]
32. Hamzah, M.H.; Bowra, S.; Cox, P. Effects of ethanol concentration on organosolv lignin precipitation and aggregation from *Miscanthus x giganteus*. *Processes* **2020**, *8*, 845. [CrossRef]
33. Brodie, G. Microwave Heating in Moist Materials. In *Advances in Induction and Microwave Heating of Mineral and Organic Materials*; Grundas, S., Ed.; InTech Open Access Publisher: Rijeka, Croatia, 2011; pp. 553–584.
34. Tsubaki, S.; Azuma, J.I. Application of microwave technology for utilization of recalcitrant biomass. In *Advances Induction and Microwave Heating of Mineral and Organic Materials*; Grundas, S., Ed.; InTech Open Access Publisher: Rijeka, Croatia, 2011; pp. 697–722.
35. Tang, C.; Shan, J.; Chen, Y.; Zhong, L.; Shen, T.; Zhu, C.; Ying, H. Organic amine catalytic organosolv pretreatment of corn stover for enzymatic saccharification and high-quality lignin. *Bioresour. Technol.* **2017**, *232*, 222–228. [CrossRef] [PubMed]
36. Pan, X.; Gilkes, N.; Gregg, D.; Mabee, W.; Xiao, Z.; Zhang, X.; Saddler, J.; Arato, C.; Pye, K. Biorefining of softwoods using ethanol organosolv pulping: Preliminary evaluation of process streams for manufacture of fuel-grade ethanol and co-products. *Biotechnol. Bioeng.* **2005**, *90*, 473–481. [CrossRef]
37. Ravindran, R.; Desmond, C.; Jaiswal, S.; Jaiswal, A.K. Optimisation of organosolv pretreatment for the extraction of polyphenols from spent coffee waste and subsequent recovery of fermentable sugars. *Bioresour. Technol. Rep.* **2018**, *3*, 7–14. [CrossRef]
38. Gümüşkaya, E.; Usta, M. Dependence of chemical and crystalline structure of alkali sulfite pulp on cooking temperature and time. *Carbohydr. Polym.* **2006**, *65*, 461–468. [CrossRef]
39. Werner, K.; Pommer, L.; Broström, M. Thermal decomposition of hemicelluloses. *J. Anal. Appl. Pyrolysis* **2014**, *110*, 130–137. [CrossRef]
40. Lv, G.-J.; Wu, S.-B.; Lou, R. Kinetic study of the thermal decomposition of hemicellulose isolated from corn stalk. *BioResources* **2010**, *5*, 1281–1291. [CrossRef]
41. Pan, X.; Gilkes, N.; Kadla, J.; Pye, K.; Saka, S.; Gregg, D.; Ehara, K.; Xie, D.; Lam, D.; Saddler, J. Bioconversion of hybrid poplar to ethanol and co-products using an organosolv fractionation process: Optimization of process yields. *Biotechnol. Bioeng.* **2006**, *94*, 851–861. [CrossRef]

sustainability

MDPI

Article

Effect of Chemical Pre-Treatment on the Catalytic Performance of Oil Palm EFB Fibre Supported Magnetic Acid Catalyst

Shamala Gowri Krishnan [1], Fei Ling Pua [2,3,*] and Zhang Fan [4]

1 College of Graduate Studies, Universiti Tenaga Nasional, Kajang 43000, Selangor, Malaysia; shamala.gowri@uniten.edu.my
2 Institute of Sustainable Energy (ISE), Universiti Tenaga Nasional, Kajang 43000, Selangor, Malaysia
3 Department of Mechanical Engineering, College of Engineering, Universiti Tenaga Nasional, Kajang 43000, Selangor, Malaysia
4 Biomass Group, Key Laboratory of Tropical Plant Resources and Sustainable Use, Xishuangbanna Tropical Botanical Garden, Chinese Academy of Sciences, 88 Xuefulu, Kunming 650223, China; zhangfan@xtbg.ac.cn
* Correspondence: gracepua@uniten.edu.my

Abstract: The development of heterogenous catalysts using renewable materials has received wide attention. A heterogenous catalyst has been a preferred choice as it evades the disadvantages of homogeneous catalysts, nevertheless, heterogenous catalysts has limited activity and a longer separation process. The current study emphasises the preparation of a new magnetic catalyst using oil palm empty fruit bunch (EFB) fibre as a carbon-based support material. The effect of different alkaline pre-treatments over the methyl ester conversion rate were investigated. The catalyst preparation parameters were studied by using the single factor optimisation approach, including the fibre loading, impregnation time, calcination temperature, and calcination time. Their effects in the esterification of oleic acid were investigated in this study. The optimisation study shows that the Na_2CO_3-treated(T)-EFBC magnetic catalyst had the highest esterification rate of 93.5% with 7 g EFB fibre loading, a 2 h impregnation time and a calcination temperature of 500 °C for 2 h. The catalyst possessed a good acidity of 3.5 mmol/g with excellent magnetism properties. This study showed that the catalysts are magnetically separable and exhibited good stability with 82.1% after five cycles. The oil palm EFB supported magnetic acid catalyst indicates it as a potential option to the existing solid catalysts that is economical and environmentally friendly for the esterification process.

Keywords: biomass; catalyst; magnetic; oil palm; oleic acid; energy

Citation: Krishnan, S.G.; Pua, F.L.; Fan, Z. Effect of Chemical Pre-Treatment on the Catalytic Performance of Oil Palm EFB Fibre Supported Magnetic Acid Catalyst. *Sustainability* **2023**, *15*, 8637. https://doi.org/10.3390/su15118637

Academic Editor: Youzhi Miao

Received: 10 April 2023
Revised: 17 May 2023
Accepted: 23 May 2023
Published: 26 May 2023

1. Introduction

The impact of global warming and the steady depletion of non-renewable energy resources, such as fossil fuels, has led many researchers worldwide to intensively investigate a potential substitution for fossil fuels [1]. With the increased attention in identifying new resources, biodiesel, a mixture of fatty acid alkyl esters, is gaining much global interest due to it being non-toxic, renewable, sustainable and environmentally benign. Traditionally, homogeneous catalysts, such as sodium hydroxide (NaOH), potassium hydroxide (KOH) and sulphuric acid (H_2SO_4), were used for biodiesel production. Despite their effective catalytic activity to produce a high biodiesel yield, these catalysts, however, are noted to be non-environmentally friendly as they are unrecyclable [2,3]. Besides having no catalyst recyclability, contributing to equipment corrosion, there is the unavailability of the final product to recover the catalyst due to its solubility nature resulting in further purification processes, which leads to the production of large amounts of wastewater [4].

The development of a new heterogeneous catalyst for the catalytic process involves the identification of a potential catalyst that can address the difficulties of using a homogeneous catalyst in biodiesel production. Further, through a heterogeneous catalytic process, there is a high possibility of obtaining a catalyst that can be recycled with the catalytic functions of the acidic sites on the surface and that does not cause any damage to the equipment [5]. Many metal oxides catalysts, such as mixed metal oxides, zeolites, graphene oxides, sulphated zirconia/titanium and ionic liquids, have been extensively used in the biodiesel production process as they have high stability and a reduced effect on the ecosystem [6]. Despite their good stability and eco-friendliness, the difficulty in the separation of catalysts from a large-scale reaction product is still unresolved.

Over the years, several researchers have focused on the use of biomass or carbon-based materials for solid catalyst synthesis. A study conducted by Tang and Niu on the preparation of carbon-based solid catalysts from bamboo for the esterification reaction reported that a high surface area of 1208 m^2/g helped to achieve a high biodiesel yield of 97.3% [7]. The study showed that biomass-based catalysts are non-toxic, biodegradable and possess a high surface area for a higher yield, thus proving that it can be a potential replacement for the existing conventional catalyst. Additionally, another study on heterogeneous catalysts, graphene oxide and sulfonated biochar for lactic acid esterification reported that the highest yield of 35% was obtained after 420 min [8]. Although many solid catalysts have been investigated in the past, the technical difficulty in separating the solid catalysts from the product either by filtration or centrifugation remains as the main drawback, whereby more time is required, and the catalyst is lost during the recovery process [9]. In regard to this, several authors have claimed that the catalytic activity of sulphate metal oxide could be enhanced by their rapid separation from the products [10,11].

Recently, the use of a magnetic catalyst has gained much interest due to its easy separation from the product by using an external magnet. Feyzi and Norouzi worked on a novel magnetic $Ca/Fe_3O_4@SiO_2$ nano-catalyst for biodiesel production and further reported that the catalytic performance of this catalyst proved to exhibit a high biodiesel yield of 97% [12]. Besides, Gardy et al. also acquired excellent catalytic activity for the methyl oleate and biodiesel production over a magnetic core-shell catalyst, with the highest yield of 88% and 98.5%, respectively [13].

Agricultural waste is often referred to as lignocellulosic material consisting of lignin, cellulose and hemicellulose [14]. Due to its limited usability and only as a possible source of electricity and heat energy, the agricultural waste had low market value. However, agricultural biomass is naturally composed of carbons that can be utilised as recyclable carbon materials. As oil palm is one of Malaysia's mainly grown crops with vast production, enormous amounts of palm waste, such as oil palm fronds, trunks, empty fruit bunches, shells and kernels, have been generated [15,16]. The oil palm empty fruit bunch (EFB), which is of particular interest in this study, is produced as waste after the harvesting process, whereby the oil extraction is done and primarily used as a power source in palm oil mill incinerators [17]. Further, EFB, which is abundant within the lignocellulosic component, can be a potential source of sustainable biomass.

This study aims to synthesise a magnetic solid catalyst using raw EFB fibres and chemically treated fibres via a one-step impregnation and calcination method. The catalyst was then used in the catalytic esterification of oleic acid and the catalytic performance was evaluated through the optimisation of several parameters during the catalyst preparation process. The synthesised catalyst was characterised by morphological analysis, elemental analysis and an acid density test. Meanwhile, the esterified product from the catalysed reactions at the optimum condition was analysed using gas chromatography (GC) analysis.

2. Materials and Methods

2.1. Materials

Oil palm empty fruit bunch (EFB) fibres were collected from Sztech Engineering Sdn. Bhd. (Shah Alam, Selangor, Malaysia). Analytical reagents, including ferric sulphate ($Fe_3(SO_4)_2$), ferrous sulphate heptahydrate ($FeSO_4.7H_2O$), sodium bicarbonate (Na_2CO_3), potassium oxide (KOH), calcium hydroxide ($Ca(OH)_2$) and sodium hydroxide (NaOH) chemicals used for the catalyst preparation were purchased from Suria Pembekal Umum

Sdn. Bhd. (Kuala Lumpur, Malaysia). Industrial oleic acid with an acid value of 199 mg KOH/g (R&M, Petaling Jaya, Selangor, Malaysia) used in the esterification reaction.

2.2. EFB Fibre Chemical Pre-Treatment

The raw EFB fibres were ground and sieved to the size range of 125–150 µm. Next, the fibres were pre-treated with 0.1 M NaOH solution, and a solid-to-liquid ratio of 50 g/L was employed. The oven-dried EFB fibres were pre-treated by stirring in the NaOH solution for 1 h with agitation at a speed of 350 rpm at room temperature. The fibres were immersed in the alkali solution for 30 min after the process. The fibres were then washed repeatedly with distilled water to remove the excess chemical until a neutral pH (~pH 7) was achieved. The washed fibres were then poured onto an aluminium tray and oven-dried for 24 h at 105 °C. The same method was used to study the effect of chemical modification using various alkalines, including KOH, $Ca(OH)_2$ and Na_2CO_3, with the same concentration (0.1 M). All treated fibres were stored for further use.

2.3. Synthesis of the Magnetic Solid Catalyst

The magnetic solid catalyst was synthesised via the one-step impregnation method. Firstly, the method was conducted by mixing the treated EFB fibres with a 0.01 M $FeSO_4.7H_2O$ and 0.02 M $Fe_3(SO_4)_2$ solution with continuous stirring at 60 °C for 2 h. A solid-to-liquid ratio of 1:10 was employed. The solid was then oven-dried overnight at 80 °C. Following that, the oven-dried samples were calcined using a vacuum furnace. Figure 1 shows the summary of the magnetic catalyst synthesis. The EFB magnetic catalyst was prepared using fibres treated differently with NaOH, KOH, $Ca(OH)_2$ and Na_2CO_3, denoted as NaOH-T-EFBC, KOH-T-EFBC, $Ca(OH)_2$-T-EFBC and Na_2CO_3-T-EFBC, respectively.

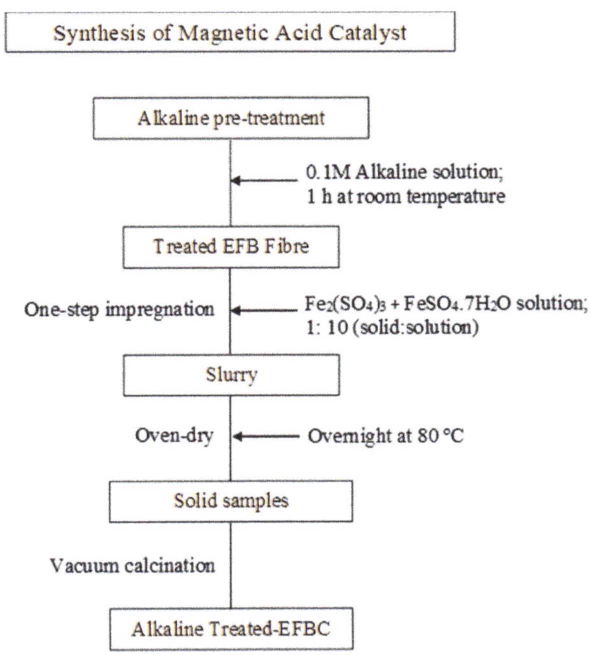

Figure 1. The summary of the magnetic acid catalyst preparation.

Later, the catalyst synthesis parameter conditions were optimised in terms of fibre loading, reaction time, calcination temperature and calcination time via the single factor optimisation method. Table 1 shows the parameters and ranges for the optimisation of the esterification reaction of oleic acid.

Table 1. The parameters and ranges for the optimisation study.

Factors	Unit	1	2	3	4	5
EFB fibre loading	g	1	3	5	7	9
Reaction time	h	1	2	3	4	5
Calcination temperature	°C	300	400	500	600	700
Calcination time	h	1	2	3	4	5

2.4. *Characterisation of the Fibre and Catalyst*

The lignin content analysis was conducted using the acid detergent lignin (ADL) method at MARDILab, Technical Services & Technology Commercialization Centre (Serdang, Selangor, Malaysia). With regards to the characterisation, the morphology and elemental composition of the samples were analysed by using a scanning electron microscopy (SEM) equipped with energy-dispersive X-ray spectroscopy (EDS) (JSM-6010PLUS/LV, Universiti Tenaga Nasional, Kajang, Malaysia). The acidity of the solid acid catalyst was determined by using the acid–base titration method. The method involved titrating with sodium hydroxide solution (0.1 M NaOH), with phenolphthalein as an indicator. The data was recorded, and the acid density was calculated using Equation (1):

$$\text{Acidic} = \frac{[\text{OH}] \times \text{VOH}}{\text{m}} \tag{1}$$

[OH] = concentration of NaOH used, VOH = volume of NaOH required and m = mass of catalyst.

2.5. *Catalytic Performance*

In total, 7 wt% of solid catalyst was mixed with 10:1 methanol and pre-heated oleic acid in a three-neck flat-bottom flask at 60 °C for 2 h at 350 rpm. The solid magnetic acid catalyst was separated by an external magnet, washed with ethanol and oven-dried at 80 °C for further use. The collected liquid was then heated at 70 °C to evaporate the excess methanol.

Next, the methyl ester conversion was examined via the base titration method using KOH. The data was collected, and the acid value of the esterified sample and oleic acid was calculated based on the ASTM D664 standard method (Equation (2)) [18], while the esterification conversion rate was calculated using Equation (3) [19]:

$$\text{Acid value} = \frac{([\text{OH}] \times \text{V} \times \text{MM})}{\text{m}} \tag{2}$$

$$\text{Conversion rate } (\%) = \left(\frac{(\text{AV}_\text{i} - \text{AV}_\text{o})}{\text{AV}_\text{i}} \times 100\right) \tag{3}$$

V = volume of KOH titrated in mL, MM = molecular mass of KOH, [OH] = concentration of KOH, m = mass of catalyst, AV_i = initial acid value and AV_o = acid value of methyl oleate.

3. Results and Discussion

3.1. *Effect of Various Chemical Pre-Treatments on the Esterification Reaction*

NaOH, KOH, $Ca(OH)_2$ and Na_2CO_3 are the essential alkalis used in the alkaline pre-treatment to investigate their efficiency in the delignification process. The pre-treatment process generally signifies the removal of lignin from lignocellulose material. The removal of complex lignin attributes to the loosening of the crystalline structure of cellulose and enhances the porosity of the biomass for catalyst preparation [20]. After alkaline pre-treatment, the lignin content changed significantly and altered the surface texture of the fibres. Lignin content is considered an important component influenced by alkaline pre-treatment because lignin forms a three-dimensional network between cellulose and hemicellulose [21]. The lignin content resulted in a range of 13.7–15.1% by using for different alkalis at a standardised concentration (Figure 2). The results showed the percentage of lignin content reduced after alkaline treatment compared to the untreated EFB fibres, with 16.5%, and this is similar to that reported by Ariffin et al. [22]. The effect of various alkaline pre-treatments resulted 14.5%, 13.7%, 14.1% and 15.1% using NaOH, KOH, $Ca(OH)_2$ and Na_2CO_3, respectively. Alkaline pre-treatment contributes to the chemically modified structure of lignocellulosic EFB, breaking up the crystalline structure and increasing the accessible surface area of EFB for a better interaction with metal sulphates in catalyst preparation [23].

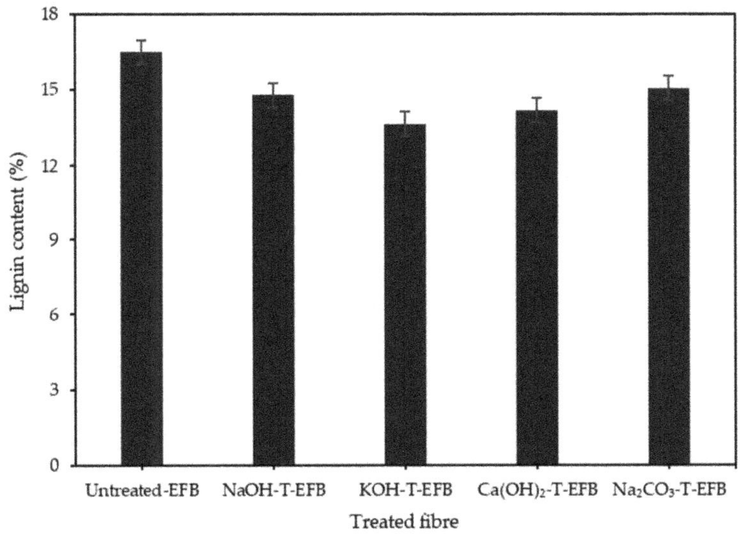

Figure 2. Effect of chemical pre-treatment on the EFB fibre delignification.

The trends in the catalysts' acidity and conversion rate over different alkaline pre-treatments and untreated-EFB are shown in Figure 3. After alkaline pre-treatment, the treated EFB fibres were used in the synthesis of magnetic solid catalysts. Five grams of treated EFB fibres were added into the chemical solution (1:10) and stirred for 2 h at 60 °C. The dried sample was calcined at 500 °C for 3 h in a vacuum furnace. The results indicated that the highest catalyst acidity was reported for catalysts treated by $Ca(OH)_2$ and Na_2CO_3, with 3.5 mmol/g, compared to the catalysts treated by NaOH and KOH, which were 3.3 mmol/g and 2.5 mmol/g, respectively. Moreover, both $Ca(OH)_2$-T-EFBC and Na_2CO_3-T-EFBC catalysts observed a higher conversion rate of 85.1% and 87.8%, respectively, compared with the catalysts treated by NaOH, KOH and untreated-EFB, which had 80.6%, 74.5% and 79.03, respectively. The highest acidity and conversion rate wsd obtained in EFB treated with Na_2CO_3 (Na_2CO_3-T-EFB), showing that Na_2CO_3 pre-treatment has a significant effect on the conversion rate. The dissociation of Na_2CO_3 in water produces carbonic acid, and sodium hydroxide that has the potential to disrupt the structure and remove silica on the EFB fibre surface [24]. Although the lignin content for Na_2CO_3-T-EFB is higher compared to other alkaline treatments, the esterification rate was exhibited to be higher using the derived catalyst. This could be due to the existence of hydrophilic and hydrophobic groups, such as hydroxyl groups and polycyclic aromatic carbons, in the carbonaceous catalyst [25]. The reactants can easily be absorbed onto the solid catalyst surface to contact with the active sites. The active sites of sulfonic acid ($-SO_3H$) and iron oxide (Fe-O) are known as Bronsted acid and Lewis acid, respectively, which contribute to a higher conversion rate [26].

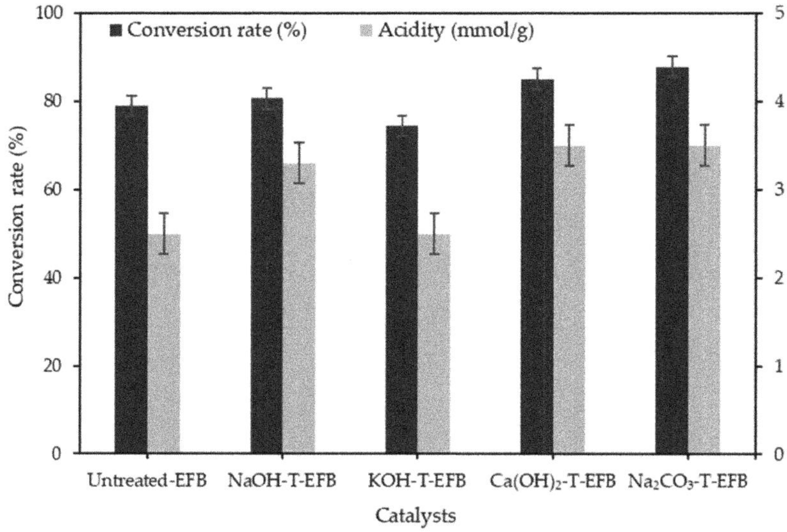

Figure 3. The relationship between chemically treated EFB fibre-derived magnetic catalysts with the acidity and esterification rate.

3.2. Single Factor Optimisation

As the Na_2CO_3-T-EFBC magnetic catalyst recorded the highest conversion rate, the Na_2CO_3 treated EFB fibres were utilised to investigate the effect of catalyst synthesis parameters, such as EFB treated fibre loading, impregnation time, calcination temperature and calcination time, on conversion rate and catalyst acidity (Figure 4a–d). The effect of EFB fibre loading on the conversion rate is demonstrated in Figure 4a. As aforementioned, the catalyst was synthesised with different ranges of EFB fibre loading by weight (1, 3, 5, 7 and 9 g). As shown in Figure 4a, the findings indicated an increase in the conversion rate from 78.32% to 91.18% as the EFB fibre loading increased from 1 g to 7 g. The increment of the esterification rate was observed to increase with the incline in the catalyst acidity from 1.7 mmol/g to 2.2 mmol/g. The findings are supported by a previous study by Jing et al., which stated that the concentration of support material or precursor is one of the key factors that affects the preparation of the catalyst by the impregnation method [27]. Further, as biomass consists of a more complex polymer structure, which is made up of functional oxygenated molecules, namely, carboxyl (-COOH) and hydroxyl (-OH) [28], the addition of carbon material enhanced the existence of weak Bronsted acid sites in the catalyst surface besides the presence of sulfonic acid ($-SO_3H$) sites [29]. Thus, the 7 g EFB fibre loading significantly increased the esterification rate with the highest acidity of 2.2 mmol/g, and the esterification reaction was favoured by the high acidity of the catalyst. However, a further increment in the amount of EFB fibre loading (9 g) was insignificant in the esterification reaction of oleic acid. This could be attributed to the limited active sites and the lack of strong acid sites for the catalytic reaction, which resulted in a slight reduction in the conversion rate to 90.81% with an acidity of 2.1 mmol/g.

In order to investigate the effect of the impregnation time in the conversion rate, different reaction times of 1, 2, 3, 4 and 5 h were executed. Figure 4b illustrates that the conversion rate increases with the increase in the impregnation time. It was observed that as the impregnation time increased from 1 to 2 h, the conversion rate was observed as 88.02% and 91.18%, respectively. During the impregnation process, the interaction through absorption between the pores of the support material with the bulk solution used requires a certain time to occur [30]. However, no significant effect was detected in the oleic acid conversion after 2 h of impregnation time, which was similar to the decline in the acidity of the catalyst after 3 to 5 h of the impregnation time. The low acidity contributed to the decrease in the conversion rate from 89.32% to 79.87% after 5 h. A previous study reported that the impregnation time highly depended on the support size, loading amount and the temperature [31]. Based on the observation, at 2 h of impregnation time, the ferric sulphate particles may be mostly adsorbed on the surface of the treated EFB fibres. Hence, at further impregnation times, the particles have limited sites available for absorption, which reasons the lower conversion rate of oleic acid after 2 h of impregnation time. Therefore, to achieve a high conversion rate and the efficient formation of active sites, an appropriate impregnation time is required; thus, the 2 h reaction time was found to be favourable for high catalytic performance [32].

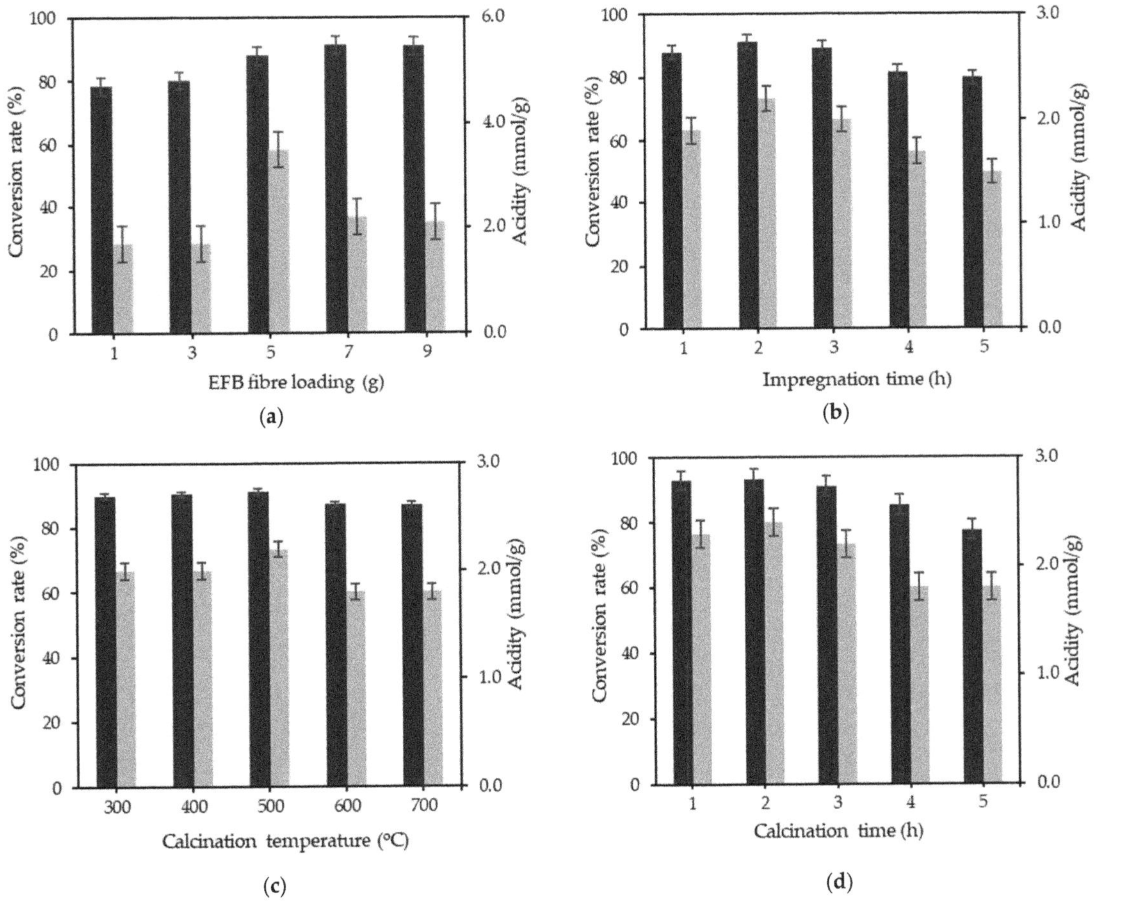

Figure 4. Optimisation studies for catalyst preparation. (**a**) Effect of EFB fibre loading (2 h impregnation time and calcination at 600 °C for 3 h). (**b**) Effect of impregnation time (EFB fibre loading 7 g and calcination at 600 °C for 3 h). (**c**) Effect of calcination temperature (EFB fibre loading 7 g, 2 h impregnation time and calcination for 3 h). (**d**) Effect of calcination time (EFB fibre loading 7 g, 2 h impregnation time and calcination at 500 °C).

Calcination temperature is an essential factor for a catalyst because of its effect on the structural and catalytic properties [33]. The effect of calcination temperature on oleic acid conversion was studied, and the results are shown in Figure 4c. The magnetic catalyst was calcined using a vacuum furnace at various calcination temperatures of 300, 400, 500, 600 and 700 °C. The findings indicated that the conversion rate gradually increases from 300 °C to 500 °C with values of 89.95% and 91.18%, respectively. This shows that calcination temperature has a significant effect on the catalytic activity as the temperature improves the formation of particles on the catalyst surface [34]. However, a gradual decrease in the conversion rate was noted at a higher temperature, which indicates that calcination at a higher temperature would contribute to sintering of particles on the surface of the support material, thus, resulting in a decrease in surface area, which eventually affects the catalytic activity [35]. Previous studies have also stated that a suitable calcination temperature was required for an effective esterification reaction because both low and high calcination temperatures were not beneficial for the conversion of oleic acid [36,37]. Hence, the suggested 500 °C was determined as an optimum temperature for the calcination process of the catalyst.

The effect of the calcination time on the esterification rate was also investigated to determine the optimum catalyst preparation parameters. The catalyst was calcined for a range of calcination times (1, 2, 3, 4 and 5 h) at a fixed temperature of 500 °C. Figure 4d presents the summary of the findings. It was found that calcination for 2 h produced an active catalyst with the maximum conversion rate of 93.46% and an acidity of 2.4 mmol/g. The high rate was attributed to the sufficient time for the formation of active components, which helps to increase the conversion rate [38]. Furthermore, the results show that the increment in calcination time was not favourable for esterification as the conversion rate gradually decreased to 77.62% after 5 h of calcination. A reason for this could be the particle sintering or agglomeration on the catalyst surface, which resulted in weak catalytic activity after a longer calcination duration [39]. Besides, a longer calcination time could deactivate the catalyst's acidic sites [40]. Thus, an optimum calcination time of 2 h was essential to synthesise an EFB supported magnetic acid catalyst. The optimum conditions for catalyst preparation are: 7 g EFB fibre loading, 2 h impregnation time and a calcination temperature of 500 °C for 2 h.

3.3. Morphology and Elemental Analysis

Additionally, scanning electron microscopy (SEM) was used to evaluate the morphology of the raw EFB fibres, treated EFB fibres and the magnetic acid catalyst. Figure 5 depicts the SEM observations of the effect of alkaline treatment on the EFB fibre surface. Figure 5a indicates the presence of white granules, known as silica bodies, and impurities on the surface of raw EFB fibres. However, these impurities and silica bodies were removed from the EFB surface after the alkaline treatment using sodium carbonate, Na_2CO_3, as illustrated in Figure 5b. Rosli et al. reported that the silica bodies are known as phytoliths, which are found on the fibre surface and could lead to poor adsorption of particles on the fibre surface [41]. After the silica bodies were removed using an alkaline treatment, a rough surface and the appearance of tiny pores were noticed. Meanwhile, the SEM image of a treated fibre impregnated with sulphated and iron oxide particles is presented in Figure 5c. It showed that the catalyst had a rough surface, with the particles embedded on the EFB surface after calcination at 500 °C for 2 h.

In addition, Table 2 presents the energy-dispersive spectrometry (EDS) results of treated EFB fibres and the Na_2CO_3-T-EFB magnetic catalyst. The treated fibres, which consist of pure carbon and oxygen, indicates that alkaline pre-treatment improves the EFB fibre surface. Meanwhile, the EDS result of the impregnated catalyst shows the existence of carbon (C), iron (Fe), oxygen (O) and sulphur (S) groups embedded on the surface of treated EFB fibres with 37.88 wt%, 21.82 wt%, 34.78 wt% and 5.52 wt%, respectively. Hence, the oxygen and sulphur content could indicate the existence of the sulfonic group (-SO_3H) in the catalyst [19]. Besides, the composition of iron and oxygen indicate that the catalyst is comprised of Fe-O particles, which further induces its magnetic properties.

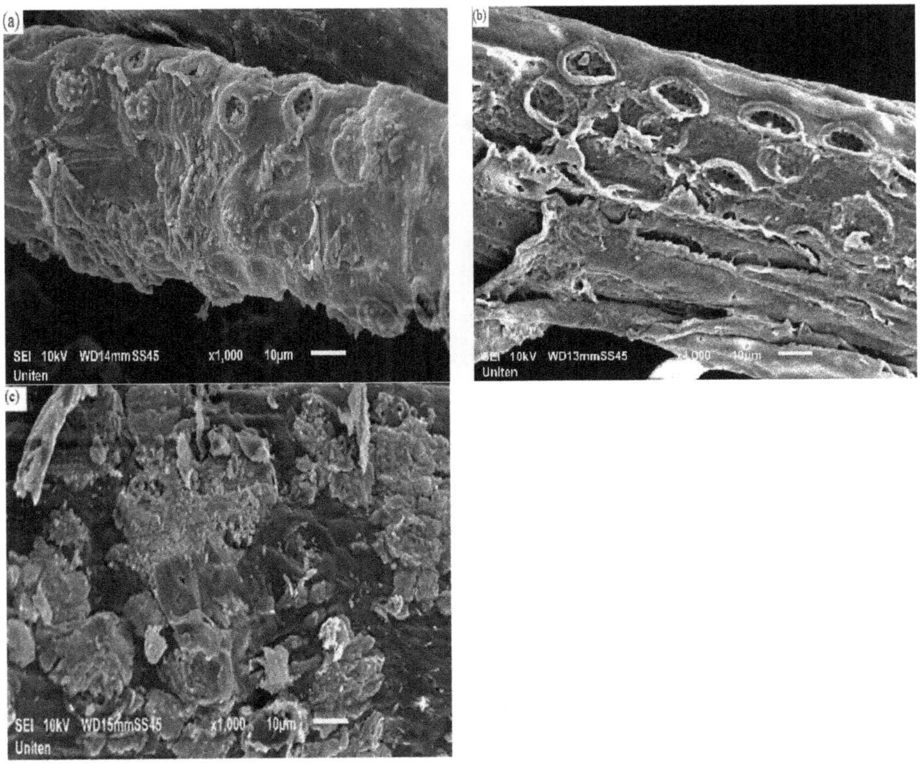

Figure 5. SEM images of (**a**) raw EFB fibre, (**b**) Na_2CO_3-T-EFB, and (**c**) Na_2CO_3-T-EFB derived magnetic acid catalyst.

Table 2. Elemental composition of treated EFB fibres and the EFB supported magnetic acid.

Samples	Elemental Composition (wt%)			
	Carbon (C)	**Oxygen (O)**	**Iron (Fe)**	**Sulphur (S)**
Treated EFB fibres	57.76	42.24	n.d [a]	n.d [a]
Na_2CO_3-T-EFB magnetic catalyst	37.88	34.78	21.82	5.52

[a] n.d = not determined.

3.4. Catalyst Reusability

The performance of the Na_2CO_3-T-EFBC magnetic catalyst was tested using the esterification reaction of oleic acid at 60 °C for 2 h, which obtained a conversion rate of 93.5% (Figure 6). After the reaction, the used catalyst was collected by using an external magnet and was washed and dried to be used in the next cycle of esterification. It was observed that the conversion rate decreased in the second cycle to 89.5%, which was possibly due to the leaching of active sites from the catalyst during the reaction. However, the conversion rate slightly dropped to 86.3% and the catalytic activity was further retained between 82.9 and 82.1% after the fourth and fifth cycles. It was noticed that the catalyst performance remains stationary after the fifth cycle. The stability of conversion rate over Na_2CO_3-T-EFBC could be associated to the higher acidity after the fifth cycle of 2.8 mmol/g.

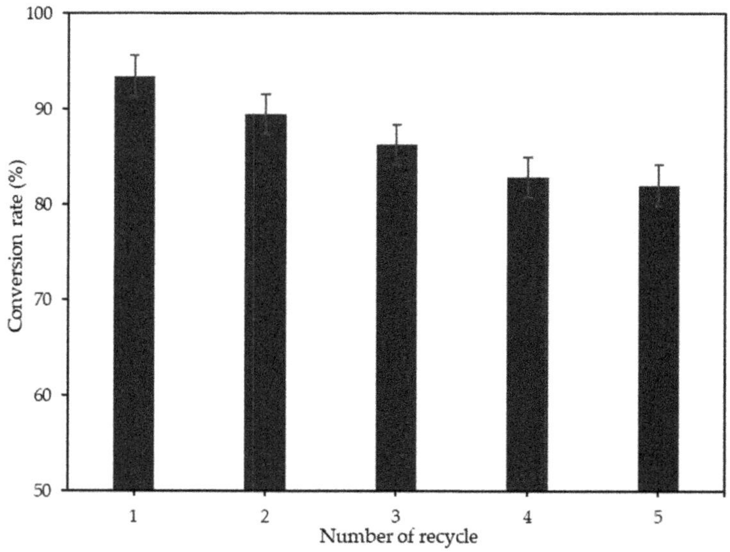

Figure 6. Reusability of the Na_2CO_3-T-EFB derived magnetic acid catalyst.

3.5. Comparison of Surface Acidity on the Catalytic Activity of Biomass Derived Solid Catalysts

The surface acidity over esterification performance of the biomass-based solid acid catalysts prepared in this study was compared with other studies in Table 3. It was observed that the Na_2CO_3-T-EFBC magnetic catalyst has a comparable or higher conversion rate at a shorter time of 2 h under a lower reaction temperature of 60 °C, which was attributed to the strong acidity on the catalyst surface. Most importantly, the magnetic catalyst prepared in this study had showed good stability compared to those in the studies reported. This proves that the Na_2CO_3-T-EFBC magnetic catalyst was exhibited as a practical catalyst for acid catalysed reactions.

Table 3. Comparison of the surface acidity over esterification performance of carbon-based solid acid catalysts prepared in this work with other studies.

Catalysts	Surface Acidity			Esterification Performance			Refs.
	By S Content (mmol SO_3H/g) [a]	By Base Titration (mmol/g)	By NH_3-TPD (mmol/g)	Conditions [b]	Conversion (%)	Reusability	
Spent coffee grounds-derived solid acid catalyst	3.36	4.22	-	CL of 10 wt%, MeOH:OA of 10:1, 80 °C/7 h	>90.0	~70% after 4 cycles	[42]
Cacao shell-derived solid acid catalyst	1.48	4.56	-	CL of 0.05 wt%, MeOH:OA of 7:1, 42 °C/4 h	78.0	48.0% after 3 cycles	[43]
EFB derived MBC02-SO_3H	-	-	0.28	CL of 5 wt%, MeOH:OA of 8:1, 150 °C/1.5 h	81.0	Not reported	[44]
Bamboo derived solid acid catalyst (S150-4)	0.82	-	-	CL of 10 wt%, MeOH:OA of 8:1, 65 °C/8 h	98.0	79.2% after 4 cycles	[45]
Dealkaline lignin derived E−260-20-SO_3H (Supercritical ethanol)	1.41	5.05	-	CL of 5 wt%, MeOH:OA of 12:1, 80 °C/7 h	95.4	≥81.9% after 5 cycles	[46]
Dealkaline lignin derived E-P400−2-SO_3H (Subcritical ethanol)	1.06	5.35	-	CL of 5 wt%, MeOH:OA of 15:1, 80 °C/5 h	95.5	≥84.6% after 3 cycles	
Na_2CO_3-T-EFBC magnetic catalyst	1.72	3.50	-	CL of 7 wt%, MeOH:OA of 10:1, 60 °C/2 h	93.5	82.1% after 5 cycles	This study

[a] Surface acidity by S content (mmol SO_3H/g) was calculated using the following formula: Surface acidity (mmol SO_3H/g) = (S content in catalyst/100) × 1000/32.07. [b] CL = catalyst loading, MeOH = methanol, OA = oleic acid.

3.6. Reaction Mechanism Using Na_2CO_3-T-EFBC

The possible reaction mechanism of the catalytic esterification of oleic acid with methanol using Na_2CO_3-T-EFBC is able to be described in several steps and is presented in Figure 7. Meanwhile, -SO_3H groups derived from the magnetic acid catalyst catalyse the esterification reaction. Three-oxygen molecules accept electrons from sulphur to form electron-withdrawing (-SO_3H, sulfonic acid) groups, which act as the Bronsted acid active sites [47]. The catalytic mechanism involves several steps: (1) The strong H^+ proton of the catalyst attacks the oxygen in the carboxyl group of oleic acid during the reaction, (2) the alcohol molecule reacts with the activated carboxyl carbon through nucleophilic substitution, (3) the intramolecular dehydration and hydrogen ion desorption lead to the formation of water molecules, and lastly, (4) the formation of methyl esters. In addition, the Lewis acid nature of iron oxide (Fe-O) particles also serve as active sites and contribute to the catalytic activity.

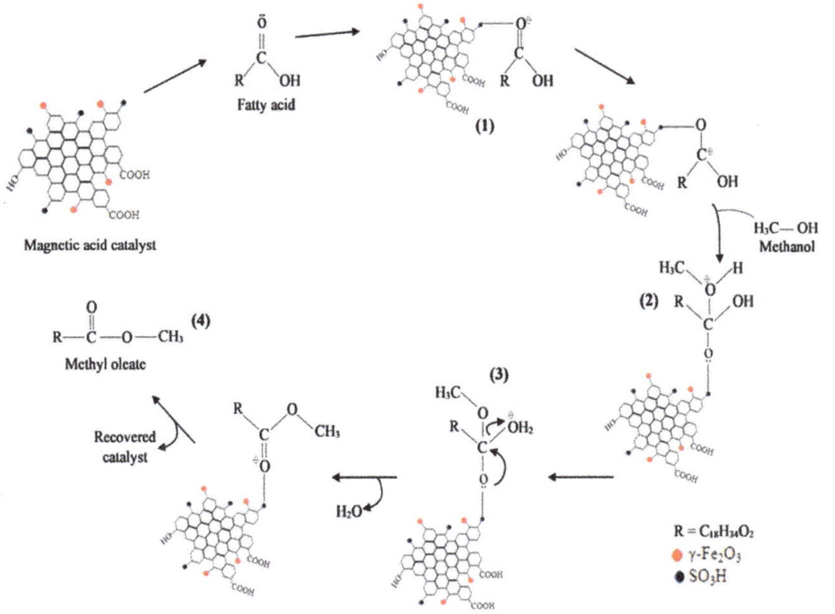

Figure 7. Proposed 4-step ((1)–(4)) reaction mechanism of the esterification of oleic acid using the Na_2CO_3-T-EFBC catalyst.

4. Conclusions

EFB fibres were utilised to produce efficient biomass-supported magnetic acid catalysts. The EFB fibre-supported magnetic acid catalyst was successfully synthesised and the performance of the catalyst was evaluated via the esterification of oleic acid. The effect of chemical pre-treatments on the catalytic performance and acidity were investigated, and the Na_2CO_3-T-EFBC magnetic catalyst showed a noticeable effect with a high conversion rate of 87.8%. Further, under the optimal catalyst preparation conditions of 7 g EFB fibres loading at 2 h impregnation time and 2 h of calcination at 500 °C, a high conversion rate of 93.46% was reported. The EFB fibre-supported magnetic catalyst revealed good properties with a high acidic value, indicating the existence of the sulfonic group, which interacts with methanol and oleic acid for the esterification reaction. Besides, the catalyst had strong magnetism, and it was recovered using an external magnet. The catalyst reusability showed good stability with 82.1% after five cycles. In conclusion, the EFB magnetic solid catalyst had good properties, such as strong magnetism and high acidity, with good catalytic performance. The findings in this study provide a potential substitution to be used as a green heterogeneous catalyst in the biodiesel industry. Thus, this assures the prominent usage of oil palm waste in wide applications and minimises waste.

Author Contributions: Conceptualization, S.G.K., F.L.P. and Z.F.; methodology, S.G.K.; validation, S.G.K.; formal analysis, S.G.K.; investigation, S.G.K.; resources, F.L.P.; data curation, S.G.K.; writing—original draft preparation, S.G.K.; writing—review and editing, S.G.K. and F.L.P.; supervision, F.L.P.; project administration, F.L.P.; funding acquisition, F.L.P. All authors have read and agreed to the published version of the manuscript.

Funding: This research was funded by Ministry of Education (MOHE), grant number FRGS/1/2018/STG07/UNITEN/02/3.

Institutional Review Board Statement: Not applicable.

Informed Consent Statement: Not applicable.

Data Availability Statement: All the data are presented inside this paper.

Acknowledgments: The authors are also grateful for the support and facilities provided by the Institute Sustainable Energy, UNITEN. This work was supported by the Ministry of Higher Education, Malaysia, through the Fundamental Research Grant Scheme (FRGS), under the project code of 20180108FRGS. This work was supported by Tenaga Nasional Berhad (TNB) and UNITEN through the BOLD Refresh Publication Fund under the project code of J510050002-IC-6 BOLDREFRESH2025-Centre of Excellence.

Conflicts of Interest: The authors declare no conflict of interest.

References

1. Rath, B.N.; Akram, V.; Bal, D.P.; Mahalik, M.K. Do Fossil Fuel and Renewable Energy Consumption Affect Total Factor Productivity Growth? Evidence from Cross-Country Data with Policy Insights. *Energy Policy* **2019**, *127*, 186–199. [CrossRef]
2. Gebremariam, S.N.; Marchetti, J.M. Economics of Biodiesel Production: Review. *Energy Convers. Manag.* **2018**, *168*, 74–84. [CrossRef]
3. Faruque, M.O.; Razzak, S.A.; Hossain, M.M. Application of Heterogeneous Catalysts for Biodiesel Production from Microalgal Oil—A Review. *Catalysts* **2020**, *10*, 1025. [CrossRef]
4. Hsiao, M.C.; Kuo, J.Y.; Hsieh, S.A.; Hsieh, P.H.; Hou, S.S. Optimized Conversion of Waste Cooking Oil to Biodiesel Using Modified Calcium Oxide as Catalyst via a Microwave Heating System. *Fuel* **2020**, *266*, 117114. [CrossRef]
5. Ullah, F.; Dong, L.; Bano, A.; Peng, Q.; Huang, J. Current Advances in Catalysis toward Sustainable Biodiesel Production. *J. Energy Inst.* **2016**, *89*, 282–292. [CrossRef]
6. Chen, M.N.; Mo, L.P.; Cui, Z.S.; Zhang, Z.H. Magnetic Nanocatalysts: Synthesis and Application in Multicomponent Reactions. *Curr. Opin. Green Sustain. Chem.* **2019**, *15*, 27–37. [CrossRef]
7. Tang, X.; Niu, S. Preparation of Carbon-Based Solid Acid with Large Surface Area to Catalyze Esterification for Biodiesel Production. *J. Ind. Eng. Chem.* **2019**, *69*, 187–195. [CrossRef]
8. Vu, T.H.T.; Nguyen, M.H.; Nguyen, M.D. Synthesis of Acidic Heterogeneous Catalysts with High Stability Based on Graphene Oxide/Activated Carbon Composites for the Esterification of Lactic Acid. *J. Chem.* **2019**, *7815697*, 1–7. [CrossRef]
9. Gardy, J.; Osatiashtiani, A.; Céspedes, O.; Hassanpour, A.; Lai, X.; Lee, A.F.; Wilson, K.; Rehan, M. A Magnetically Separable SO_4/Fe-Al-TiO_2 Solid Acid Catalyst for Biodiesel Production from Waste Cooking Oil. *Appl. Catal. B* **2018**, *234*, 268–278. [CrossRef]
10. Kristiani, A.; Sembiring, K.C.; Aulia, F.; Abimanyu, H. Sulfated Zirconia Catalyst for Hydrolysis of Palm Oil Lignocellulosic Wastes. *Energy Procedia* **2015**, *65*, 8–13. [CrossRef]
11. Iwanow, M.; Gärtner, T.; Sieber, V.; König, B. Activated Carbon as Catalyst Support: Precursors, Preparation, Modification and Characterization. *Beilstein J. Org. Chem.* **2020**, *16*, 1188–1202. [CrossRef]
12. Feyzi, M.; Norouzi, L. Preparation and Kinetic Study of Magnetic Ca/Fe_3O_4@SiO_2 Nanocatalysts for Biodiesel Production. *Renew Energy* **2016**, *94*, 579–586. [CrossRef]
13. Gardy, J.; Nourafkan, E.; Osatiashtiani, A.; Lee, A.F.; Wilson, K.; Hassanpour, A.; Lai, X. A Core-Shell SO_4/Mg-Al-Fe_3O_4 Catalyst for Biodiesel Production. *Appl. Catal. B* **2019**, *259*, 118093. [CrossRef]
14. Loh, S.K. The Potential of the Malaysian Oil Palm Biomass as a Renewable Energy Source. *Energy Convers. Manag.* **2017**, *141*, 285–298. [CrossRef]
15. Thushari, I.; Babel, S. Sustainable Utilization of Waste Palm Oil and Sulfonated Carbon Catalyst Derived from Coconut Meal Residue for Biodiesel Production. *Bioresour. Technol.* **2018**, *248*, 199–203. [CrossRef]
16. Jafri, N.H.S.; Jimat, D.N.; Azmin, N.F.M.; Sulaiman, S.; Nor, Y.A. The Potential of Biomass Waste in Malaysian Palm Oil Industry: A Case Study of Boustead Plantation Berhad. *IOP Conf. Ser. Mater. Sci. Eng.* **2021**, *1192*, 12028. [CrossRef]
17. Dolah, R.; Karnik, R.; Hamdan, H. A Comprehensive Review on Biofuels from Oil Palm Empty Bunch (Efb): Current Status, Potential, Barriers and Way Forward. *Sustainability* **2021**, *13*, 10210. [CrossRef]
18. Zhang, Q.; Wu, J.; Ma, P.; Cai, J.; Zhang, Y. Acid Value Determination and Pre-Esterification of Crude Euphorbia Lathyris L. Oil. *World J. Eng. Technol.* **2015**, *3*, 70–75. [CrossRef]

19. Zhou, Y.; Niu, S.; Li, J. Activity of the Carbon-Based Heterogeneous Acid Catalyst Derived from Bamboo in Esterification of Oleic Acid with Ethanol. *Energy Convers. Manag.* **2016**, *114*, 188–196. [CrossRef]
20. Az-Zahraa, B.; Zakaria, S.; Daud, M.F.B.; Jaafar, S.N.S. Removal of Oil Palm Trunk Lignin in Ammonium Hydroxide Pretreatment. *AIP Conf. Proc.* **2018**, *1940*, 20017. [CrossRef]
21. Li, Y.; Li, F.; Yang, Y.; Ge, B.; Meng, F. Research and Application Progress of Lignin-Based Composite Membrane. *J. Polym. Eng.* **2021**, *41*, 245–258. [CrossRef]
22. Ariffin, H.; Hassan, M.; Umi Kalsom, M.; Abdullah, N.; Shirai, Y. Effect of Physical, Chemical and Thermal Pretreatments on the Enzymatic Hydrolysis of Oil Palm Empty Fruit Bunch (OPEFB) (Kesan Prarawatan Fizikal, Kimia Dan Termal Terhadap Hidrolisis Enzimitik Tandan Kosong Kelapa Sawit). *J. Trop. Agric. Food Sci.* **2008**, *36*, 259–268.
23. Ismail, S.; Saharuddin, M.Q.; Zahari, M.S.M. Upgraded Seawater-Alkaline Pre-Treatment of Lignocellulosic Biomass for Bio-Methane Production. *Energy Procedia* **2017**, *138*, 372–379. [CrossRef]
24. Chin, S.X.; Chia, C.H.; Zakaria, S.; Ahmad, S.H.; Tasirin, S.M. Combination of Gamma Irradiation and Sodium Carbonate Pretreatment on Oil Palm Empty Fruit Bunch (EFB) for High Acidic Hydrolysis Yield. *Sains Malays.* **2017**, *46*, 167–173. [CrossRef]
25. Guo, F.; Xiu, Z.L.; Liang, Z.X. Synthesis of Biodiesel from Acidified Soybean Soapstock Using a Lignin-Derived Carbonaceous Catalyst. *Appl. Energy* **2012**, *98*, 47–52. [CrossRef]
26. Wang, Y.T.; Yang, X.X.; Xu, J.; Wang, H.L.; Wang, Z.B.; Zhang, L.; Wang, S.L.; Liang, J.L. Biodiesel Production from Esterification of Oleic Acid by a Sulfonated Magnetic Solid Acid Catalyst. *Renew Energy* **2019**, *139*, 688–695. [CrossRef]
27. Jing, H.; Wang, X.; Liu, Y.; Wang, A. Preparation of Magnetic Nanocomposites of Solid Acid Catalysts and Their Applicability in Esterification. *Chin. J. Catal.* **2015**, *36*, 244–251. [CrossRef]
28. Jain, A.; Balasubramanian, R.; Srinivasan, M.P. Hydrothermal Conversion of Biomass Waste to Activated Carbon with High Porosity: A Review. *Chem. Eng. J.* **2016**, *283*, 789–805. [CrossRef]
29. Chellappan, S.; Aparna, K.; Chingakham, C.; Sajith, V.; Nair, V. Microwave Assisted Biodiesel Production Using a Novel BrØnsted Acid Catalyst Based on Nanomagnetic Biocomposite. *Fuel* **2019**, *246*, 268–276. [CrossRef]
30. Eterigho, E.J.; Alueshima, B.M.; Ejejigbe, S.E. Optimization of Process Parameters for the Synthesis of Locally Sourced Alumina-Supported Eggshell Catalyst. *Lect. Notes Eng. Comput. Sci.* **2018**, *2238*, 523–527.
31. Hussain, Z.; Kumar, R. Synthesis and Characterization of Novel Corncob-Based Solid Acid Catalyst for Biodiesel Production. *Ind. Eng. Chem. Res.* **2018**, *57*, 11645–11657. [CrossRef]
32. Liu, K.; Wang, R.; Yu, M. An Efficient, Recoverable Solid Base Catalyst of Magnetic Bamboo Charcoal: Preparation, Characterization, and Performance in Biodiesel Production. *Renew Energy* **2018**, *127*, 531–538. [CrossRef]
33. Abdedayem, A.; Guiza, M.; Ouederni, A. Copper Supported on Porous Activated Carbon Obtained by Wetness Impregnation: Effect of Preparation Conditions on the Ozonation Catalyst's Characteristics. *Comptes Rendus Chim.* **2015**, *18*, 100–109. [CrossRef]
34. Amadine, O.; Essamlali, Y.; Fihri, A.; Larzek, M.; Zahouily, M. Effect of Calcination Temperature on the Structure and Catalytic Performance of Copper-Ceria Mixed Oxide Catalysts in Phenol Hydroxylation. *RSC Adv.* **2017**, *7*, 12586–12597. [CrossRef]
35. Ding, Y.; Zhao, C.; Li, Y.; Ma, Z.; Lv, X. Effect of Calcination Temperature on the Structure and Catalytic Performance of the Cu-Mcm-41 Catalysts for the Synthesis of Dimethyl Carbonate. *Quim. Nova* **2018**, *41*, 1156–1161. [CrossRef]
36. Lu, J.; Zhao, Z.; Xu, C.; Duan, A.; Zhang, P. Effects of Calcination Temperature on the Acidity and Catalytic Performances of HZSM-5 Zeolite Catalysts for the Catalytic Cracking of n-Butane. *J. Nat. Gas Chem.* **2005**, *14*, 213–220. [CrossRef]
37. Zainol, M.M.; Amin, N.A.S.; Asmadi, M. Synthesis and Characterization of Porous Microspherical Ionic Liquid Carbon Cryogel Catalyst for Ethyl Levulinate Production. *Diam. Relat. Mater.* **2019**, *95*, 154–165. [CrossRef]
38. Zainol, M.M.; Amin, N.A.S.; Asmadi, M. Effects of Thermal Treatment on Carbon Cryogel Preparation for Catalytic Esterification of Levulinic Acid to Ethyl Levulinate. *Fuel Process. Technol.* **2017**, *167*, 431–441. [CrossRef]
39. Singh, S.; Kumar, R.; Setiabudi, H.D.; Nanda, S.; Vo, D.V.N. Advanced Synthesis Strategies of Mesoporous SBA-15 Supported Catalysts for Catalytic Reforming Applications: A State-of-the-Art Review. *Appl. Catal. A Gen.* **2018**, *559*, 57–74. [CrossRef]
40. He, B.; Fu, X.; Lian, X.; Jiang, S.; Xu, P.; Deng, X.; He, C.; Chen, C. Catalytic Deacidification of Vacuum Gas Oil by Zno/Al2o3 and Its Modification with Fe_2O_3. *Catalysts* **2019**, *9*, 499. [CrossRef]
41. Rosli, N.S.; Harun, S.; Jahim, J.M.; Othaman, R. Chemical and Physical Characterization of Oil Palm Empty Fruit Bunch. *Malays. J. Anal. Sci.* **2017**, *21*, 188–196. [CrossRef]
42. Agapay, R.C.; Liu, H.C.; Ju, Y.H.; Go, A.W.; Angkawijaya, A.E.; Nguyen, P.L.T.; Truong, C.T.; Quijote, K.L. Synthesis and Initial Evaluation of Solid Acid Catalyst Derived from Spent Coffee Grounds for the Esterification of Oleic Acid and Methanol. *Waste Biomass Valorization* **2021**, *12*, 4387–4397. [CrossRef]
43. Bureros, G.M.A.; Tanjay, A.A.; Cuizon, D.E.S.; Go, A.W.; Cabatingan, L.K.; Agapay, R.C.; Ju, Y.H. Cacao Shell-Derived Solid Acid Catalyst for Esterification of Oleic Acid with Methanol. *Renew Energy* **2019**, *138*, 489–501. [CrossRef]
44. Jenie, S.N.A.; Kristiani, A.; Sudiyarmanto; Khaerudini, D.S.; Takeishi, K. Sulfonated Magnetic Nanobiochar as Heterogeneous Acid Catalyst for Esterification Reaction. *J. Environ. Chem. Eng.* **2020**, *8*, 103912. [CrossRef]
45. Zhang, B.; Gao, M.; Geng, J.; Cheng, Y.; Wang, X.; Wu, C.; Wang, Q.; Liu, S.; Cheung, S.M. Catalytic Performance and Deactivation Mechanism of a One-Step Sulfonated Carbon-Based Solid-Acid Catalyst in an Esterification Reaction. *Renew Energy* **2021**, *164*, 824–832. [CrossRef]

46. Huang, M.; Luo, J.; Fang, Z.; Li, H. Biodiesel Production Catalyzed by Highly Acidic Carbonaceous Catalysts Synthesized via Carbonizing Lignin in Sub- and Super-Critical Ethanol. *Appl. Catal. B* **2016**, *190*, 103–114. [CrossRef]
47. Rechnia-Gorący, P.; Malaika, A.; Kozłowski, M. Acidic Activated Carbons as Catalysts of Biodiesel Formation. *Diam. Relat. Mater.* **2018**, *87*, 124–133. [CrossRef]

sustainability MDPI

Article

From Waste Biomass to Cellulosic Ethanol by Separate Hydrolysis and Fermentation (SHF) with *Trichoderma viride*

Małgorzata Hawrot-Paw * and Aleksander Stańczuk

Department of Renewable Energy Engineering, West Pomeranian University of Technology in Szczecin, Pawla VI 1, 71-459 Szczecin, Poland
* Correspondence: malgorzata.hawrot-paw@zut.edu.pl

Abstract: Advanced biofuels can reduce fossil fuel use and the number of harmful compounds released during combustion, by reducing the use of fossil fuels. Lignocellulosic materials, especially waste biomass, are suitable substrates for the production of advanced biofuels. Among the most expensive steps in the production of ethanol is enzyme-based hydrolysis. Using microorganisms can reduce these costs. This study investigated the effectiveness of hydrolyzing three waste lignocellulosic biomass materials (barley straw, oak shavings, spent grains) into ethanol, after biological pretreatment with *Trichoderma viride* fungi. The number of fermentable sugars obtained from each substrate was subjected to preliminary study, and the correlation between the temperature and fungal activity in the decomposition of lignocellulosic materials was determined. Ethanol was produced by the separate hydrolysis and fermentation (SHF) method. It was found that not all lignocellulosic biomass is suitable to decomposition and hydrolysis in the presence of *T. viride*. Regardless of the process temperature, the average enzymatic activity of fungi (activity index) ranged from 1.25 to 1.31. 94 mL of distillate, with a 65% (v/v) ethanol concentration produced by the hydrolysis and fermentation of the sugars released from the barley straw.

Keywords: waste biomass; lignocellulose; biological treatment; *Trichoderma viride*; ethanol

Citation: Hawrot-Paw, M.; Stańczuk, A. From Waste Biomass to Cellulosic Ethanol by Separate Hydrolysis and Fermentation (SHF) with *Trichoderma viride*. *Sustainability* **2023**, *15*, 168. https://doi.org/10.3390/su15010168

Academic Editors: Attila Bai, Steven Lim, Shuit Siew Hoong, Santi Chuetor and Pang Yean Ling

Received: 26 October 2022
Revised: 15 December 2022
Accepted: 16 December 2022
Published: 22 December 2022

1. Introduction

The greenhouse effect, mainly caused by transport emissions, is driving the search for solutions to curb global temperature rise and environmental pollution. Conventional fuels can be substituted with biofuels. There are three main generations of biofuels, based on the substrates that are used for their production [1]. First-generation biofuels (1G) are produced from ingredients that are also intended for food and fodder; they are called conventional biofuels [2]. In Directive 2015/1513 of the European Parliament and the Council of the European Union of 9 September 2015, the share of first-generation biofuels was limited to 7%, in favor of advanced biofuels. In December 2018, the RED II Directive entered into force, according to which the target for 2030 is 14% biofuels used in transport, with first-generation biofuels no longer being included in renewable energy sources. Therefore, this favors the sector of second-generation (2G) biofuels, the production of which involves non-food energy plants, as well as waste lignocellulosic and third-generation (3G) biomass, which are produced from algae biomass. Compared to the production of first-generation fuels, the production of second-generation biofuels is more complex, but the use of residues, inter alia, from agriculture, forestry, industrial processes, and municipal management, allows for the production of biofuels with respect for the environment, under the conditions of sustainable development [3,4], without competing with food production [5].

Lignocellulosic materials require appropriate preparation; however, they are a very attractive, also for economic reasons, source of the sugars necessary for fermentation [6]. The ethanol obtained in this process, after dehydration, can be used as an additive to gasoline, but it can also be an independent fuel. The first step in the production of ethanol is pretreatment, which separates lignin from cellulose and hemicellulose [7,8]. For this purpose, the options include physical methods (milling, microwave, mechanical extrusion, pyrolysis, pulse electric field), chemical methods (acid and alkali pretreatment, organosolv, ionic liquids, ozonolysis), physico-chemical methods (ammonia fiber expansion (AFEX), steam explosion, carbon dioxide explosion, liquid hot water (LHW), wet oxidation, SPORL treatment), or biological methods based on the enzymatic activity of bacteria and fungi [9], which have two types of exogenous enzyme systems. One of them is responsible for the degradation of polysaccharides, while the other is the ligninolytic system [10]. Fungi break down lignin more efficiently than bacteria [11]. Strains belonging to the genus *Trichoderma* have a significant role in these processes [12,13]. Biological pretreatment is an environmentally safe and low-energy method [14].

The next stage is enzymatic hydrolysis, incl. using cellulases and xylanases [15], the aim of which is to depolymerize polysaccharides isolated from biomass and release monomeric sugars [16]. These are five-carbon and six-carbon sugars, mainly glucose, used by ethanol-fermentation microorganisms [17]. Hydrolysis conditions and biomass types affect the conversion of lignocellulosic biomass to fermentable sugars [18]. In the hydrolysis process, not only commercial enzymes but also microbial cells can be used. The advantage of using microorganisms is the enzymatic activity closely adapted to the specific type of biomass [19], and the technology itself is classified as one of the future-oriented and promising methods favoring the increase in the level of saccharification [20]. The hydrolysis can be carried out separately from the fermentation process or simultaneously. Five basic methods are most often mentioned: separate hydrolysis and fermentation—SHF, simultaneous saccharification and fermentation—SSF [21], separate hydrolysis and co-fermentation—SHCF [22], simultaneous saccharification and co-fermentation—SSCF [23], and consolidated bioprocessing—CBP [24]. The hexose fermentation process can be carried out by bacteria and various types of mold fungi; however, on a commercial scale, mainly due to the efficiency of sugar conversion to ethanol, fast growth rate, efficient production of ethanol, and tolerance to its high concentration in the environment, *Saccharomyces cerevisiae* yeasts are mainly used [25].

After the fermentation process is completed, the solution is distilled and rectified, which not only increases the ethanol content but also purifies the biofuel [26]. The last stage of bioethanol production is its dehydration [27].

The economic feasibility of ethanol production from lignocellulosic feedstocks is currently a major challenge. The hydrolysis of lignocellulosic biomass applies commercial enzymes with cellulolytic and hemicellulolytic activity. The utilization of such enzymes increases production costs; therefore, it is necessary to search for simple and low-cost alternatives, especially for fuel utilization of the waste biomass. The aim of this study was to estimate the potential of *Trichoderma viride* to degrade different lignocellulosic materials, considering the optimum conditions for enzymes biosynthesis, and to estimate the efficiency of the enzymatic hydrolysis process based on ethanol yield. It was hypothesized that, under appropriate conditions, using *T. viride* for the pretreatment and hydrolysis of lignocellulosic substrates would enable simple sugars to be obtained, which are the ethanol precursors during the ethanol fermentation process. Due to the biological treatment of biomass, chemical compounds and a detoxification process are not required. Different species of *Trichoderma* have been used during the pretreatment of lignocellulosic feedstocks or as a source of the enzymes required for hydrolysis; however, to our best knowledge, this is the first time that whole *T. viride* cells have been used simultaneously for both processes.

2. Materials and Methods

2.1. Feedstock for Ethanol Fermentation

The materials that were used in the research include brewing spent grain (from the brewery), barley straw, and oak shavings (both from the university research center). The biomass was ground using a laboratory mill to a particle size of 1–2 mm and then stored at room temperature before pretreatment.

2.2. Biological Biomass Pretreatment

The possibility of obtaining fermentable sugars from biomass was determined during the enriched culture, which was carried out in Erlenmeyer conical flasks with a capacity of 250 mL. The flasks contained 100 mL of liquid medium with the following composition: (g/L) peptone 1.0 (BTL), $(NH_4)_2SO_4$ 1.4 (cz.d.a., Chempur), KH_2PO_4 2.0 (cz.d.a., Chempur), urea 0.3 (cz.d.a., Chempur), $CaCl_2$ 0.3 (cz.d.a., Chempur), $MgSO_4 \cdot 7H_2O$ 0.3 (cz.d.a., Chempur), (mg/L) $FeSO_4 \cdot 7H_2O$ 5.0 (cz.d.a., Chempur), $MnSO_4 \cdot H_2O$ 1.6 (cz.d.a., Chempur), $ZnSO_4 \cdot 7H_2O$ 1.4 (cz.d.a., Chempur), and $CoCl_2$ 2.0 (cz.d.a., Chempur). Next, 1 g of substrate was added to each of the flasks. The flasks with their contents were sterilized at a temperature of approx. 100 °C. Then, one disc (size 5 mm) cut from a 7-day *T. viride* culture was added to each flask (Figure 1). The flasks were set in a thermostat at 25 °C. The experiment was carried out for 28 days. The samples subjected to enzymatic hydrolysis were subjected to the qualitative test for monomer sugar content using Benedict's test. The solutions from each culture flask were taken into test tubes, Benedict's reagent was added, and the whole was mixed and placed for 5 min in a bath of boiling water. The color change and the presence of sediment after the samples had cooled down were used to determine the presence of sugars and their concentration. Analyses were performed in three replications after 7, 14, 21, and 28 days of culture.

Figure 1. *Trichoderma viride* colony on potato dextrose agar (PDA).

2.3. Production of Cellulase—Index Activity

The cellulolytic activity of *T. viride* was assessed on a culture medium with a 1% addition of carboxymethyl cellulose (CMC; Sigma-Aldrich, St. Louis, MI, USA). The experiment was prepared in accordance with the previously described methodology. Conical flasks were inoculated with *T. viride* culture discs and then incubated for 7 days at 25, 30, and 35 °C, respectively. After one week, 0.1 mL of the culture solution was withdrawn from each flask and transferred to a Petri dish with solidified medium. The Petri dishes were incubated at 25, 30, and 35 °C for the next 7 days, and then the activity index (I_A) was determined. A 1% aqueous solution of Congo red (Sigma-Aldrich St. Louis, MI, USA) was poured into the plates and left for 15 min. Then, the excess solution was poured off, and 1M NaCl solution was introduced to the dishes. Positive reaction was indicated by the formation of a clear zone (halo) around the colony. The decolorization was carried out over a period of 20 min. The cellulolytic activity index was determined by comparing the hydrolysis zone visible in the form of brightening around the colony (A) to the diameter of the colony (B), according to the following formula:

$$I_A = A/B$$

where A—diameter of the hydrolysis zone (mm), and B—diameter of the colony (mm).

2.4. Experimental Procedure of SHF Method

The lignocellulosic biomass was subjected to physical and mechanical pretreatment. In a laboratory mill, 200 g of the substrate was ground to 1–2 mm particle size. Then, thermal treatment was applied. The substrate was placed in 2 L of water, boiled for about 30 min, cooled to 25 °C. and poured into a 3L polyethylene container. Ten discs were cut from a 7-day *T. viride* culture (5 mm in diameter). The experiment was prepared in three containers to obtain enough solution for the fermentation process. The containers were kept in a thermostat at 25 °C (optimum temperature for *T. viride* cellulolytic enzymes) for 21 days and then sterilized at approx. 100 °C. The contents of the containers were filtered through a mesh filter, separating the solid from the liquid fraction used for ethanol fermentation. After filtration, 3.4 L of solution were obtained; therefore, fermentation of monomeric sugars present in the hydrolysates was carried out in two fermentation containers with a capacity of 3 L, with the participation of *Saccharomyces cerevisiae* distillery yeast (Turbo Pure Yeast, MAXX Johnnie Cotton). The pH was set at 5.0 using 10 N sulfuric acid (cz.d.a, Chempur). The containers were placed in a thermostat at 30 °C (optimum temperature for *S. cerevisiae* activity) and incubated for 14 days. The wort was distilled at a temperature of 78.32 °C in an apparatus equipped with a heating mantle with a power regulator, a dephlegmator with a receiver, a cooler, and a thermometer that allowed for control of the temperature of the vapors. The ethanol content of the distillate was determined with an alcohol meter and is given as a percentage by volume (*v*/*v*). The experimental flow chart for all stages is shown in Figure 2.

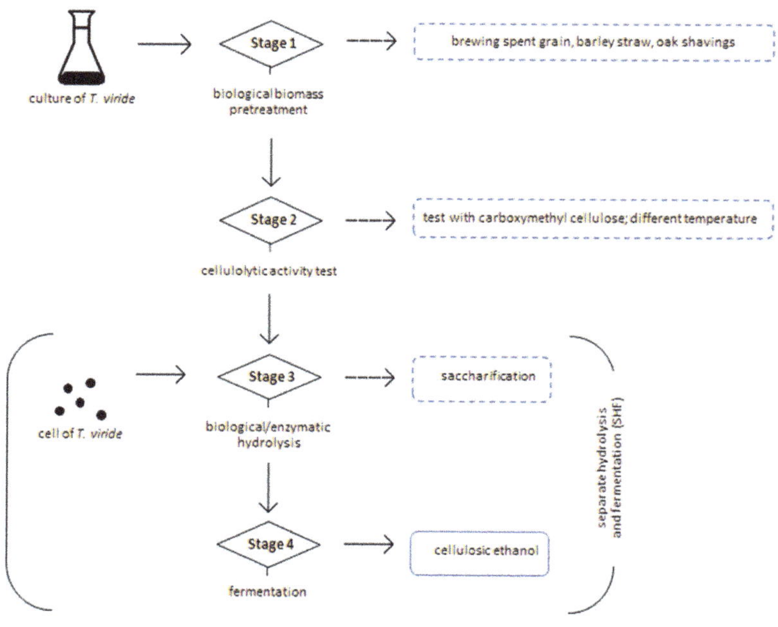

Figure 2. The experimental procedure flowchart.

2.5. Statistical Analysis

The results concerning the cellulolytic activity of the *T. viride* strain were analyzed statistically. Software for Windows (Statistica version 13.3; Dell Inc., Tulsa, OK, USA) and one-way analysis of variance were used. The standard deviation was also determined.

3. Results and Discussion

3.1. Susceptibility of Lignocellulosic Biomass to Biological Pretreatment

The production of second-generation biofuels is important both for environmental and economic reasons [28]. Lignocellulosic biomass, especially waste biomass, is not only a widely available but also a relatively cheap substrate for second-generation biofuels' production [29]. Its annual global energy potential is estimated at 100–270 EJ [30].

These substrates contain polysaccharides, which are the source of ethanol-fermentable monomeric sugars.

Different biomass types have different amounts of cellulose and hemicellulose, which may affect the efficiency of the biofuel production using them [31]. This composition may also be important in the processes of the biological treatment of lignocellulosic biomass, which is confirmed by the conducted research. Not all lignocellulose substrates could be used as a source of sugars for the alcohol fermentation obtained with the participation of enzymes secreted by *T. viride*. During the first 14 days of incubation, no reducing sugars were found in any of the cultures. After 21 days, a positive result was obtained in one of the samples of barley straw. After 28 days of incubation, the presence of fermentable sugars in the flasks containing oak sawdust was additionally noted. The lignin content in oak wood is 28% [32]. Lignin can bind to cellulolytic enzymes and reduce their effectiveness [33,34]. The efficiency of lignin degradation depends on pretreatment method. Singhvi and Kim [35], in a study of hydrogen production, treated raw maize cobs with peroxidase mimicking $CeFe_3O_4$ and noted lignin degradation at the level of 43.26%. Wi et al. [36] carried out a study on the conversion of lignocellulose to biofuel. The authors used hydrogen peroxide and acetic acid (HPAC) for pretreatment, removing up to 97.2% of lignin from the substrate. Biological methods are also effective. Su et al. [37] used the fungus *Myrothecium verrucaria* to biological treatment of corn stover and reported a 42.30% reduction in lignin content. Bari et al. [38] analyzed the effect of fungi on the degradation of beech wood and recorded a decrease in lignin content from 22.23% to 9.67% for *Pleurotus ostreatus* and 9.47% for *Trametes versicolor* after 120 days. Ghorbani et al. [39], during delignification of rice straw by Trichoderma viride, reported an increase in lignin-removal efficiency of up to 74% under optimal conditions. Ladeira Azar et al. [40], in their research on enzymatic hydrolysis of low-lignin bagasse from sugarcane, confirmed that even slight differences in lignin content can increase sugar yield. There is definitely less of it in barley straw. This straw contains 31–45% cellulose, 27–38% hemicellulose, and 14–19% lignin [41]. Earlier studies confirm the usefulness of this biomass for the production of cellulosic ethanol [19]. During simultaneous microbial saccharification and fermentation, ethanol with a concentration of 15% (v/v) was obtained.

The sugar content in the flasks was relatively low and did not exceed 0.1–0.3 g/100 mL. No reducing sugars were found in culture solutions in which the substrate was spent grains (BSG). The composition of this material is variable, which may make their valorization difficult [42]. In a study by Mishra et al. [43], BSG contained 24.5% cellulose, 9.8% hemicellulose, and 15.8% lignin. In the presented research, this substrate was not used by *T. viride* as a source of carbon and energy. Barley straw was selected for the next stage of the research.

3.2. Cellulolytic Activity of Trichoderma Viride

The temperature had no statistically significant effect on the extracellular *T. viride* hydrolase production. According to measurements of the hydrolysis zones (Figure 3), the activity index (IA) values varied from 1.18 to 1.38, as shown in Table 1.

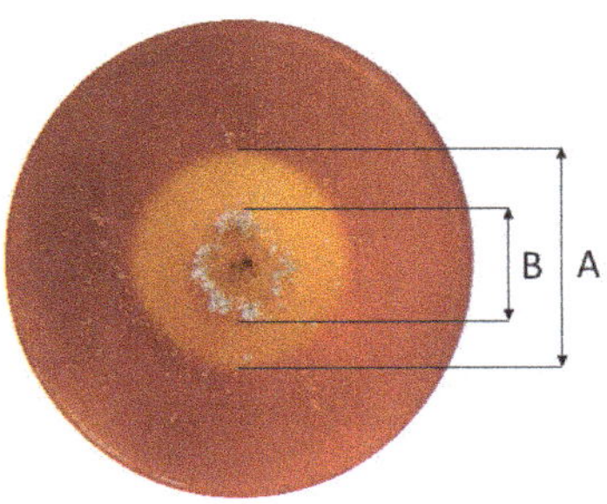

Figure 3. Scheme for determining the index activity (A—diameter of clear zone, B—diameter of fungal colony).

Table 1. Enzymatic activity index of *T. viride*.

	Temperature		
	25 °C	30 °C	35 °C
Index Activity (I_A)	1.31	1.33	1.25
	1.25	1.20	1.18
	1.38	1.23	1.38
Mean ± SD	1.31 ± 0.7	1.25 ± 0.7	1.27 ± 0.10

Bhattacharya et al. [44] investigated the relationship between the cellulase production efficiency of *T. viride* and temperature and found that the 30 °C is the optimum temperature for the production of these enzymes. This was also confirmed by the results of previous experiments on the evaluation of the potential enzymatic activity of *T. viride* for the decomposition of various types of straw [45]. For barley straw, the activity index increased from 1.45 at 25 °C to 1.82 at 30 °C. According to Malik et al. [46] an increase in temperature above 30 °C reduces the production of enzymes by *T. viride*.

3.3. Separate Hydrolysis and Fermentation

Under optimal conditions, it is possible to obtain the sufficient amount of reducing sugars necessary for the proper course of the next stage of the experiment—ethanol fermentation. Therefore, according to the results obtained, the straw material was hydrolyzed at a temperature of 25 °C. The solution after hydrolysis (3.4 L) was subjected to the ethanol fermentation process, and the result was 94 mL of distillate with an ethanol concentration of 65% (*v*/*v*). The available studies also confirmed the effectiveness of obtaining ethanol in the process of separate hydrolysis and fermentation. Mitra et al. [47] investigated ethanol production from lignocellulosic-starch biomass. Fermentation was carried out by SHF or SSF with batch feed. Substrates were pretreated with steam or dilute sulphuric acid (DSA). The highest ethanol yield (42.6 g/L) was obtained for separate hydrolysis and fermentation with DSA. Mendez et al. [48] carried out an SHF process for sugarcane bagasse and obtained a total ethanol concentration of 6.5% (*v*/*v*), including 33 g/L ethanol from the cellulose fraction and 18 g/L from the hemicellulose fraction. Different results were presented by Dahnum et al. [21]. When optimizing bioethanol production from an empty fruit bunch, they obtained 4.74% ethanol by the SHF process and 6.05% ethanol by the SSF process. This indicates that the efficiency of these processes also depends on other factors, including the type of substrate used during ethanol fermentation.

For the production of 2G ethanol, different types of waste biomass are used, e.g., rice; wheat and corn straw [49]; corn and cotton stalks and sugar cane processing waste [50]; and coffee pomace [51]. The amount of sugar obtained, precursors for the fermentation process, and high ethanol concentration depend significantly on the type of lignocellulosic biomass (Table 2).

Table 2. Example ethanol production from different types of lignocellulosic biomass.

Lignocellulosic Biomass	Pretreatment	Ethanol Yield	References
barley straw	biological	65% (*v*/*v*)	present study
barley straw	ionic liquids	18.5 g/L	Lara-Serrano et al. [52]
maize residues	-	10.22% (*v*/*v*)	Cutzu and Bardi [53]
wheat straw	deep eutectic solvent (des)	15.42 g/L	Xian et al. [54]
sugar cane pulp	dilute acid	15.5%	Saka and Afolabi [55]
wheat straw	steam explosion	21.3 g/L	Tomás-Pejó et al. [56]
corncob	alkaline	32.32 g/L	Boonchuay et al. [57]
sugarcane bagasse	acid	56.1 g/L	Unrean and Ketsub [58]
sugarcane bagasse	mixed H_2O_2 + NaOH	66 g/L	Irfan et al. [59]

Table 2. *Cont.*

Lignocellulosic Biomass	Pretreatment	Ethanol Yield	References
rapeseed straw	acid	39.9 g/L	López-Linares et al. [60]
pineapple leaves waste	hydrothermal	9 g/L	Saini et al. [61]
sugarcane leaves and tops	combination of oxidative alkali–peroxide	13.7 g/L	Dodo et al. [62]

The amount of substrate is also important. In a study by Paschos et al. [63], the concentration of ethanol increased from 39.55 to 46.62 g/L when the amount of barley straw biomass subjected to hydrolysis and the SSF process was increased from 15% to 20%. A similar relation was reported by Zhang et al. [64] for corn stover. With 15% solids, they obtained 24.7 g/L of ethanol, while 31.0 g/L, 39.3 g/L, and 40.6 g/L were obtained with 20%, 25%, and 30% solids, respectively. On the other hand, these authors observed a correlation between increasing solids content and decreasing ethanol yields, compared to theoretical values.

Conversion efficiency can be changed by using microorganisms other than yeast. Tivari et al. [65], from rice bran with *Bacillus cereus* strain McR-3, obtained a maximum of $10.50 \pm 0.10\%$. The efficiency of bioethanol production depended on temperature and the appropriate pH. Da Silva et al. [66] used alkaline-pretreated Carnauba straw for ethanol production. During simultaneous saccharification and fermentation (SSF), they used two strains of *Saccharomyces cerevisie*; however, the highest yield above 7.5 g/L was recorded for the *Kluyveromyces marxianus* strain. Takano and Hoshino [67] carried out simultaneous saccharification and fermentation of rice straw. They used a mixture of commercial cellulases for hydrolysis and a mutant of *Mucor circinelloides* J (a xylose-fermenting fungus) for fermentation. Under aerobic conditions, 30.5 g/L ethanol was obtained after 36 h; however, the straw was also pretreated with alkaline. Waghmare et al. [68] used *Saccharomyces cerevisiae* (KCTC 7296) and *Pachysolen tannophilus* (MTCC 1077) to improve the yield from 2.113% and 1.095% for the single strains to 2.348% for the co-culture, respectively.

Our study demonstrated that biological methods could be used to produce a high concentration of ethanol from lignocellulosic material. These methods do not require chemicals, can be conducted at milder temperature conditions, and obtain strains that actively degrade and hydrolyze lignocellulosic substrates, which would enable large-scale ethanol production to be cost-effective [11].

4. Conclusions

The efficiency of the pretreatment method and the type of feedstock utilized both determine the amount of ethanol produced from lignocellulosic substrates. Pretreatment and hydrolysis with the participation of fungi do not require chemical and energy supply, as they are effective and environmentally friendly; however, not every waste biomass is a suitable material for obtaining fermentable sugars in biological processes. Among the lignocellulosic materials used in the research, the susceptibility to bioconversion with the participation of *T. viride* was demonstrated by oak chips and barley straw. The enzymatic activity of fungi did not depend on the temperature, and the determined values of the activity index ranged from 1.25 to 1.31. Ninety-four mL of distillate was obtained from barley straw, subjected to pretreatment and hydrolysis with the *T. viride* strain, and then fermented with the *Saccharomyces cerevisiae* strain. The ethanol concentration in the distillate was 65%.

Author Contributions: Conceptualization, M.H.-P.; methodology, M.H.-P.; validation, M.H.-P.; formal analysis, M.H.-P. and A.S.; investigation, M.H.-P. and A.S.; writing—original draft preparation, M.H.-P.; writing—review and editing, M.H.-P.; visualization, M.H.-P.; supervision, M.H.-P. All authors have read and agreed to the published version of the manuscript.

Funding: This research received no external funding.

Institutional Review Board Statement: Not applicable.

Informed Consent Statement: Not applicable.

Data Availability Statement: The data presented in this study are available on request from the corresponding author.

Conflicts of Interest: The authors declare no conflict of interest.

References

1. Lee, R.A.; Lavoie, J.-M. From first- to third-generation biofuels: Challenges of producing a commodity from a biomass of increasing complexity. *Anim. Front.* **2013**, *3*, 6–11. [CrossRef]
2. Baudry, G.; Delrue, F.; Legrand, J.; Pruvost, J.; Vallée, T. The challenge of measuring biofuel sustainability: A stakeholder-driven approach applied to the French case. *Renew. Sustain. Energy Rev.* **2017**, *69*, 933–947. [CrossRef]
3. Heo, S.; Choi, J. Potential and environmental impacts of liquid biofuel from agricultural residues in Thailand. *Sustainability* **2019**, *11*, 1502. [CrossRef]
4. Jeswani, H.K.; Chilvers, A.; Azapagic, A. Environmental Sustainability of Biofuels: A Review. *Proc. R. Soc. A* **2020**, *476*, 20200351. [CrossRef] [PubMed]
5. Shahare, V.V.; Kumar, B.; Singh, P. Biofuels for sustainable development: A global perspective. In *Green Technologies and Environmental Sustainability*; Singh, R., Kumar, S., Eds.; Springer: Berlin/Heidelberg, Germany, 2017; pp. 67–89. [CrossRef]
6. Singh, D.P.; Trivedi, R.K. Biofuel from wastes an economic and environmentally feasible resource. *Energy Procedia* **2014**, *54*, 634–641. [CrossRef]
7. Putro, J.N.; Soetaredjo, F.E.; Lin, S.Y.; Ju, Y.H.; Ismadi, S. Pretreatment and conversion of lignocellulose biomass into valuable chemicals. *RSC Adv.* **2016**, *6*, 46834–46852. [CrossRef]
8. Hasanov, I.; Raud, M.; Kikas, T. The Role of Ionic Liquids in the Lignin Separation from Lignocellulosic Biomass. *Energies* **2020**, *13*, 4864. [CrossRef]
9. Aftab, M.N.; Iqbal, I.; Riaz, F.; Karadag, A.; Tabatabaei, M. *Different Pretreatment Methods of Lignocellulosic Biomass for Use in Biofuel Production In Biomass for Bioenergy-Recent Trends and Future Challenges*; Abomohra, A.E.-F., Ed.; IntechOpen: London, UK, 2019. [CrossRef]
10. Sanchez, C. Lignocellulosic residues: Biodegradation and bioconversion by fungi. *Biotechnol. Adv.* **2009**, *27*, 185–194. [CrossRef]
11. Tsegaye, B.; Balomajumder, C.; Roy, P. Microbial delignification and hydrolysis of lignocellulosic biomass to enhance biofuel production: An overview and future prospect. *Bull. Natl. Res. Cent.* **2019**, *43*, 51. [CrossRef]
12. Neethu, K.; Rubeena, M.; Sajith, S.; Sreedevi, S.; Priji, P.; Unni, K.N.; Sarath Josh, M.K.; Jisha, V.N.; Pradeep, S.; Benjamin, S. A novel strain of *Trichoderma viride* shows complete lignocellulolytic activities. *Adv. Biosci. Biotechnol.* **2012**, *3*, 1160–1166. [CrossRef]
13. Rubeena, M.; Neethu, K.; Sajith, S.; Sreedevi, S.; Priji, P.; Unni, K.N.; Sarath Josh, M.K.; Jisha, V.N.; Pradeep, S.; Benjamin, S. Lignocellulolytic activities of a novel strain of *Trichoderma harzianum*. *Adv. Biosci. Biotechnol.* **2013**, *4*, 214–221. [CrossRef]
14. Wagner, A.; Lackner, N.; Mutschlechner, M.; Prem, E.; Markt, R.; Illmer, P. Biological pretreatment strategies for second-generation lignocellulosic resources to enhance biogas production. *Energies* **2018**, *11*, 1797. [CrossRef]
15. Hu, J.; Arantes, V.; Saddler, J.N. The enhancement of enzymatic hydrolysis of lignocellulosic substrates by the addition of accessory enzymes such as xylanase: Is it an additive or synergistic effect? *Biotechnol. Biofuels* **2011**, *4*, 36. [CrossRef]
16. Yang, B.; Dai, Z.; Ding, S.Y.; Wyman, C.E. Enzymatic hydrolysis of cellulosic biomass. *Biofuels* **2011**, *2*, 421–449. [CrossRef]
17. Bhaumik, P.; Dhepe, P.L. Conversion of biomass into sugars. In *Biomass Sugars for Non-Fuel Applications*; Murzin, D., Simakova, O., Eds.; Royal Society of Chemistry: London, UK, 2015; pp. 1–53.
18. Loow, Y.-L.; Wu, T.Y.; Jahim, J.M.; Mohammad, A.W.; Teoh, W.H. Typical conversion of lignocellulosic biomass into reducing sugars using dilute acid hydrolysis or alkaline pretreatment. *Cellulose* **2016**, *23*, 1491–1520. [CrossRef]
19. Hawrot-Paw, M.; Koniuszy, A.; Zając, G.; Szyszlak-Bargłowicz, J.; Jaklewicz, J. Production of 2nd generation bioethanol from straw during simultaneous microbial saccharification and fermentation. *Arch. Environ. Prot.* **2020**, *46*, 47–52.
20. Sindhu, R.; Binod, P.; Pandey, A. Biological pretreatment of lignocellulosic biomass–An overview. *Bioresour. Technol.* **2016**, *199*, 76–82. [CrossRef]
21. Dahnum, D.; Tasum, S.O.; Triwahyuni, E.; Nurdin, M.; Abimanyu, H. Comparison of SHF and SSF processes using enzyme and dry yeast for optimization of bioethanol production from empty fruit bunch. *Energy Procedia* **2015**, *68*, 107–116. [CrossRef]
22. Erdei, B.; Franko, B.; Galbe, M.; Zacchi, G. Separate hydrolysis and co-fermentation for improved xylose utilization in integrated ethanol production from wheat meal and wheat straw. *Biotechnol. Biofuels* **2012**, *5*, 12. [CrossRef]
23. Qin, L.; Zhao, X.; Li, W.-C.; Zhu, J.-Q.; Liu, L.; Li, B.-Z.; Yuan, Y.-J. Process analysis and optimization of simultaneous saccharification and co-fermentation of ethylenediamine-pretreated corn stover for ethanol production. *Biotechnol. Biofuels* **2018**, *11*, 118. [CrossRef]
24. Olguin-Maciel, E.; Singh, A.; Chable-Villacis, R.; Tapia-Tussell, R.; Ruiz, H.A. Consolidated Bioprocessing, an Innovative Strategy towards Sustainability for Biofuels Production from Crop Residues: An Overview. *Agronomy* **2020**, *10*, 1834. [CrossRef]
25. Binood, P.; Sindhu, R.; Pandey, A. The Alcohol Fermentation Step: The Most Common Ethanologenic Microorganisms Among Yeasts, Bacteria and Filamentous Fungi. In *Lignocellulose Conversion*; Faraco, V., Ed.; Springer: Berlin/Heidelberg, Germany, 2013. [CrossRef]

26. Bušić, A.; Mardetko, N.; Kundas, S.; Morzak, G.; Belskaya, H.; Šantek, M.I.; Komes, D.; Novak, S.; Šantek, B. Bioethanol production from renewable raw materials and its separation and purification: A review. *Food Technol. Biotechnol.* **2018**, *56*, 289–311. [CrossRef] [PubMed]
27. Kumar, S.; Singh, N.; Prasad, R. Anhydrous ethanol: A renewable source of energy. *Renew. Sustain. Energy Rev.* **2010**, *14*, 1830–1844. [CrossRef]
28. Ferdes, M.; Dinca, M.N.; Moiceanu, G.; Zabava, B.S.; Paraschiv, G. Microorganisms and Enzymes Used in the Biological Pretreatment of the Substrate to Enhance Biogas Production: A Review. *Sustainability* **2020**, *12*, 7205. [CrossRef]
29. Anwar, Z.; Gulfraz, M.; Irshad, M. Agro-industrial lignocellulosic biomass a key to unlock the future bioenergy: A brief review. *J. Rad. Res. Appl. Sci.* **2014**, *7*, 163–173. [CrossRef]
30. Li, Y.; Khanal, S.K. *Bioenergy: Principles and Applications*; Wiley Blackwell: Hoboken, NJ, USA, 2017.
31. Ahorsu, R.; Medina, F.; Constantí, M. Significance and Challenges of Biomass as a Suitable Feedstock for Bioenergy and Biochemical Production: A Review. *Energies* **2018**, *11*, 3366. [CrossRef]
32. Fiserova, M.; Opalena, E.; Illa, A. Comparative study of hemicelluloses extraction from beech and oak wood. *Wood Res.* **2013**, *58*, 543–554.
33. Maeda, R.N.; Serpa, V.I.; Rocha, V.A.L.; Mesquita, R.A.A.; Santa Anna, L.M.M.; Machado de Castro, A.; Driemeier, C.E.; Pereira Jr., N.; Polikarpov, I. Enzymatic hydrolysis of pretreated sugar cane bagasse using *Penicillium funiculosum* and *Trichoderma harzianum* cellulases. *Process Biochem.* **2011**, *46*, 1196–1201. [CrossRef]
34. Vermaas, J.V.; Petridis, L.; Qi, X.; Schulz, R.; Lindner, B.; Smith, J.C. Mechanism of lignin inhibition of enzymatic biomass deconstruction. *Biotechnol. Biofuels* **2015**, *8*, 217. [CrossRef]
35. Singhvi, M.; Kim, B.S. Green hydrogen production through consolidated bioprocessing of lignocellulosic biomass using nanobiotechnology approach. *Bioresour. Technol.* **2022**, *365*, 128108. [CrossRef]
36. Wi, S.G.; Cho, E.J.; Lee, D.-S.; Lee, S.J.; Lee, Y.J.; Bae, H.-J. Lignocellulose conversion for biofuel: A new pretreatment greatly improves downstream biocatalytic hydrolysis of various lignocellulosic materials. *Biotechnol. Biofuels* **2015**, *8*, 228. [CrossRef]
37. Su, Y.; Yu, X.; Sun, Y.; Wang, G.; Chen, H.; Chen, G. Evaluation of Screened Lignin-degrading Fungi for the Biological Pretreatment of Corn Stover. *Sci. Rep.* **2018**, *8*, 5385. [CrossRef]
38. Bari, E.; Taghiyari, H.R.; Naji, H.R.; Schmidt, O.; Ohno, K.M.; Clausen, C.A.; Bakar, E.S. Assessing the destructive behaviors of two white-rot fungi on beech wood. *Int. Biodeterior. Biodegrad.* **2016**, *114*, 129–140. [CrossRef]
39. Ghorbani, F.; Karimi, M.; Biria, D.; Kariminia, H.R.; Jeihanipour, A. Enhancement of fungal delignification of rice straw by *Trichoderma viride* sp. to improve its saccharification. *Biochem. Eng. J.* **2015**, *101*, 77–84. [CrossRef]
40. Ladeira Ázar, R.I.S.; Bordignon-Junior, S.E.; Laufer, C.; Specht, J.; Ferrier, D.; Kim, D. Effect of Lignin Content on Cellulolytic Saccharification of Liquid Hot Water Pretreated Sugarcane Bagasse. *Molecules* **2020**, *25*, 623. [CrossRef]
41. Saini, J.K.; Saini, R.; Tewari, L. Lignocellulosic agriculture wastes as biomass feedstocks for second-generation bioethanol production: Concepts and recent developments. *3 Biotech* **2015**, *5*, 337–353. [CrossRef]
42. Pinheiro, T.; Coelho, E.; Romaní, A.; Domingues, L. Intensifying ethanol production from brewer's spent grain waste: Use of whole slurry at high solid loadings. *New Biotechnol.* **2019**, *53*, 1–8. [CrossRef]
43. Mishra, P.K.; Gregor, T.; Wimmer, R. Utilising brewer's spent grain as a source of cellulose nanofibres following separation of protein-based biomass. *BioResources* **2017**, *12*, 107–116. [CrossRef]
44. Bhattacharya, S.; Das, A.; Patnaik, A.; Bokade, P.; Rajan, S.S. Submerged fermentation and characterization of carboxymethyl cellulase from a rhizospheric isolate of *Trichoderma viride* associated with Azadirachta indica. *J. Sci. Ind. Res.* **2014**, *73*, 225–230.
45. Hawrot-Paw, M.; Izwikow, M. Cellulolytic activity of *Trichoderma viride* with regard to selected lignocellulosic waste materials. *J. Ecol. Eng.* **2016**, *17*, 119–122. [CrossRef]
46. Malik, S.K.; Mukhtar, H.; Farooqi, A.A.; Haq, I. Optimization of process parameters for the biosynthesis of cellulases by *Trichoderma viride*. *Pak. J. Bot.* **2010**, *42*, 243–4251.
47. Mithra, M.G.; Jeeva, M.L.; Sajeev, M.S.; Padmaja, G. Comparison of ethanol yield from pretreated lignocellulo-starch biomass under fed-batch SHF or SSF modes. *Heliyon* **2018**, *4*, e00885. [CrossRef] [PubMed]
48. Mendez, J.; de França Passos, D.; Wischral, D.; Modesto, L.F.; Pereira Jr, N. Second-generation ethanol production by separate hydrolysis and fermentation from sugarcane bagasse with cellulose hydrolysis using a customized enzyme cocktail. *Biofuels* **2021**, *12*, 1225–1231. [CrossRef]
49. Kim, S.; Dale, B.E. Global potential bioethanol production from wasted crops and crop residues. *Biomass Bioenerg.* **2004**, *26*, 361–375. [CrossRef]
50. Yasin, M. Performance evaluation of mobile straw baler. *J. Agric. Res.* **2012**, *50*, 261–270.
51. Phuong, D.V.; Quoc, L.P.T.; Tan, P.V.; Duy, L.N.D. Production of bioethanol from Robusta coffee pulp (*Coffea robusta* L.) in Vietnam. *Foods Raw Mater.* **2019**, *7*, 10–17. [CrossRef]
52. Lara-Serrano, M.; Angulo, F.S.; Negro, M.J.; Morales-delaRosa, S.; Campos-Martín, J.M.; Fierro, J.L. Second-Generation Bioethanol Production Combining Simultaneous Fermentation and Saccharification of IL-Pretreated Barley Straw. *ACS Sustain. Chem. Eng.* **2018**, *6*, 7086–7095. [CrossRef]
53. Cutzu, R.; Bardi, L. Production of bioethanol from agricultural wastes using residual thermal energy of a cogeneration plant in the distillation phase. *Fermentation* **2017**, *3*, 24. [CrossRef]

54. Xian, X.; Fang, L.; Zhou, Y.; Li, B.; Zheng, X.; Liu, Y.; Lin, X. Integrated Bioprocess for Cellulosic Ethanol Production from Wheat Straw: New Ternary Deep-Eutectic-Solvent Pretreatment, Enzymatic Saccharification, and Fermentation. *Fermentation* **2022**, *8*, 371. [CrossRef]
55. Saka, A.; Afolabi, A.S. Production and characterization of bioethanol from sugarcane bagasse as alternative energy sources. In Proceedings of the World Congress on Engineering, London, UK, 1–3 July 2015; pp. 876–880.
56. Tomás-Pejó, E.; Fermoso, J.; Herrador, E.; Hernando, H.; Jiménez-Sánchez, S.; Ballesteros, M.; González-Fernández, C.; Serrano, D.P. Valorization of steam-exploded wheat straw through a biorefinery approach: Bioethanol and bio-oil co-production. *Fuel* **2017**, *199*, 403–412. [CrossRef]
57. Boonchuay, P.; Techapun, C.; Leksawasdi, N.; Seesuriyachan, P.; Hanmoungjai, P.; Watanabe, M.; Takenaka, S.; Chaiyaso, T. An integrated process for xylooligosaccharide and bioethanol production from corncob. *Bioresour. Technol.* **2018**, *256*, 399–407. [CrossRef]
58. Unrean, P.; Ketsub, N. Integrated lignocellulosic bioprocess for co-production of ethanol and xylitol from sugarcane bagasse. *Ind. Crops Prod.* **2018**, *123*, 238–246. [CrossRef]
59. Irfan, M.; Nadeem, M.; Syed, Q. Ethanol production from agricultural wastes using *Sacchromyces cervisae*. *Braz. J. Microbiol.* **2014**, *45*, 457–465. [CrossRef]
60. López-Linares, J.C.; Romero, I.; Cara, C.; Ruiz, E.; Moya, M.; Castro, E. Bioethanol production from rapeseed straw at high solids loading with different process configurations. *Fuel* **2014**, *122*, 112–118. [CrossRef]
61. Saini, R.; Chen, C.-W.; Patel, A.K.; Saini, J.K.; Dong, C.-D.; Singhania, R.R. Valorization of Pineapple Leaves Waste for the Production of Bioethanol. *Bioengineering* **2022**, *9*, 557. [CrossRef]
62. Dodo, C.M.; Mamphweli, S.; Okoh, O. Bioethanol production from lignocellulosic sugarcane leaves and tops. *J. Energy S. Afr.* **2017**, *28*, 1–11. [CrossRef]
63. Paschos, T.; Louloudi, A.; Papayannakos, N.; Kekos, D.; Mamma, D. Potential of barley straw for high titer bioethanol production applying pre-hydrolysis and simultaneous saccharification and fermentation at high solid loading. *Biofuels* **2020**, *13*, 467–473. [CrossRef]
64. Zhang, J.; Chu, D.; Huang, J.; Yu, Z.; Dai, G.; Bao, J. Simultaneous saccharification and ethanol fermentation at high corn stover solids loading in a helical stirring bioreactor. *Biotechnol. Bioeng.* **2010**, *105*, 718–728. [CrossRef]
65. Tiwari, S.; Jadhav, S.K.; Tiwari, K.L. Bioethanol production from rice bran with optimization of parameters by *Bacillus cereus* strain McR-3. *Int. J. Environ. Sci. Technol.* **2015**, *12*, 3819–3826. [CrossRef]
66. da Silva, F.L.; de Oliveira Campos, A.; dos Santos, D.A.; Batista Magalhães, E.R.; de Macedo, G.R.; dos Santos, E.S. Valorization of an agroextractive residue—Carnauba straw—For the production of bioethanol by simultaneous saccharification and fermentation (SSF). *Renew. Energy* **2018**, *127*, 661–669. [CrossRef]
67. Takano, M.; Hoshino, K. Bioethanol production from rice straw by simultaneous saccharification and fermentation with statistical optimized cellulase cocktail and fermenting fungus. *Bioresour. Bioprocess.* **2018**, *5*, 16. [CrossRef]
68. Waghmare, P.R.; Khandare, R.V.; Jeon, B.; Govindwar, S.P. Enzymatic hydrolysis of biologically pretreated sorghum husk for bioethanol production. *Biofuel Res. J.* **2018**, *5*, 846–853. [CrossRef]

sustainability

MDPI

Article

Review on Phytoremediation Potential of Floating Aquatic Plants for Heavy Metals: A Promising Approach

Yean Ling Pang [1,2,*], Yen Ying Quek [1], Steven Lim [1,2] and Siew Hoong Shuit [1,2]

[1] Department of Chemical Engineering, Lee Kong Chian Faculty of Engineering and Science, Universiti Tunku Abdul Rahman, Kajang 43000, Malaysia
[2] Centre for Photonics and Advanced Materials Research, Universiti Tunku Abdul Rahman, Kajang 43000, Malaysia
* Correspondence: pangyl@utar.edu.my or pangyeanling@hotmail.com; Tel.: +60-39-086-0288; Fax: +60-39-019-8868

Abstract: Water pollution due to heavy metals has become a serious environmental concern due to their hazardous properties. Since conventional water remediation techniques are generally ineffective and non-environmentally friendly, phytoremediation has gained increasing attention from worldwide researchers and scientists due to its cost-effectiveness and environmental friendliness. Hence, this review first discussed soil and water remediations. Phytoremediation can be divided into five techniques to remove heavy metals from the polluted environment, namely, phytostabilization (phytosequestration), phytodegradation (phytotransformation), phytofiltration (rhizofiltration), phytoextraction (phytoaccumulation), and phytovolatilization. Four common floating aquatic plants (accumulator plants), such as duckweed (*Lemna minor*), water lettuce (*Pistia stratiotes*), water hyacinth (*Eichhornia crassipes*), and watermoss (*Salvinia*) were discussed in detail due to their great capability in absorbing the metal ions by their roots and further translocating the metal ions to the aerial parts. Furthermore, the parameter studies, such as optimum pH and temperature of the water, exposure duration, initial metal concentration, water salinity, and the addition of chelating agents, were evaluated. The absorption kinetics of the plants was discussed in detail. In short, phytoremediation is a promising green and sustainable water remediation approach. However, further research is necessary to enhance its practicability and performance at large-scale implementation.

Keywords: phytoremediation; hyperaccumulator plants; heavy metals pollution; accumulation; bioremediation

check for updates

Citation: Pang, Y.L.; Quek, Y.Y.; Lim, S.; Shuit, S.H. Review on Phytoremediation Potential of Floating Aquatic Plants for Heavy Metals: A Promising Approach. *Sustainability* **2023**, *15*, 1290. https://doi.org/10.3390/su15021290

Academic Editor: Antonio Galvagno

Received: 29 October 2022
Revised: 11 December 2022
Accepted: 26 December 2022
Published: 10 January 2023

1. Introduction

Due to rapid urbanization and industrialization, about 40% of the world's population is facing water scarcity problems [1,2]. Some contributing factors to these issues are climate change, food necessity, and the inefficient utilization of natural resources. In addition, a study revealed that the death of 1.8 million people was linked to water pollution in 2015 [3]. The progressive revision and assessment of water resource policy at all levels are indeed necessary. Water pollution has been linked to severe human health issues such as infectious diseases, nervous system damage, and even death [4]. Moreover, water pollutants are tremendously harmful to aquatic life. Therefore, the rehabilitation of wastewater is the only remediation solution to cope with the greater demand for water for industrial and agricultural use.

Nowadays, most of the surface waters are not achieving the optimal standard owing to a variety of stressors that affect freshwater quality, namely point source and non-point source pollution, the change in land use and climate, which further magnifies the challenge of supplying water security [5]. In detail, point sources of water pollution are those discharged from a single and identifiable origin. Contrarily, non-point sources of pollution are those pollutants eliminated from various sources and diverse non-identifiable sources [6].

In the United Kingdom, one of the primary stressors on water quality is an excessive nutrient released from a diffuse source of water pollution [5]. On the other hand, in China, other issues of heavy metal pollution are notable. The interactions between various stressors in time and space could lead to additional effects [5]. For instance, an increase in land-use change on account of vigorous agricultural activities and a potential rise in storm frequency might escalate the distribution of nutrients, such as phosphorus and nitrogen, and fine sediment to receiving water bodies. Eventually, the rapid expansion of industrial and residential activities would negatively impact the water quality of rivers, lakes, and oceans [4,7].

Phytoremediation is a promising green technology in wastewater remediation by using plants and microorganisms to eliminate, translocate, immobilize, or degrade the contaminants from the environment [8–12]. In other words, phytoremediation employs the fact that a living plant can act as a photosynthetic-driven pump proficient in eliminating pollutants such as metals and metalloids from the environment and water effectively [9,13]. Notably, aquatic plants play an essential role as a natural absorber in phytoremediation for heavy metals and contaminants with their extensive root systems, making them the best selection for the uptake of pollutants through their shoots and roots [1]. Phytoremediation technology has received global attention among scientists and administration bodies due to its effectiveness in lowering unparalleled environmental pollution via an environmentally friendly pathway. Figure 1 demonstrates the increasing trend for phytoremediation based on the data extracted from Scopus by using the keywords of "phytoremediation and phytoremediation and water" from 2002 to 2022.

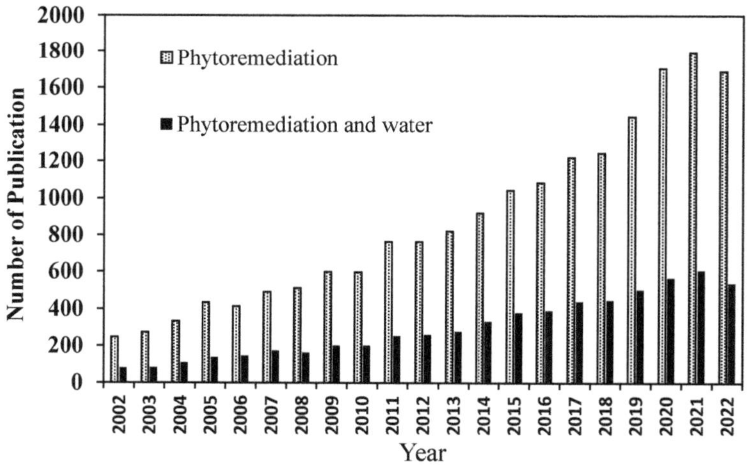

Figure 1. Increasing use of aquatic plants in phytoremediation.

This review intends to examine the phytoremediation of heavy metals within different accumulator plants. It focuses on the emerging phytoremediation results for the removal of heavy metals from contaminated water and the main mechanisms occurring between the heavy metals and the accumulator plants. The phytoremediation parameters and kinetic studies were also updated and discussed. Lastly, this review also proposed some useful prospects and challenges for further development to boost the development of phytoremediation that is capable of improving the efficiency of the process.

2. Types of Phytoremediation

Phytoremediation can be applied to remove heavy metals in contaminated soil or contaminated wastewater through the bioaccumulation process. The utilization of accumulator plants is the crucial element of phytoremediation, and its selection is based on the bioaccumulation capacity of the accumulator plants on the targeted heavy metal. Some studies have ascertained the bioaccumulation capability through a series of experiments.

2.1. Soil Remediation

Conventional soil remediation approaches such as chemical oxidation and solvent extraction used to treat the contaminated soil with heavy metals and residues are generally cost-intensive and destructive to soil nature. Additionally, these methods require the transportation of contaminated substances to the treatment site, introducing additional risks of secondary pollution [10]. Lately, phytoremediation has received attention in remediating soil contamination sites due to its profitability, environmental friendliness, and durable application. Phytoremediation employs plants and microorganisms to eliminate, isolate, or degrade toxic substances away from the environment. The reliable mechanisms for soil remediation are phytoextraction and phytostabilization [10]. Phytoextraction uses plants to extract and capture the contaminant, whereas phytostabilization contains the contaminant. Other workable phytoremediation mechanisms include rhizofiltration and phytovolatilization. Rhizofiltration works by absorbing and adsorbing the contaminant. On the other hand, phytovolatilization works by absorbing the contaminant from the medium via plant roots and discharging them into the atmosphere [14].

Bioavailability expresses the degree of contaminants readily absorbed by plants through exposure to them [14]. Plants will only absorb or uptake metals that are bioavailable to them. Metal bioavailability is vital in determining the success of phytoremediation by plants. Low metal bioavailability is the primary factor that restricts the phytoextraction of metal contaminants [15]. In addition, soil microbes play a crucial role in catalyzing redox reactions, altering the metal bioavailability in soil and the tendency for root uptake.

The factors affecting the metal bioavailability in soil are the pH, microorganisms, root exudates, soil organic matter, and competitive cations [10]. The acidity and alkalinity of the soil determines the metal solubility and mobility in the soil. At acidic or low pH conditions, plants liberate more metals into the soil solution to compete with hydrogen ions (H^+), thus enhancing the metal bioavailability. Under alkaline or neutral pH conditions, the immobilization of metals such as lead (Pb) and chromium (Cr) would happen. Therefore, the metals are not bioavailable to plants [16]. In addition, soil microorganism such as the strain of *Xanthomonas maltophyla* was proven to accelerate the precipitation of Cr^{6+} to trivalent chromium ions (Cr^{3+}) from a state of high mobility to low mobility and less toxic compounds [17].

Once taking up the heavy metals, the metals concentrate in the root tissues through immobilization or further translocate towards the aerial part of the plant via xylem vessels [14]. In shoots, the metal accumulation usually happens in vacuoles, which are cellular organelles that have a low metabolism. The hyperaccumulator plants are usually equipped with vital metal tolerance mechanisms, namely metal detoxification and metal exclusion, to cope with the toxic effects of metal ions at elevated concentrations [17]. For metal exclusion, the excluders prevent metal absorption by roots and preclude further translocation and accumulation in plant shoots.

The phytoextraction mechanism to eliminate soil contaminants includes five necessities: (1) mobilization of metal ions in the rhizosphere, (2) uptake of metal ions through plant roots, (3) translocation of metal ions internally to the shoots, stems, and leaves of the plant, (4) heavy metal tolerance, and (5) metal sequestration in plant tissues [18]. Among these requirements, heavy metal tolerance is an essential requirement for phytoremediation since higher plant tolerance to metal stress indicates that a higher number of metals could be accumulated in the plant tissues with the lowest detrimental impacts on plant health [18]. Metals such as cadmium (Cd) are easier to be absorbed from the soil via phytoextraction [16].

The potential metal tolerance in a plant relies on some mechanisms, such as metal binding in the plant cell wall, active transportation of metal ions into the plant vacuoles, the formation of metal complexes, and the chelation of metal ions with peptides and proteins.

Apart from the physiological processes dominating the plant tolerance, another crucial factor in predicting the phytoextraction potential is the yearly production of biomass, including the dry weight of shoots and the net composition of metal harvested [14].

2.2. Water Remediation

It has been reported that the application of aquatic plants in the remediation of wastewater was initiated about 300 years ago [19]. Several plant species have been examined and evaluated for their efficiencies in concentrating organic and inorganic contaminants from the water via hydroponic, constructed wetlands, or natural habitats. However, in wetland systems, the precipitated inorganic contaminants from the water often enter the sediments, leading to a complex recovery. Opposingly, floating plant systems could eliminate contaminants via biomass harvesting [20].

Among the diverse floating aquatic plant species, phytoremediators such as *Azolla*, *Eichhornia*, *Lemna*, *Spirodela*, *Wolfia*, and *Wolfialla* demonstrated high efficiency in removing water pollutants through bioaccumulation in their plant tissues [19]. Explicitly, *Lemna minor*, *Eichhornia crassipes*, and *Pistia stratiotes* are specifically employed to eliminate metal ions present in the aquatic system [20]. For instance, *Eichhornia crassipes* can biodegrade inorganic pollutants by concentrating various metal ions, such as copper (Cu), Cd, Cr, lead (Pb), and zinc (Zn).

Additionally, it could remove other contaminants such as total dissolved solids (TDS), total suspended solids (TSS), biological oxygen demand (BOD), and chemical oxygen demand (COD) from industrial wastewater [21]. The harvesting process is comparatively simple due to its floating and not-rooted structure [22]. Furthermore, the treatment by plants such as *Lemna minor* and *Pistia stratiotes* successfully reduced the TDS, BOD, COD, chloride, and sulphate in the wastewater.

3. Classification of Phytoremediation Technique

Several mechanisms are involved in phytoremediation to uptake heavy metals by using aquatic plants from the polluted water bodies and subsequently transforming them into a non-toxic form. These mechanisms are phytofiltration, phytoextraction, phytodegradation, phytostabilization, and phytovolatilization, as shown in Figure 2 [23]. The application of phytoremediation typically starts with the recognition and screening of suitable aquatic plants with great potency to concentrate metals, dissolved nutrients, and other pollutants [19,24]. Each mechanism has its unique characteristics, applications, and uptake routes, which are discussed in the following subsections. For example, in a water environment, rhizofiltration works well by overcoming the inherent biological limitation found in phytoextraction. Specifically, phytoextraction is more suitable for treating polluted soils at shallow depths. Table 1 shows the phytoremediation strategies that are applicable for the removal of different categories of contaminants present in the water bodies.

Table 1. Phytoremediation mechanisms for the removal of pollutants present in the aquatic environment [19].

Pollutants	Mechanisms	Descriptions
Inorganic	Phytoextraction	Eliminate the contaminants in the form of harvestable plant biomass.
	Phytostabilization	Minimize the contaminants mobility.
	Phytoaccumulation	Hyperaccumulation due to hypertolerance.
	Rhizofiltration	Roots filter water via absorption or adsorption.
Organic	Phytodegradation	Degrade the contaminants in the plant.
	Phytostimulation	Stimulate the microbial activity to degrade the contaminants.
	Phytoassimilation	Transport and metabolize the contaminants in plants.
	Phytovolatilization	Extract the contaminants from media and liberate them through air.
	Phytotransformation	Degrade contaminants into a simpler form.

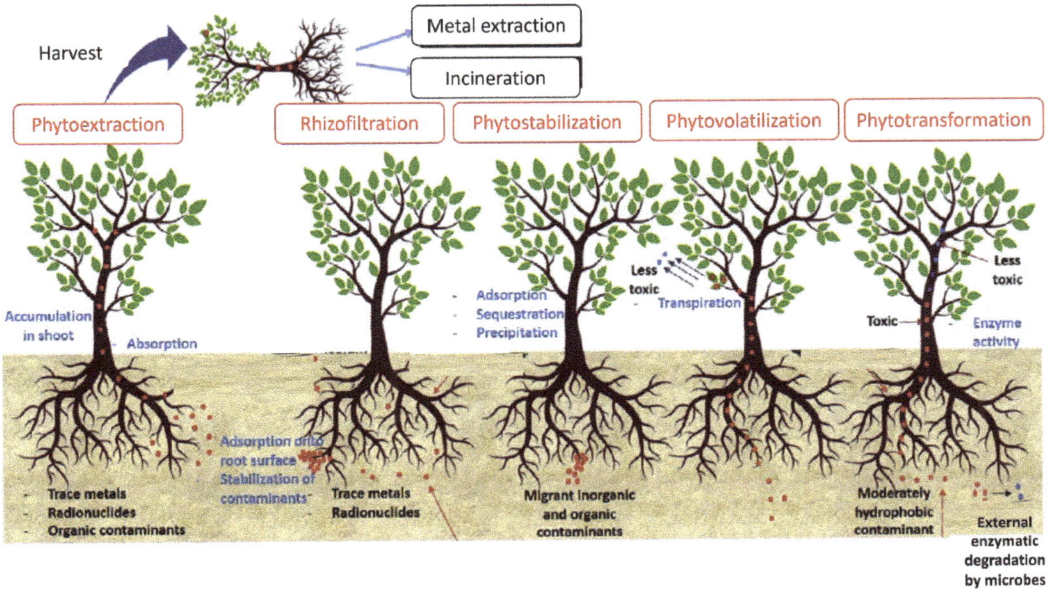

Figure 2. Phytoremediation mechanisms for various types of heavy metals mitigation. Reproduced with permission from [25]. Copyright 2020 Elsevier Ltd.

Rhizo- implies root; thus, rhizofiltration/phytofiltration involves the adsorption or absorption of pollutants in the solution adjacent to the plant roots. The working mechanism in rhizofiltration is similar to that of phytoextraction. However, the plants are mainly used to address the polluted water. The plant species for pollutant removal are cultivated in greenhouses, in which their roots are in contact with water [26]. It could only happen once the widespread fibrous root system has been successfully established. The plant species that grew in the polluted area will uptake the water containing pollutants by their roots. Once the plant roots reach the pollutant saturation limit, they will be harvested. Plants with dense root systems are preferred for rhizofiltration to concentrate the maximum concentration of the contaminants with the larger root adsorption area [20]. Some rootless or floating macrophytes demonstrate high efficiency and potential for the rhizofiltration of metal ions such as Cr, Pb, and Zn from an aqueous system. Nevertheless, the performance of rhizofiltration will depend on the types of metal and the plant's metabolism [21]. The plants selected for rhizofiltration must be resistant to metal, tolerant to hypoxia, and have a large surface area for absorption [14].

Meanwhile, phytoextraction could also be referred to as phytoaccumulation. The phytoextraction mechanism involves the heavy metal uptake from the soil through the root of crop species and further translocates into the aerial part of the plant [1]. The translocation process is regulated by employing leaf transpiration and root pressure. According to Shaari et al. [16], one of the strategies to improve metal solubility in the soil is through the addition of chelating agents. Upon completion of the phytoextraction, the plant will be harvested and disposed of with care [20]. In addition, the plant can be burned for energy generation and further recovery or recycle important metal from the ash. The hyperaccumulators selected should have high efficiency to accumulate high concentrations of crucial micronutrients and uptake considerable quantities of non-crucial metals, such as Cd [27]. Hyperaccumulators occupy environments rich in metals because of their greater necessity for metals than normal accumulator and non-accumulator plants. Additionally, plants with high biomass production are efficient in pollutant uptake. Lastly, plants with high translocation factors, elongated roots, and simple harvesting processes are suitable for phytoextraction [14].

On the other hand, phytostabilization/phytosequestration uses specific plant species to deactivate contaminants contained in the groundwater and soil. As its name implies, phytostabilization could be referred to as in-place immobilization or phytoimmobilization [1]. This process happens when the roots of plants absorb, accumulate, adsorb, or precipitate the pollutants to restrict their movement. Microbes restrict the mobility of contaminants by deliberating chelating substances, avoiding the upward migration of the contaminants to the groundwater, and lessen the bioavailability of metal in the food chain [26]. For example, *Arbuscular mycorrhizal* symbiosis could stabilize the metals in soil [20]. Furthermore, phytostabilization can restore vegetation cover at contaminated sites by employing metal-tolerant plants, especially when the natural vegetation is unable to sustain in the soils contaminated with metals or physically disturbing surface materials. Accordingly, plants could limit the migration of pollutants via wind destruction, delivery of the affected soil surface, and leaching of soil pollutants to groundwater. Apart from that, plants that are inefficient in translocating the absorbed metals from the root to the aerial part are preferable for phytostabilization [14].

In phytovolatilization, the plant absorbs contaminants from the soil and converts them into various volatile compounds, then discharges the metal in gaseous form into the adjacent environment or atmosphere through the plant's stomata via transpiration [1]. Phytovolatilization, primarily concerned with remediating organic acids, is also useful in treating mercury (Hg) and selenium (Se). For the phytovolatilization of Se, with a long half-life of 327,000 years, the suitable plant species are Indian mustard and canola. Phytovolatilization offers numerous advantages, including the unnecessary auxiliary management of vegetation, less soil corrosion, absence of soil interruption, unreturned harvesting, and dumping of plant biomass. Furthermore, the presence of bacteria in the plant's rhizosphere assists in biotransforming the contaminant and eventually bolsters the phytovolatilization rate. However, phytovolatilization is still an arguable method since it discharges toxic metals and returns them to the atmosphere [20]. The research undertaken by Jeevanantham et al. [23] suggested that the heavy metals taken up by plants could transform into a water-soluble and non-toxic form while being transported from the root to the leaves of the plant and being compartmented in the vacuole, followed by the volatilization of metal ions to the atmosphere. The accumulation of metal begins in the epidermis of leaves, followed by the accumulation in the mesophyll of leaves. However, a hyperaccumulator usually inhibits the accumulation of metals in mesophyll with its high evaporation rate of modified soluble metals. Therefore, no adverse effects could result on the plants.

Lastly, phytodegradation, also known as phytotransformation, involves the breakdown of contaminants into simpler molecules through the enzymatic metabolic activity in plants with their corresponding microorganisms [21]. In certain circumstances, selected plants possess the capability to uptake toxic compounds, followed by the detoxification and metabolization of toxic compounds as nutrients [27]. The detoxification of toxic compounds usually involves three stages, namely bioactivation, conjugation, and compartmentalization. Each phase requires various types of enzymes, such as oxygenizes and nitro reductases, classified by the properties and distribution of their reaction products. The enzymes generated by plants could catalyze and speed up the degradation process. The phytodegradation process could be classified into internal and external processes [23]. In particular, two mechanisms that work for the degradation are the plant's enzymatic activity and photosynthetic oxidation. In terms of external degradation, pollutants absorbed by plants will be hydrolyzed into smaller sizes, whereas, for internal degradation, the organic pollutants absorbed are further broken down into smaller sizes by plant enzymes and are eventually used as metabolites.

4. Accumulator Aquatic Plants

The selected plant species must have a high potential to take up various pollutants, rapid growth, be easy to cultivate, and be simple to harvest [8]. Moreover, an ideal plant species applied in phytoremediation should meet a few criteria: (1) plants that show a high accumulation rate for heavy metals even at low concentrations, (2) plants that are easy to harvest, (3) plants that have resistance towards pests and diseases, (4) plants that are capable to uptake several types of heavy metals, and (5) plants that display environmentally friendly and economical application [28]. In reality, aquatic plants such as water lettuce, water hyacinth, and vetiver grass have demonstrated excellent capability to eliminate different pollutants such as heavy metals, TDS, TSS, BOD, COD, and nutrients present in wastewater.

4.1. Duckweed (Lemna minor)

Duckweed, with the scientific name *Lemna minor*, is the fastest-growing and smallest plant species on the planet. The five aquatic genera under the *Lemnaceae* family are *Lemna*, *Landoltia*, *Spirodela*, *Wolffia*, and *Wolffiella* [1]. The *Lemnaceae* family has been the focus of recent research for phytoremediation due to their rapid growth, rapid biomass productivity, phytoplankton, microbial minimization, and high metal and nutrient accumulation capability [19]. These plant species usually appeared in the form of small leaf-like structures known as fronds. They can propagate under various environmental conditions, specifically in the pH ranges from 3.5 to 10.5 and the temperature range between 7 and 35 °C [27]. Furthermore, duckweed can be cultivated in different seasons owing to its cold tolerance characteristic. However, the growth of duckweed species needs special environmental considerations owing to its high sensitivity towards various contaminants. Moreover, diverse duckweed species have various metal tolerances depending on the ambient water conditions such as temperature, pH, metal concentration, and electrical conductivity.

Based on the recent study conducted by Rezania et al. [29], the findings revealed that duckweed has been widely used to recover the nutrients and heavy metals released from agricultural and domestic wastewaters. Referring to another study, the results demonstrated that *Lemna gibba* was more efficient than *Salvinia minima* and *Azolla caroliniana*, thus, was appraised as a hyperaccumulator plant. Additionally, *Lemna minor* could accumulate high concentrations of Cd, Cu, nickel (Ni), magnesium (Mn), Zn, arsenic (As), and uranium (U). *Wolffia globose* possesses a great tolerance to concentrate 400 mg As per kilogram of its dry weight and subsequently removes it effectively [19]. Nevertheless, the burning of metal-contaminated duckweed has become an issue for safe disposal [1,30]. In addition, their degradation through carbonization, incineration, hydrolyzation, or anaerobic digestion is necessary to avoid successive contamination in the environment.

4.2. Water Lettuce (Pistia stratiotes)

Water lettuce, with the scientific name *Pistia stratiotes*, is a free-floating macrophyte that can absorb and concentrate pollutants in the plant body [31]. *Pistia stratiotes*, which belongs to the Araceae family, is a floatable aquatic plant with a hanging root structure submerged in water [29]. Water lettuce is abundant in many regions, such as the tropical and subtropical areas of Asia and America, because of its simple growth requirements and ability to adapt to an extensive range of growth environments. Based on the study conducted by Gupta, Roy, and Mahindrakar [31], a considerable portion of iron (Fe), Mg, Mn, Cd, calcium (Ca), and cobalt (Co) were adsorbed or deposited on the outer root surfaces of water lettuce, whereas more aluminum (Al), Cu, Cr, Ni, and Pb were absorbed by the plant roots. *Pistia stratiotes* are also an effective phytoremediator plant species in treating Mn-polluted wastewater [32]. The advantages of water lettuce are that it is fast-growing, able to cover large water surfaces, and that it requires an uncomplicated harvesting process. Referring to Lu et al. [33], water lettuce possesses an excellent capability in accumulating metal ions from the water bodies with a high concentration factor greater than 10^2. By taking its bioconcentration factor as an indicator, this plant was regarded as a hyperaccumulator for Cu, Cr, Fe, Mn, Ni, Pb, and Zn. Therefore, it was feasible to apply it for surface water remediation.

4.3. Water Hyacinth (*Eichhornia crassipes*)

Water hyacinth, with the scientific name of *Eichhornia crassipes*, is a rooted macrophyte belonging to the families of *Pontederiaceae* and *Eichhornia*. Water hyacinth usually grows largely in polluted water systems and eutrophic lakes [34]. This fast-growing free-floating perennial aquatic weed appears with upright and rounded leaves with a dark blue root system, and it has been demonstrated to be highly competent in remediating domestic wastewater due to its highly resistant features [19]. It is one of the most prevailing invasive vascular plants in the aquatic system due to its tolerance to high concentrations of heavy metals, acetic acids, formaldehyde, formic acids, oxalic acids, and phenols. In addition, it can rapidly adapt to various aquatic physiochemical surroundings, such as those caused by drought and moist sediment conditions. Furthermore, it can uptake tremendous quantities of contaminants, especially heavy metals and nutrients. Various researchers have claimed that water hyacinth exhibited a modest accumulation efficiency towards Cd and Zn. Meanwhile, the plants were efficient in treating waters containing toxic Cr^{6+}. Moreover, water hyacinth was highly efficient in eliminating nitrogen and potassium from the aquatic system. The plant's pollutant removal efficiency was closely related to its maximum growth. The optimal growth conditions for water hyacinth were a pH of 6 to 8, a temperature between 10 and 40 °C, and a water salinity below 5 mg/L [29].

4.4. Watermoss (*Salvinia*)

Salvinia, which belongs to the Salviniaceae family, is a floating aquatic plant with a fast growth rate and a high tolerance towards metal toxicities [29]. *Salvinia* species is a popular plant for heavy metal remediation due to its inherent capacity to absorb and concentrate high compositions of different heavy metals. Essentially, the roots of the plants have shown an unreasonably high potential to accumulate metal ions, such as Cr, Ni, and Pb, higher than their leaves [34,35]. In particular, *Salvinia natans* is a hyperaccumulator for some specific heavy metals, and its leaves could accumulate more heavy metals compared to the other parts of the plant. Table 2 summarizes the heavy-metal removal efficiencies demonstrated by different aquatic plants in phytoremediation applications.

Table 2. Summary of heavy metals removal efficiency by floating aquatic plants in phytoremediation.

Aquatic Plants	Conditions	Heavy Metals Removal Efficiency	References
Duckweed (*Lemna minor*)	Sampling time: 25 days Temperature: 7 to 20 °C Initial concentration (ppb): 16.31 As, 1.47 Cd, 67.37 Cr, 25.84 Cu, 0.36 Hg, 347.8 Ni, 23.37 Pb, 49.59 Zn Framework: industrial wastewater	90.95% As, 97.79% Cd, 90.25% Cr, 98.46% Cu, 82.84% Hg, 98.08% Ni, 99.91% Pb, 98.00% Zn	[36]
	Sampling time: 7 days Temperature: 13 to 20 °C Relative humidity: 70% Photoperiod: 16 h light, 8 h dark Concentration: 10^{-6} mol/L metal solutions	95% Cd, 93% Pb, 81.2% Zn, 86.5% Cu	[36]
Water lettuce (*Pistia stratiotes*)	Sampling time: 15 days Initial concentration (mg/L): 0.08–0.46 Cu, 0.03–1.36 Ni, 0.09–0.86 Pb, 0.26–1.31 Zn Framework: field	39.72–72.58% Cu, 28.96–68.79% Ni, 43.02–76.66% Pb, 26.99–79.57% Zn	[3]
	Sampling time: 30 days Initial concentration (mg/L): 22.17 Al, 5.03 As, 0.028 Cd, 2.84 Cr, 0.16 Cu, 14.70 Fe, 20.37 Mn, 5.25 Pb, 2.01 Zn Framework: steel industry effluent	73% Al, 74% As, 82.8% Cd, 62.8% Cr, 78.6% Cu, 61% Fe, 39.5% Mn, 73% Pb, 65.2% Zn	[37]

Table 2. *Cont.*

Aquatic Plants	Conditions	Heavy Metals Removal Efficiency	References
Water hyacinth (*Eichhornia crassipes*)	Sampling time: 15 days Temperature: 25 ± 5 °C Humidity: 72 ± 15% Initial concentration (mg/L): 1.12 Fe, 0.62 Cu, 1.41 Ni, 0.77 Pb, 1.42 Zn Framework: landfill leachate	87.56% Fe, 87.09% Cu, 81.56% Ni, 84.41% Pb, 90.18% Zn	[38,39]
	Initial concentration (mg/L): 0.24 Pb, 1.20 Pb, 4.97 Hg, 3.34 Ni Framework: industrial wastewater	97.50% Cd, 95.10% Ni, 99.90% Hg, 83.40% Pb	[29]
Watermoss (*Salvinia*)	Sampling time: 28 days Initial concentration (mg/L): 0–12.39 Framework: field	72–91% Cd, 80% Cu, 72–91% Ni, 72–91% Zn	[3]
	Sampling time: 12 days Temperature: 25 °C Humidity: 70–75% Photoperiod: 16 h light, 8 h dark Initial concentration: 1.0 mg/dm^3 Cr, 1.0 mg/dm^3 Hg	74% Cr, 93% Hg	[40]

5. Phytoremediation Mechanisms Using Accumulator Aquatic Plants

In general, heavy metal accumulation in plants involves the uptake of metals into the plant tissue and the liberation of the absorbed metals back to the external medium. In aquatic ecosystems, the adsorption of heavy metals onto the sediment takes place. However, these adsorbed metals could be freed from the sediment and remobilized in the water system if the sediment is disturbed or the water chemistry changes [41]. The potential sediment disturbances could be due to bioturbation, resuspension, the presence of organic matter, and an alteration in water salinity, thereby manipulating the equilibrium concentration between the metal ions in the water and the sediment [42]. Heavy metal remobilization is unwanted since this would contaminate clean areas when conveyed by the water current. Free-floating plants take up metals from the water by their roots. Despite uptake, the metals could be released back into the water and soil environments from the plant tissue. Aside from that, metals could be liberated into the air in a gaseous form from the surfaces of the leaves.

For instance, water lettuce must be harvested periodically not only to maintain growth density at an optimal level but also to remove metals and nutrients efficiently from water bodies. This is because the pollutants taken up by the plants will be released into the aquatic environment following the death and decomposition of the plants [33]. In the cases where a higher metal removal rate is desirable, the plant biomass should be harvested more frequently and on time. Moreover, the residence time of metals in different mediums will directly manipulate the metal concentration available in that particular medium [43]. Most metals tend to accumulate in the soil and sediment, which are known as sinks. Contrarily, metals usually have a shorter retention time in water and air since these mediums only usually serve as transport mediums. Explicitly, the retention time of metals in water varies depending on the type of heavy metals. For instance, Pb^{2+} exhibits a shorter residence time compared to Zn^{2+} [43].

5.1. *Absorption, Adsorption, and Efflux of Metals by Plants*

Plants can function as accumulators and excluders in phytoremediation. Accumulator plants can survive without being affected by a large number of metals concentrated within their aerial tissues owing to their ability to biodegrade and biotransform the metals into non-toxic forms. Opposingly, excluders limit the metal uptake into their plant biomass due to the presence of barriers [43]. As a result, excluders demonstrate low metal uptake even at high external metal concentrations. Normal plants commonly accumulate metals in quantities that do not exceed their short-term metabolic needs. Lower metal concentrations between 10 and 15 ppm are sufficient for the basic functioning of normal plants. Nevertheless, an exception is hyperaccumulators, which can absorb and tolerate thousands of ppm of metal concentrations within their tissues. The reason for this is that hyperaccumulators possess more than one detoxification mechanism for preventing metal toxicity, such as metal storage into vacuoles, metal chelation, and metal efflux [44].

According to Huynh, Chen, and Tran [45], there are two different mechanisms of heavy metal absorption in plants, namely root absorption and foliar absorption. Concerning root absorption, plant roots absorb heavy metals into the apoplast while absorbing water. The presence of –COOH groups in the pectin of the plant roots allows the exchange of cations within the cell membrane [46]. In turn, it becomes a transportation means for heavy metals to move into the cell wall from the external medium through diffusion or mass flow, where absorption actively takes place. The total concentration of metal uptake could be bound to the anions in the cell wall, transported apoplastically, and into the cells [43]. The distribution of the absorbed metals among these three locations relies on the types of metal species and the genotype of the plants.

Since water hyacinths possess dense and fibrous root systems, aerobic bacteria are well established in these aquatic environments. These bacteria gather the nutrients and inorganic pollutants that serve as food for plant nourishment [45]. Hence, the plants grow faster and are harvestable as phytoremediation plant biomass after storing the heavy metals in their tissues. Aside from root absorption, foliar absorption might occur in the plants, where the passive absorption of heavy metals occurs through stomata cells and cuticle fissures on the plant's leaves [46]. A high density of stomata cells stimulates greater ion uptake capacity as most of the uptake process is initiated in the ectodesmata. However, cuticle fissures could only act as weak ion exchangers owing to their non-esterified cutin polymers and cationic pectin substances. Specifically, the penetration of ions occurs from a low-charge density outer surface to high-charge density cell walls through the cuticle. Correspondingly, cation absorption is more likely to happen over anion absorption in this mechanism [43].

Furthermore, heavy metals could be adsorbed by plants with the aid of the bacteria attached to the feathery and fibrous roots. Meanwhile, the ionic imbalance could happen within the cell membrane [45]. In aquatic macrophytes, the usual metal transportation mechanism is rhizofiltration, in which the metal is contained, immobilized, and accumulated within the plant's roots [34]. The roots exudate within the rhizosphere, allowing the adsorption of metals on the root surfaces of the plant.

A different discovery was reported by Lissy and Madhu [46]; they suggested that phytoextraction was the process accounting for the uptake of heavy metals from a contaminated aquatic system. Despite metal absorption, metal efflux could happen. Metal efflux is a process of releasing the metal from the vacuole to the cytoplasm, from the cytoplasm to the apoplast, and seepage from the apoplast to the external medium. Various liberation and seepages are probably non-metabolic processes [43]. In addition, the efflux of metals from the cuticular layer of leaves might happen when metal ions are exchanged with hydrogen ions (H^+) during acid rainfall. Metal ions, such as Hg in a gaseous form, might also be liberated to the atmosphere through open stomata.

5.2. Bioconcentration, Translocation, and Distribution of Metals

The two important parameters for evaluating the heavy metal uptake by aquatic plants are the concentration factor (CF) and the bioconcentration factor (BCF). The CF is an indicator that assesses the total metal accumulation by plants through absorption and adsorption, whereas the BCF is an index that accounts for the metal absorption by plants from the external medium [15]. The BCF values are usually smaller than the CF values, and the difference between both values is small if the absorption of metal ions is dominant in plants. Furthermore, BCF values greater than 1000 are commonly regarded as a sign of great phytoremediation potential [47]. The BCF is a more suitable indicator for distinguishing the hyperaccumulator from the normal plants since the concentration of metals accumulated in the plant through absorption is more significant [33].

Furthermore, Greger [43] found that the majority of metals tend to bind to the cell walls during their transportation. The findings revealed that there was about 75 to 90% metal uptake by the plant's roots while only 10 to 25% were further translocated in the shoots. For instance, the distribution of Cd was lower in higher parts of the plants, following the descending order: dense fibrous roots > storage roots > stems > leaves. Vesely et al. [48] also found that more Pb was accumulated in water lettuce roots compared to that in the leaves. Additionally, a higher accumulation of Pb in the roots of water hyacinth than that in the stems and leaves was reported by [49]. On top of that, the addition of chelating agents could increase the metal bioavailability in the soil and facilitate the transportation of metal ions within the plant [14]. For example, the introduction of ethylenediamine tetraacetic acid promotes plant Cd uptake.

The translocation factor (TF) is the ratio between the concentration of metal ions accumulated in the plant shoot to that in the plant root. Ideally, a hyperaccumulator plant should have a TF value greater than one [47]. A TF value larger than one indicates that the heavy metals absorbed by the plants have been translocated effectively to the aerial parts of the plant [31]. In contrast, a TF value lesser than one implies that the heavy metals tend to accumulate and store in the plant's roots with less translocation to the aerial parts. A low TF value might be due to the exclusion strategy and restriction of metal movement towards the plant's aerial parts [47].

Generally, the TF value increased with an increased contamination level in the tributary. The translocation mechanism is crucial for the plant as it could prevent the excessive accumulation of toxic metal ions in the plant's roots [42]. The detoxification of metal ions might happen within the leaves of the plants through evapotranspiration. Evapotranspiration is an evaporation process of water from the plant leaves, promoting the absorption of nutrients and other substances from the medium into the plant's roots. Meantime, it accounts for the movement of heavy metals into the plant's shoots [13].

Another study reported on the metal distribution within the shoots and roots of plants, indicated by the root/shoot (R/S) ratio. The R/S ratio implies the metal concentration accumulated in the plant root over the shoot. For exemplification, approximately 80% of Cr, Cu, Fe, and Ni accumulated in the plant root with an R/S ratio equal to or greater than 6, while Fe has an R/S ratio greater than 17 [33]. Concretely, plant roots are the final destination of the absorbed metals since the roots can concentrate a greater amount of metal ions than their shoots. However, hyperaccumulator plants should have a shoot-to-root ratio of greater than one, reflecting the effective transportation of metals from the plant's roots to the harvestable parts of plants. Nevertheless, non-accumulator plants have a shoot-to-root ratio much smaller than one [13].

5.3. Phytotoxicity of Heavy Metals in Plants

Undesirable effects on the plant's growth and development were observed due to the accumulation of toxic metals in their roots, stems, and leaves. Bioactive metals could be classified into two groups: redox-active and non-redox-active metals. Redox-active metals, such as Cr, Cu, Mn, and Fe, could directly disrupt the plant cell homeostasis, break DNA strands, defragment proteins or cell membranes, destroy photosynthesis pigments, and cause cell death. Opposingly, non-redox active metals could impose oxidative stress on plants [50]. Moreover, Kumar, Singh, and Chopra [31] reported that phytoremediation using water lettuce to remediate sugar mill effluent containing Cd, Cu, Cr, Fe, Pb, Mn, and Zn, induced the yellowing of the plant's leaves as well as chlorosis and necrosis.

The phytotoxic responses of various plants to heavy metals are presented in Table 3. For example, Mishra and Tripathi [34] reported that the exposure of water hyacinth to Cr ions at 10.0 and 20.0 mg/L concentrations could result in the yellowing of plant leaves, chlorosis, and root exfoliating. In addition, the chlorophyll, protein, and sugar content in the plants were found to reduce along with the escalating metal concentration and exposure time. They also found that Cr demonstrated a higher degree of metal toxicity as compared with Zn. In addition, Hasan, Talat, and Rai [51] revealed that Cd was more toxic than Zn. The increase in Cd concentration affected the relative growth rate and demonstrated a negative growth rate when the Cd concentration in the growth culture medium was 4.0 ppm and above. A similar declining plant growth trend was observed as increasing Zn concentration from 2.0 to 12.0 ppm but without showing a negative growth rate.

Table 3. Phytotoxicity of heavy metals in plants.

Plants	Heavy Metals	Concentration	Experimental Layout and Duration	Phytotoxic Responses	References
Water hyacinth	Cr	10.0 to 20.0 mg/L	15 L experimental tanks filled with 10 L of tap water and investigated up to day 11	Yellowing of leaves, leaf chlorosis, and growth retardation.	[34]
	Zn	2.0 to 12.0 ppm	2 L container filled with 1 L tap water and investigated up to day 16	Growth reduction, leaf chlorosis, metabolism disruption.	[51]
	Cd	1.0 to 4.0 ppm		Growth reduction, growth retardation, new root growth inhibition, root function disruption, leaf chlorosis.	
Duckweed	Cd Cu	>10 mM >50 μM	10 L plastic reactors with 5 L of lake water and investigated up to day 15	Pigment degradation and photosynthesis restriction.	[52]
Water lettuce	Pb	1 to 2 mmol/L	60 L PE containers filled with 10 L of Hoagland nutrient solution and investigated up to day 8	Chlorophyll synthesis inhibition, chlorophyll reduction, loss of photosynthesis activity.	[48]
	Ni	1.0 and 10.0 ppm	Unknown size for hydroponic tubs filled with 10% Hoagland's solution and investigated up to day 6	Plant wilting, chlorosis in leaves, chlorophyll reduction, carotenoid reduction, water loss, browning of root tips, and root damage.	[53]

At a Cd concentration of 3.3 ppm, the metal toxicity led to the retardation of plant growth by hindering the growth of new roots and disrupting the function of the roots. The leaf chlorosis was fast, implying the decaying of plant tissue due to acute metal toxicity. This could eventually hinder the metabolism of plants [51]. Furthermore, exposure to excessive Cr concentration could result in a loss of photosynthesis pigments, protein, and sugar in plants. For instance, the presence of Cr in duckweed could result in a slower growth rate due to the restriction in photosynthesis [52]. Kumar, Singh, and Chopra [31] also found that higher heavy metal concentrations in wastewater could restrict aquatic plant growth and limit plant metabolism and physiological processes.

Nevertheless, the exposure of water hyacinth to Zn could cause oxidative impairments and alter the metalloenzymes of the plant. Moreover, the loss of chlorophyll could interfere with photosynthesis because of the interrupted chloroplasts. The reduction of sugar might slow down photochemical activities and chlorophyll initiation. Eventually, the loss of protein content resulting from the production of protein complexes might impede enzymatic activity [34]. The study presented by Buta et al. [49] suggested that the chlorophyll contents declined after six days of exposure to multi-metallic systems. Generally, the carotenoid content in plants decreased in all plants. For water lettuce, the uptake of Zn and Cu could restrict the biosynthesis of chlorophyll and carotenoids, resulting in an obvious discoloration of the plant leaves [54].

6. Phytoremediation Parameters and Kinetic Studies

There are several influencing factors that enhance or inhibit the performance of phytoremediation of heavy metals, as shown in Table 4. The selection of appropriate types of plants is the key to the success of phytoremediation technology; additionally, the factors such as solution pH, solution temperature, exposure duration, water salinity, initial metal concentration and chelating agents' concentration can also directly influence phytoremediation efficiency.

Table 4. Factors affecting on the phytoremediation performance in previous studies.

Plant Species	Heavy Metals	Influence/Enhancement Factor and Details	Significant Results	Reference
Thlaspi caerulescens	Cd, Zn	pH	The soluble metal form of both Cd and Zn was greatly increased with decreasing pH.	[55]
Eichhornia crassipes	Cd, As, Pb, Zn, and Cu	Temperature	The ideal water temperature for growth is between 28 °C and 30 °C. Temperatures exceeding 33 °C stifle further development.	[45]
Elodea canadensis, Potamogeton natans	Cu, Zn, Cd, Pb		The metal concentration and accumulations increased with increasing water temperature.	[56]
		Salinity	The metal concentration increased with decreasing salinity.	[56]
Eichchornia crassipes	Zn, Cd	Exposure duration	The overall metal uptake by the plant increased with the duration of the exposure time.	[51]
		Initial solution concentration	The uptake of heavy metals increased with an increase in the initial solution concentration.	[51]
Sasa argenteostriata (Regel) E.G. Camus	Pb	Chelating agents (Ethylenediaminetetraacetic acid (EDTA) and nitrilotriacetic acid (NTA))	The combined application of EDTA and NTA brought the accumulation of Pb availability to a more reasonable level than EDTA alone.	[57]
Zea mays L.	Cd	Chelating agents (ethylenediamine tetraacetic acid (EDTA), diethylenetriacetic acid (NTA), tetrasodium N, N-diacetate (GLDA), aspartate dibutyric acid ether (AES), and iminodisuccinic acid (IDSA))	Total Cd extraction followed the order AES (6 mmol kg^{-1}) > GLDA > NTA > EDTA > IDSA (3 mmol kg^{-1})	[58]

6.1. Effect of Solution pH

The uptake of metal ions by aquatic macrophytes is reliant on the solution's pH. According to Obinnaa and Ebere [3], the metal uptake was usually higher at a lower solution pH of about 4, thus reducing the metal concentration in the external culture medium. Notably, the pH of the medium would alter the metal speciation and metal bioavailability [43]. In low pH or acidic environments, most heavy metal ions exist as free positively charged species because of the higher H^+ concentration in the water, implying that more metals are soluble and bioavailable to biota. Therefore, plants could absorb the heavy metal ions easily, resulting in higher metal uptake. According to Soltan and Rashed [59], the pH of the water medium decreased with escalating metal concentrations from 50 to 100 mg^{-1} owing to the ionic exchange potential and the discharge of proton from the water hyacinth root while accumulating the metal ions.

Based on the study conducted by Singh, Gupta, and Tiwari [60], the author suggested that plants showed a better accumulation of Pb^{2+} at pH 6 than that at pH 9. As evidence, 89% metal removal was attained at pH 6, while only 56% metal removal was achieved at pH 9. Different findings were reported by Uysal and Taner [61], in which the highest Pb^{2+}

uptake by plants occurred at pH 4.5, followed by a decreasing metal accumulation within pH ranges from 4.5 to 6 and a constant uptake rate within pH 6 to 8.

In addition to that, the pH level affects the growth of plants. Generally, the plant cytoplasm environment was best maintained at pH 7 to ensure optimal plant growth and survival. In addition, Hardy and Raber [62] found that the Cd^{2+} uptake by plants increased within the pH ranges between 2 and 4. At pH 2, the acidic growth environment caused the blenching of the plant's roots and plant death, inhibiting metal uptake. In other words, the heavy metal uptake by plants reduced significantly when the pH decreased from 4 to 2 due to the fact that there were fewer anionic sites available for the ion exchange and more competitive metal binding between the protons and the metal ions to the plant's cell walls [63]. However, it was noteworthy that the presence of other contaminants in the medium could affect the metal-uptake efficiency [64]. For example, a solution with pH ranges between 6 and 9 might only be feasible for the remediation of wastewater without heavy metal contamination.

6.2. *Effect of Solution Temperature*

Solution temperature is another crucial ecological factor affecting the performance of metal uptake by aquatic macrophytes. The uptake of most metal ions by plants relies upon the temperature of the medium. This is because the change in temperature might influence the solubility and kinetic energy of the metal ions [65]. Based on the findings presented by Singh, Gupta, and Tiwari [60], the removal percentage of Pb^{2+} by plants increased by 22% when raising the temperatures from 20 °C to 28 °C. This finding was in agreement with Rai [66], illustrating an increasing metal uptake trend by plants with increasing temperatures. Uysal and Taner [61] also revealed that the Pb^{2+} accumulation by plants was the highest at 30 °C and the lowest at 15 °C. However, at a temperature beyond 30 °C, the metal uptake reduced again. The sudden decrease in the metal accumulation by plants might be due to the stress effects imposed on plants at the higher temperature of 35 °C, thus lowering the metal-uptake efficiency.

Similarly, Giri [67] reported that the removal of As and Cr metal ions by plants gradually decreased when increasing the temperature from 25 °C to 45 °C. The author suggested that the fast absorption rate and maximum metal removal by plants had taken place at 25 °C, which might be owing to the wide availability of metal binding sites on the plant roots during the initial ion exchange process. Hence, it was deduced that the metal absorption processes of plants were regulated by an exothermic process. Additionally, Rakhshaee, Khosravi, and Ganji [63] revealed that the metal uptake by plants increased with increasing temperatures from 10 to 25 °C.

Apart from affecting the metal-uptake efficiency, the solution's temperature influences the growth of plants. The behavior of aquatic accumulators varies depending on the temperature. Temperatures between 20 °C and 30 °C could result in the optimal cultivation of most aquatic plants. Conversely, a temperature equivalent to or lower than 10 °C could hinder the metabolic activities of most aquatic plants [64]. Additionally, it would hinder the growth of plants and inactivate microbial activity, leading to a low metal removal efficiency by plants [65]. Instead, a minimum temperature of 15 °C should be maintained to ensure optimal pollutant removal by microbes. The study also suggested that the optimal water temperature for the growth of water hyacinth was between 28 and 30 °C, while the optimum air temperature was between 21 and 30 °C. Nevertheless, at greater than 33 °C, it would suppress the successive growth of the plant [45]. Unfavorable culture medium temperatures restricted the growth of plants and caused plants to cease the uptake. As a result, the plants were incapable of accumulating the metals [68]. However, in some cases, plants can grow at colder temperatures.

6.3. *Effect of Exposure Duration*

Various exposure durations of aquatic plants to the metal concentrations in the culture medium resulted in various metal uptake performances. Lu et al. [33] reported that the total metal accumulation in the roots and shoots of the plant generally increased with increasing exposure durations. In addition, Soltan and Rashed [59] discovered that the plants cultivated in 100 mg/L of metal solution portrayed a declining metal uptake trend at increasing exposure durations due to the wilting of the plants resulting from the high toxicity of the metal accumulated in the plant tissue. Consequently, the metal uptake by plants via diffusion and osmosis reduced significantly with increasing exposure times.

Furthermore, Hardy and Raber [62] found that the Cd^{2+} uptake rate by plants was fast at the first 4 h but decreased linearly for the subsequent 72 h, implying that the percentage of metal uptake declined with increasing exposure durations. The trend of Cd^{2+} and Zn^{2+} absorption by the plants as the function of exposure time at various concentrations has been studied by [51]. The uptake of Cd^{2+} by the plants took place in two stages at higher metal concentrations of 4.0 and 6.0 ppm. A greater uptake efficiency was observed during the second stage, from the 6th to the 16th day, implying that the metal uptake rate increased with increasing exposure times. However, at lower Cd^{2+} concentrations of 1.0 and 2.0 ppm, the metal-uptake rate reduced with increasing exposure duration. The Zn^{2+} uptake trend by plants only showed a single stage of biphasic at any exposure concentration.

6.4. *Effect of Water Salinity*

Another crucial parameter affecting the metal uptake by plants is water salinity. The salt concentration of water affects the growth and reproductive potential of aquatic plants. Different plants had varying degrees of salinity tolerance, which regulates the capability of the plants to remove pollutants from the water environment. It revealed that floating macrophytes, such as water hyacinth and water lettuce, were likely influenced by the low water salinity at about 2.50% [69]. In the case of high salt concentrations in the water, it slowed down the transpiration rates and reduced the total dry weight of the plants [64]. Moreover, high water salinity might induce the complexation of metal–chloride, making the metal uptake process more complex, hence reducing the metal uptake by plants [43]. Correspondingly, the aquatic plants would die owing to the decreased osmotic potential levels as the water molecules had a lower potential to flow from a less solute region to a high solute region.

6.5. *Effect of Initial Metal Concentration*

The initial concentration of metal in the culture medium would also manipulate the metal-uptake efficiency of the plant's roots and leaves. The heavy metal uptake by plants usually increases with increased initial metal concentrations [59]. A similar observation was reported by Lu et al. [33], suggesting that the removal capacity of metal was higher when the aquatic plants were cultivated in wastewater with higher metal contamination levels. For instance, Pb^{2+} accumulated in the roots and leaves of plants increased with the escalating Pb^{2+} concentration in the growth medium [48].

Moreover, Uysal and Taner [61] found that the amount of metal accumulated in the plants increased with escalating initial metal concentrations ranging between 0 and 50 mg/L. However, it decreased when increasing the metal concentration from 50 to 100 mg/L. At higher initial metal concentrations, the plants wilted due to the metal toxicity imposed on the plant tissues. The decreased metal accumulation might also be attributable to the transpiration of the metal ions in the roots to the surrounding solution, imposing adverse effects on the survival of the plants. The common phytotoxicity effect on the plants was truncated plant growth resulting from hindered photosynthesis [59]. Furthermore, the higher concentrations of metal ions in the medium imposed inhibitory consequences on plant metabolism, alternately minimize plant growth, cause leaf necrosis, and destroy the plant's physiological systems [67]. In general, the high correlation coefficient of 0.9801 confirmed the positive relationship between the metal uptake by plants and the initial metal concentration in the culture medium.

The metal concentrations in the sediment and the metal accumulated in the roots of the plant reflected a positive linear correlation for most heavy metals, such as Co, Mn, Ni, and Sn, with R^2 values of 0.3559, 0.4216, and 0.7616, respectively. This implied that the plant could accumulate and remove more heavy metals at higher contamination levels [42]. However, the linear correlation between the metal concentration in the sediment and that accumulated in the plant's roots was weaker for the accumulation of Cu and Pb, with R^2 values of 0.3338 and 0.3011, respectively. For instance, when the plant was treated with escalating Pb concentrations from 30 to 50 mg/L, it demonstrated a declining metal accumulation rate [60].

Nevertheless, the accumulation of Cd, Sb, and Zn in the plant's roots became constant with escalating concentrations of metals in the medium upon reaching its absorption limit. This was because the plant might restrict the metal uptake by immobilizing or activating selective barriers in the plasma membrane when reaching the uptake limit of metal accumulation [42]. A similar finding was reported by Greger [43], where the metal uptake was not linearly correlated to the increasing initial metal concentration. The over-accumulation of metal ions in the plant tissue saturated the limited binding sites, subsequently reducing the metal removal rate. Therefore, it could be inferred that the metal uptake from the soil, sediment, and water was the greatest at lower external metal concentrations due to the less competitive ionic exchange process.

6.6. *Effect of Other Metals Concentration*

The presence of other metals in the culture medium also affected the metal uptake by aquatic plants because of the metal binding competition at the plant cell wall [43]. For example, the absorption of Cd^{2+} by the plant's roots declined when other cations of escalating ionic radii or valency are present. Specifically, the Cd^{2+} uptake by plants declined when increasing the concentration of Zn^{2+} in the medium [62]. Apart from that, a slower rate of metal uptake in the multi-metallic system than the single metal system was portrayed due to the competition between Cd^{2+} and Zn^{2+} metals for the similar metal exchange sites of plants, limiting the metal-uptake efficiency during metabolism. Moreover, Zn^{2+} could protect against Cd^{2+}, giving rise to the loss of potassium ions at the plant-membrane level [51].

6.7. *Effect of Chelating Agent Addition*

Another crucial factor affecting phytoremediation is the bioavailability of metals in the soil that facilitates the transportation of metal ions within the plant [14]. Most metals were not readily bioavailable to the plant due to their high binding abilities to the soil [70]. According to Prasad [71], heavy metals or metalloids, such as As, Cd, Cu, Ni, Se, and Zn, were more bioavailable to plants. In contrast, Cr and Pb were the least bioavailable metals, implying that these metals were more resistant to phytoextraction. Hence, the addition of a chelating agent became an effective way to enhance the metal bioavailability in the soil and improve the metal uptake by the plants. Upon the introduction of a chelating agent to the soil, it formed water-soluble metal–chelant complexes, which were further taken up by the plants via the apoplastic route [72]. The formation of complexes, in turn, restricted the precipitation of the heavy metal, increased their mobility, enhanced the bioavailability of the heavy metals, and promoted metal desorption [73].

Chelating agents could be generally classified into synthetic and organic types. Synthetic chelating agents include ethylenediamine tetraacetic acid (EDTA), diethylene triamine pentaacetic acid, and ethylene glycol tetra-acetic acid [14]. According to Dhaliwal et al. [74], the addition of a chelating agent to the soil improved the phytoextraction of Cd^{2+} uptake by plants. The results showed that the metal uptake was increased by about 15% with increasing amounts of EDTA in the soil from 1 to 2 mg/kg levels. This might be attributed to the higher bioavailability and mobility of the metal ions in the soil, thus boosting the metal uptake and translocation by the plants. Additionally, a dosage of EDTA at 2.7 mmol/kg enhanced phytoextraction [75].

EDTA was regarded as an efficient chelate, yet it had unfavorable toxic effects on plants, soils, and ecosystems, which might introduce risks to the environment [76]. Accordingly, natural, biodegradable, non-toxic, and environmental-friendly organic chelators such as citric, acetic, oxalic, and malic acids have been proposed to overcome the toxic effects that resulted from the use of EDTA. Shinta, Zaman, and Sumiyati [75] suggested that citric acid was more effective than EDTA as it acidified and lowered the pH level of the soil, created a microbial community in the soil, and promoted the growth of plant roots. As a result, the plants showed higher metal absorption due to faster growth. On the other hand, Souza et al. [15] recommended that organic chelating agents might be employed during the plant harvesting process to enhance the metal desorption and metal bioavailability in soil.

6.8. Kinetics of Phytoremediation

The kinetic model of phytoremediation is important to determine the efficiency, effectiveness, and natural behavior of aquatic plants during heavy metal removal from water or soil [77]. In addition, the kinetic model is useful for investigating the mass transfer rate of metals from a medium to plant tissues. On top of that, the kinetic study provides insightful information regarding the design and optimization of biological treatment technology at a large scale. Based on the findings reported by Naaz [78], the bioaccumulation kinetics of the heavy metals in the entire plant of water hyacinth were investigated. The results showed that the experimental data could fit into a linearized first-order kinetic equation with minimal adjustments, as presented in Equation (1). Moreover, Ingole and Bhole [79] proposed linearized first-order kinetics for heavy metal removal by the plant, as shown in Equation (2). Notably, the uptake rate constant (k) for the plants is an important parameter for evaluating the metal uptake performance by plants. The linear relationship observed from the plot of log (C_t) versus time (t) confirmed the first-order behavior of heavy metals uptake by plants.

First-order kinetic equation with slight adjustment:

$$\log(C_t) = -\frac{k}{2.303}t + \log(C_0) \tag{1}$$

First-order kinetic equation:

$$\log(C_t) = -kt + \log(C_0) \tag{2}$$

where

C_0 = initial concentration of metal in water, mg/L
C_t = concentration of metal in water at time t, mg/L
k = first-order uptake rate constant, day^{-1}
t = sampling time, days

The kinetic parameters of several heavy metal removals by accumulator plants are summarized in Tables 5 and 6. According to Singh et al. [77], the first-order kinetics of heavy metals uptake by the plants demonstrated the best-fit results with a high determination coefficient (R^2) greater than 0.82 and a rate constant larger than 0.023 mg/L·day. Similarly, Ingole and Bhole [79] found that the heavy metal uptake by the plant fitted well to the first-order behavior. The plant demonstrated the highest uptake rate of 0.1027 day^{-1} during the removal of Pb compared to other metals such as Ni, Hg, Zn, As, and Cr. Overall, it attained high R^2 values larger than 0.789, confirming the fitness to the straight-line plot of log (C_t) and t. Apart from that, Rakhshaee, Khosravi, and Ganji [63] revealed that the removal of heavy metals by living plants corresponded to the first-order kinetic, following the descending sequence of the first-order kinetic constant: $Zn^{2+} > Ni^{2+} > Pb^{2+} > Cd^{2+}$. The highest removal rate of 0.94 min^{-1} was attained for the removal of Zn^{2+} by the living plant, while the lowest of 0.118 min^{-1} was achieved for the removal of Cd^{2+}.

Table 5. Summary of kinetic model and equilibrium isotherm model.

Plant	Heavy Metal	Research Highlight	References
Dendrocalamus asper	Cu	The removal rate of Cu from the contaminated source had an order of 2.71 and a kinetic constant of 0.0013 $ppm^{-1.71}$ day^{-1}	[80]
Bambusa merilliana, *Bambusa blumeana*, *Dendrocalamus asper*	Cu	The zero-order model has well described the uptake of metal ions per mass of plant with a correlation value R^2 of 0.954 and a rate constant of 3.136 mg/(kg·day).	[81]
Eichhornia sp., *Pistia* sp.	Cr	Pseudo-first-order (0.910) and pseudo-second model (0.665) are more suitable for bioaccumulation kinetic in *Pistia* sp. rather than *Eichhornia* sp.	[82]
Eichhornia crassipes, *Lemma valdniana*	As	Pseudo-first-order gave a good correlation for both plants, with a correlation value R^2 > 0.8 for all the concentrations involved.	[83]

Table 6. Kinetic parameters of heavy metals removal by plants.

Heavy Metal	k (day^{-1})	R^2	References
Cd	0.0625	0.930	[77]
Cu	0.0700	0.890	
Fe	0.0800	0.920	
Mn	0.0825	0.870	
Pb	0.0575	0.980	
Zn	0.0875	0.890	
As	0.0693	0.825	[79]
Cr	0.0548	0.968	
Hg	0.0879	0.885	
Ni	0.0937	0.950	
Pb	0.1027	0.789	
Zn	0.0749	0.990	

According to Kamalu et al. [84], Richards's pseudo-first-order (PFO) and pseudo-second-order (PSO) models had been adopted through the verification with experimental results. The kinetic model of a common plant hyperaccumulator was established by studying the pathways starting from its rhizosphere to the atmosphere via the stem. By solving the two systems of phloem and xylem ordinary differential equations for the upward and downward transportation of the metal through the plant xylem and the phloem, the kinetic models for both PFO and PSO were developed, as illustrated in Equations (3) and (4), respectively. By deriving Equation (4), the PSO kinetic equation generated a dumb-bell shape profile and eventually optimized the model, as displayed in Equation (5).

PFO kinetic model:

$$q = q_m - (q_m - q_0)e^{k_1(t_0 - t)} \tag{3}$$

PSO kinetic model:

$$q = \frac{q_0 - q_m(q_m - q_0)e^{k_2(t_0-t)}}{1 - (q_m - q_0)e^{k_2(t_0-t)}} \tag{4}$$

Derivation of PSO kinetic model:

$$D_q = \frac{{k_2}^2(q_m - q_0)^2 e^{k_2(t_0-t)}}{1 - (q_m - q_0)e^{k_2(t_0-t)}} \tag{5}$$

where

q = metal concentration at time t, mg/L

q_m = maximum concentration of absorbed metal, mg/L

q_0 = initial metal concentration, mg/L

k_1 = PFO kinetic rate constant, day^{-1}

k_2 = PSO kinetic rate constant, mg/L·day
t = sampling time, day
t_0 = initial sampling time, day

The results showed that the phytoremediation process followed the PSO relationship of Richard's model, achieving high R^2 values ranging between 0.9979 and 0.9991, implying that the prediction obtained from the model were highly consistent with the experimental data. Opposingly, the phytoremediation process showed a low degree of compatibility with the PFO kinetic model [84]. Hence, it can be inferred that the natural phenomenal process of phytoremediation demonstrated a sigmoidal profile. Moreover, the concentration of metal uptake by the plants via the xylem tissue generally decreased with time. On the other hand, the uptake of heavy metals via the phloem exhibited an increasing trend or a free-fall profile along with time, which means that the sigmoidal profile might set in at longer exposure times due to its natural behavior.

7. Conclusions

Water plays an irreplaceable role in sustaining the life of living beings, including humans, animals, and plants. Increasing water contamination with heavy metals has become a serious concern nowadays due to rapid industrial development, growing population, and frequent anthropogenic activities. Therefore, phytoremediation using floating plants is regarded as a promising green technology to remediate the heavy-metal-contaminated water in an environmentally-friendly and cost-effective way. This research study investigated various process parameters affecting the phytoremediation behavior by aquatic macrophytes, such as the solution pH, solution temperature, exposure duration, water salinity, initial metal concentration, presence of other metals concentration, and the addition of chelating agent. The findings revealed that the optimal solution pH and temperature were pH 4 and 30 °C. In addition, it was found that the metal uptake by aquatic macrophytes generally enhanced with increasing temperature, exposure duration, and initial metal concentration in the growth culture medium. Additionally, the metal uptake was enhanced with the addition of chelating agents on the soil. However, it decreased with increasing water salinity and the presence of other metals concentration.

The kinetics of phytoremediation were studied to determine the effectiveness and natural behavior of plants during heavy metal uptake. The finding revealed that aquatic plants obeyed the first-order kinetic model, illustrating a linear relationship between the metal uptake concentration with time. On the other hand, some plants demonstrated good fitness to the PSO of Richard's model, reflecting a natural sigmoidal profile of heavy metal uptake by plants. Apart from that, various post-harvested biomass disposal methods were studied. These disposal methods included composting, compaction, direct disposal, leaching, pyrolysis, incineration, and nanoparticle synthesis. Each method had different working principles and demonstrated various performances and limitations. Among these methods, incineration was regarded as the most feasible technology due to its ability to significantly reduce biomass volume, save transportation costs, and decompose almost all the organic matter in the contaminated plant biomass.

In a nutshell, all of the research objectives were attained. The aquatic plants, especially water hyacinth, demonstrated great efficiency in absorbing various metal ions. Most of the metal absorption processes by aquatic plants followed the first-order kinetic model. In short, phytoremediation is a promising green water remediation technique. However, further research and studies are required to enhance its feasibility and practicability at large-scale implementation.

Author Contributions: Resources, Y.L.P.; writing—review and editing, Y.L.P.; writing—original draft, Y.Y.Q.; data curation, S.H.S.; supervision and visualization, S.L. All authors have read and agreed to the published version of the manuscript.

Funding: This research was supported in part with Ministry of Education (MOE) Malaysia that provided the Fundamental Research Grant Scheme (FRGS/1/2022/TK05/UTAR/02/34), Kurita Asia Research Grant (21Pmy076) provided by Kurita Water and Environment Foundation and Universiti Tunku Abdul Rahman (UTAR) Research Fund (IPSR/RMC/UTARRF/2020-C2/P01).

Institutional Review Board Statement: Not applicable.

Informed Consent Statement: Not applicable.

Data Availability Statement: The data presented in this study are available upon request from the corresponding author.

Conflicts of Interest: The authors declare no conflict of interest.

References

1. Ali, S.; Abbas, Z.; Rizwan, M.; Zaheer, I.E.; Yavaş, İ.; Ünay, A.; Abdel-DAIM, M.M.; Bin-Jumah, M.; Hasanuzzaman, M.; Kalderis, D. Application of floating aquatic plants in phytoremediation of heavy metals polluted water: A review. *Sustainability* **2020**, *12*, 1927. [CrossRef]
2. Cauberghe, V.; Vazquez-Casaubon, E.; Van de Sompel, D. Perceptions of water as commodity or uniqueness? The role of water value, scarcity concern and moral obligation on conservation behavior. *J. Environ. Manag.* **2021**, *292*, 112677. [CrossRef]
3. Obinhaa, I.B.; Ebere, E.C. A Review: Water pollution by heavy metal and organic pollutants: Brief review of sources, effects and progress on remediation with aquatic plants. *Anal. Methods Environ. Chem. J.* **2019**, *2*, 5–38. [CrossRef]
4. Afroz, R.; Masud, M.M.; Akhtar, R.; Duasa, J.B. Water pollution: Challenges and future direction for water resource management policies in malaysia. *Environ. Urban. ASIA* **2014**, *5*, 63–81. [CrossRef]
5. Fletcher, J.; Willby, N.; Oliver, D.M.; Quilliam, R.S. Phytoremediation using aquatic plants. In *Phytoremediation: In-Situ Applications*; Shmaefsky, B.R., Ed.; Springer International Publishing: Cham, Switzerland, 2020; pp. 205–260.
6. Gupta, A. *Water Pollution—Sources, Effects and Control*; Pointer Publishers: Jaipur, India, 2016.
7. Anawar, H.; Chowdhury, R. Remediation of polluted river water by biological, chemical, ecological and engineering processes. *Sustainability* **2020**, *12*, 7017. [CrossRef]
8. Rezania, S.; Ponraj, M.; Talaiekhozani, A.; Mohamad, S.E.; Md Din, M.F.; Taib, S.M.; Sabbagh, F.; Sairan, F.M. Perspectives of phytoremediation using water hyacinth for removal of heavy metals, organic and inorganic pollutants in wastewater. *J. Environ. Manag.* **2015**, *163*, 125–133. [CrossRef]
9. Prasad, M.N.V. Prospects for manipulation of molecular mechanisms and transgenic approaches in aquatic macrophytes for remediation of toxic metals and metalloids in wastewaters. In *Transgenic Plant Technology for Remediation of Toxic Metals and Metalloids*; Elsevier: Amsterdam, The Netherlands, 2018; pp. 395–428.
10. Sarwar, N.; Imran, M.; Shaheen, M.R.; Ishaque, W.; Kamran, M.A.; Matloob, A.; Rehim, A.; Hussain, S. Phytoremediation strategies for soils contaminated with heavy metals: Modifications and future perspectives. *Chemosphere* **2017**, *171*, 710–721. [CrossRef]
11. Rahman, S.U.; Yasin, G.; Nawaz, M.F.; Cheng, H.; Azhar, M.F.; Riaz, L.; Javed, A.; Lu, Y. Evaluation of heavy metal phytoremediation potential of six tree species of Faisalabad city of Pakistan during summer and winter seasons. *J. Environ. Manag.* **2022**, *320*, 115801. [CrossRef]
12. Yasin, G.; Ur Rahman, S.; Yousaf, M.T.B.; Azhar, M.F.; Zahid, D.M.; Imtiaz, M.; Hussain, B. Phytoremediation potential of E. camaldulensis and M. alba for copper, cadmium, and lead absorption in urban areas of Faisalabad City, Pakistan. *Int. J. Environ. Res.* **2021**, *15*, 597–612. [CrossRef]
13. Tangahu, B.V.; Sheikh Abdullah, S.R.; Basri, H.; Idris, M.; Anuar, N.; Mukhlisin, M. A review on heavy metals (As, Pb, and Hg) uptake by plants through phytoremediation. *Int. J. Chem. Eng.* **2011**, *2011*, 939161. [CrossRef]
14. Laghlimi, M.; Baghdad, B.; Hadi, H.E.; Bouabdli, A. Phytoremediation mechanisms of heavy metal contaminated soils: A review. *Open J. Ecol.* **2015**, *5*, 14. [CrossRef]
15. Souza, L.A.; Piotto, F.A.; Nogueirol, R.C.; Azevedo, R.A. Use of non-hyperaccumulator plant species for the phytoextraction of heavy metals using chelating agents. *Sci. Agric.* **2013**, *70*, 290–295. [CrossRef]
16. Shaari, N.E.M.; Tajudin, M.T.F.M.; Khandaker, M.M.; Majrashi, A.; Alenazi, M.M.; Abdullahi, U.A.; Mohd, K.S. Cadmium toxicity symptoms and uptake mechanism in plants: A review. *Braz. J. Biol.* **2022**, *84*, e252143. [CrossRef]
17. Lasat, M.M. Phytoextraction of toxic metals: A review of biological mechanisms. *J. Environ. Qual.* **2002**, *31*, 109–120. [CrossRef]
18. Verma, R.K.; Sankhla, M.S.; Jadhav, E.B.; Parihar, K.; Awasthi, K.K. Phytoremediation of heavy metals extracted from soil and aquatic environments: Current advances as well as emerging trends. *Biointerface Res. Appl. Chem.* **2022**, *12*, 5486–5509.
19. Ansari, A.A.; Naeem, M.; Gill, S.S.; AlZuaibr, F.M. Phytoremediation of contaminated waters: An eco-friendly technology based on aquatic macrophytes application. *Egypt. J. Aquat. Res.* **2020**, *46*, 371–376. [CrossRef]

20. Sharma, S.; Tiwari, S.; Hasan, A.; Saxena, V.; Pandey, L.M. Recent advances in conventional and contemporary methods for remediation of heavy metal-contaminated soils. *3 Biotech* **2018**, *8*, 216. [CrossRef]
21. Lakshmi, K.S.; Sailaja, V.H.; Reddy, M.A. Phytoremediation—A promising technique in waste water treatment. *Int. J. Sci. Res. Manag.* **2017**, *5*, 5480–5489. [CrossRef]
22. Dixit, A.M.; Dixit, S.; Goswami, C.S. Process and plants for wastewater remediation: A review. *Sci. Rev. Chem. Commun.* **2011**, *1*, 71–77.
23. Jeevanantham, S.; Saravanan, A.; Hemavathy, R.V.; Kumar, P.S.; Yaashikaa, P.R.; Yuvaraj, D. Removal of toxic pollutants from water environment by phytoremediation: A survey on application and future prospects. *Environ. Technol. Innov.* **2019**, *13*, 264–276. [CrossRef]
24. Shen, X.; Dai, M.; Yang, J.; Sun, L.; Tan, X.; Peng, C.; Ali, I.; Naz, I. A critical review on the phytoremediation of heavy metals from environment: Performance and challenges. *Chemosphere* **2022**, *291*, 132979. [CrossRef]
25. Gunarathne, V.; Gunatilake, S.R.; Wanasinghe, S.T.; Atugoda, T.; Wijekoon, P.; Biswas, J.K.; Vithanage, M. 7—Phytoremediation for E-waste contaminated sites. In *Handbook of Electronic Waste Management*; Prasad, M.N.V., Vithanage, M., Borthakur, A., Eds.; Butterworth-Heinemann: Oxford, UK, 2020; pp. 141–170.
26. Ahalya, N.; Ramachandra, T.V. *Phytoremediation: Processes and Mechanisms*; Academic Press: Cambridge, MA, USA, 2004; Volume 18, pp. 33–38.
27. Akpor, O.; Otohinoyi, D.; Olaolu, T.; Aderiye, J. Pollutants in wastewater effluents: Impacts and remediation processes. *Int. J. Environ. Res. Earth Sci.* **2014**, *3*, 50–59.
28. Phearkeo, O. A Study on Removal of Heavy Metals from Wastewater by Floating Plants. Master's Thesis, Master of Science (Engineering and Technology), Sirindhorn International Institute of Technology Thammasat University, Bangkok, Thailand, 2016.
29. Rezania, S.; Taib, S.M.; Md Din, M.F.; Dahalan, F.A.; Kamyab, H. Comprehensive review on phytotechnology: Heavy metals removal by diverse aquatic plants species from wastewater. *J. Hazard. Mater.* **2016**, *318*, 587–599. [CrossRef] [PubMed]
30. Ubuza, L.J.A.; Padero, P.C.S.; Nacalaban, C.M.N.; Tolentino, J.T.; Alcoran, D.C.; Tolentino, J.C.; Ido, A.L.; Mabayo, V.I.F.; Arazo, R.O. Assessment of the potential of duckweed (*Lemna minor* L.) in treating lead-contaminated water through phytoremediation in stationary and recirculated set-ups. *Environ. Eng. Res.* **2020**, *25*, 977–982. [CrossRef]
31. Kumar, V.; Singh, J.; Pathak, V.V.; Ahmad, S.; Kothari, R. Experimental and kinetics study for phytoremediation of sugar mill effluent using water lettuce (*Pistia stratiotes* L.) and its end use for biogas production. *3 Biotech* **2017**, *7*, 330. [CrossRef]
32. Hua, J.; Zhang, C.; Yin, Y.; Chen, R.; Wang, X. Phytoremediation potential of three aquatic macrophytes in manganese-contaminated water. *Water Environ. J.* **2012**, *26*, 335–342. [CrossRef]
33. Lu, Q.; He, Z.L.; Graetz, D.A.; Stoffella, P.J.; Yang, X. Uptake and distribution of metals by water lettuce (*Pistia stratiotes* L.). *Environ. Sci. Pollut. Res.* **2011**, *18*, 978–986. [CrossRef]
34. Mishra, V.K.; Tripathi, B.D. Accumulation of chromium and zinc from aqueous solutions using water hyacinth (*Eichhornia crassipes*). *J. Hazard. Mater.* **2009**, *164*, 1059–1063. [CrossRef]
35. Dhir, B.; Sharmila, P.; Pardha Saradhi, P. Photosynthetic performance of *Salvinia natans* exposed to chromium and zinc rich wastewater. *Braz. J. Plant Physiol.* **2008**, *20*, 61–70. [CrossRef]
36. Tufaner, F. Post-treatment of effluents from UASB reactor treating industrial wastewater sediment by constructed wetland. *Environ. Technol.* **2020**, *41*, 912–920. [CrossRef]
37. Basile, A.; Sorbo, S.; Conte, B.; Castaldo Cobianchi, R.; Trinchella, F.; Capasso, C.; Carginale, V. Toxicity, accumulation, and removal of heavy metals by three aquatic macrophytes. *Int. J. Phytoremediat.* **2012**, *14*, 374–387. [CrossRef] [PubMed]
38. Aurangzeb, N.; Nisa, S.; Bibi, Y.; Javed, F.; Hussain, F. Phytoremediation potential of aquatic herbs from steel foundry effluent. *Braz. J. Chem. Eng.* **2014**, *31*, 881–886. [CrossRef]
39. Abbas, Z.; Arooj, F.; Ali, S.; Zaheer, I.E.; Rizwan, M.; Riaz, M.A. Phytoremediation of landfill leachate waste contaminants through floating bed technique using water hyacinth and water lettuce. *Int. J. Phytoremediat.* **2019**, *21*, 1356–1367. [CrossRef] [PubMed]
40. Saha, P.; Shinde, O.; Sarkar, S. Phytoremediation of industrial mines wastewater using water hyacinth. *Int. J. Phytoremediat.* **2017**, *19*, 87–96. [CrossRef]
41. Bennicelli, R.; Stępniewska, Z.; Banach, A.; Szajnocha, K.; Ostrowski, J. The ability of *Azolla caroliniana* to remove heavy metals (Hg(II), Cr(III), Cr(VI)) from municipal waste water. *Chemosphere* **2004**, *55*, 141–146. [CrossRef]
42. Du, Y.; Wu, Q.; Kong, D.; Shi, Y.; Huang, X.; Luo, D.; Chen, Z.; Xiao, T.; Leung, J.Y.S. Accumulation and translocation of heavy metals in water hyacinth: Maximising the use of green resources to remediate sites impacted by e-waste recycling activities. *Ecol. Indic.* **2020**, *115*, 106384. [CrossRef]
43. Greger, M. Metal availability and bioconcentration in plants. In *Heavy Metal Stress in Plants: From Molecules to Ecosystems*; Prasad, M.N.V., Hagemeyer, J., Eds.; Springer: Berlin/Heidelberg, Germany, 1999; pp. 1–27.
44. Chaudhary, K.; Jan, S.; Khan, S. Chapter 23—Heavy metal ATPase (HMA2, HMA3, and HMA4) genes in hyperaccumulation mechanism of heavy metals. In *Plant Metal Interaction*; Ahmad, P., Ed.; Elsevier: Amsterdam, The Netherlands, 2016; pp. 545–556.
45. Huynh, A.T.; Chen, Y.-C.; Tran, B.N.T. A small-scale study on removal of heavy metals from contaminated water using water hyacinth. *Processes* **2021**, *9*, 1802. [CrossRef]
46. PN, M.; Madhu, G. Removal of heavy metals from waste water using water hyacinth. *ACEEE Int. J. Transp. Urban Dev.* **2011**, *1*, 48–52.

47. Das, S.; Goswami, S.; Talukdar, A.D. A study on cadmium phytoremediation potential of water lettuce, *Pistia stratiotes* L. *Bull. Environ. Contam. Toxicol.* **2014**, *92*, 169–174. [CrossRef]
48. Vesely, T.; Neuberg, M.; Trakal, L.; Szakova, J.; Tlustoa, P. Water lettuce *Pistia stratiotes* L. response to lead toxicity. *Water Air Soil Pollut.* **2012**, *223*, 1847–1859. [CrossRef]
49. Buta, E.; Paulette, L.; Mihăiescu, T.; Buta, M.; Cantor, M. The influence of heavy metals on growth and development of eichhornia crassipes species, cultivated in contaminated water. *Not. Bot. Horti Agrobot.* **2011**, *39*, 135–141. [CrossRef]
50. Emamverdian, A.; Ding, Y.; Mokhberdoran, F.; Xie, Y. Heavy metal stress and some mechanisms of plant defense response. *Sci. World J.* **2015**, *2015*, 756120. [CrossRef] [PubMed]
51. Hasan, S.H.; Talat, M.; Rai, S. Sorption of cadmium and zinc from aqueous solutions by water hyacinth (*Eichchornia crassipes*). *Bioresour. Technol.* **2007**, *98*, 918–928. [CrossRef] [PubMed]
52. Miretzky, P.; Saralegui, A.; Cirelli, A.F. Aquatic macrophytes potential for the simultaneous removal of heavy metals (Buenos Aires, Argentina). *Chemosphere* **2004**, *57*, 997–1005. [CrossRef]
53. Singh, K.; Pandey, S.N. Effect of nickel-stresses on uptake, pigments and antioxidative responses of water lettuce, *Pistia stratiotes* L. *J. Environ. Biol.* **2011**, *32*, 391–394.
54. Hegazy, A.K.; Kabiel, H.F.; Fawzy, M. Duckweed as heavy metal accumulator and pollution indicator in industrial wastewater ponds. *Desalinat. Water Treat.* **2009**, *12*, 400–406. [CrossRef]
55. Wang, A.S.; Angle, J.S.; Chaney, R.L.; Delorme, T.A.; Reeves, R.D. Soil pH effects on uptake of Cd and Zn by *Thlaspi caerulescens*. *Plant Soil* **2006**, *281*, 325–337. [CrossRef]
56. Fritioff, Å.; Kautsky, L.; Greger, M. Influence of temperature and salinity on heavy metal uptake by submersed plants. *Environ. Pollut.* **2005**, *133*, 265–274. [CrossRef]
57. Yang, Y.; Liao, J.; Chen, Y.; Tian, Y.; Chen, Q.; Gao, S.; Luo, Z.; Yu, X.; Lei, T.; Jiang, M. Efficiency of heterogeneous chelating agents on the phytoremediation potential and growth of *Sasa argenteostriata* (Regel) E.G. Camus on Pb-contaminated soil. *Ecotoxicol. Environ. Saf.* **2022**, *238*, 113603. [CrossRef]
58. Yang, Q.; Yang, C.; Yu, H.; Zhao, Z.; Bai, Z. The addition of degradable chelating agents enhances maize phytoremediation efficiency in Cd-contaminated soils. *Chemosphere* **2021**, *269*, 129373. [CrossRef]
59. Soltan, M.E.; Rashed, M.N. Laboratory study on the survival of water hyacinth under several conditions of heavy metal concentrations. *Adv. Environ. Res.* **2003**, *7*, 321–334. [CrossRef]
60. Singh, D.; Gupta, R.; Tiwari, A. Potential of duckweed (*Lemna minor*) for removal of lead from wastewater by phytoremediation. *J. Pharm. Res.* **2012**, *5*, 1578–1582.
61. Uysal, Y.; Taner, F. Effect of pH, temperature, and lead concentration on the bioremoval of lead from water using *Lemna minor*. *Int. J. Phytoremediat.* **2009**, *11*, 591–608. [CrossRef] [PubMed]
62. Hardy, J.K.; Raber, N.B. Zinc uptake by the water hyacinth: Effects of solution factors. *Chemosphere* **1985**, *14*, 1155–1166. [CrossRef]
63. Rakhshaee, R.; Khosravi, M.; Ganji, M.T. Kinetic modeling and thermodynamic study to remove Pb(II), Cd(II), Ni(II) and Zn(II) from aqueous solution using dead and living Azolla filiculoides. *J. Hazard. Mater.* **2006**, *134*, 120–129. [CrossRef]
64. Dijoo, Z.K.; Ali, R.; Hameed, M. Role of free-floating aquatic macrophytes in abatement of the disturbed environs. In *Bioremediation and Biotechnology, Volume 4: Techniques for Noxious Substances Remediation*; Bhat, R.A., Hakeem, K.R., Eds.; Springer International Publishing: Cham, Switerland, 2020; pp. 259–274.
65. Kumar, S.; Deswal, S. Phytoremediation capabilities of Salvinia molesta, water hyacinth, water lettuce, and duckweed to reduce phosphorus in rice mill wastewater. *Int. J. Phytoremed.* **2020**, *22*, 1097–1109. [CrossRef]
66. Rai, P.K. Heavy metal phytoremediation from aquatic ecosystems with special reference to macrophytes. *Crit. Rev. Environ. Sci. Technol.* **2009**, *39*, 697–753. [CrossRef]
67. Giri, A.K. Removal of Arsenic (III) and Chromium (VI) from the Water Using Phytoremediation and Bioremediation Techniques. Ph.D. Thesis, Degree of Doctor of Philosophy, Department of Chemistry National Institute of Technology Rourkela, Rourkela, India, July 2012.
68. Shah, M.; Hashmi, H.N.; Ghumman, A.R.; Zeeshan, M. Performance assessment of aquatic macrophytes for treatment of municipal wastewater. *J. S. Afr. Inst. Civ. Eng.* **2015**, *57*, 18–25. [CrossRef]
69. Haller, W.T.; Sutton, D.L.; Barlowe, W.C. Effects of salinity on growth of several aquatic macrophytes. *Ecology* **1974**, *55*, 891–894. [CrossRef]
70. Liang, L.; Liu, W.; Sun, Y.; Huo, X.; Li, S.; Zhou, Q. Phytoremediation of heavy metal-contaminated saline soils using halophytes: Current progress and future perspectives. *Environ. Rev.* **2016**, *25*, 269–281. [CrossRef]
71. Prasad, M.N.V. Phytoremediation of metal-polluted ecosystems: Hype for commercialization. *Russ. J. Plant Physiol.* **2003**, *50*, 686–701. [CrossRef]
72. Wuana, R.A.; Okieimen, F.E. Heavy metals in contaminated soils: A review of sources, chemistry, risks, and best available strategies for remediation. In *Heavy Metal Contamination of Water and Soil: Analysis, Assessment, and Remediation Strategies*; International Scholarly Research Network: Londoin, UK, 2014; pp. 1–50.
73. Oladoye, P.O.; Olowe, O.M.; Asemoloye, M.D. Phytoremediation technology and food security impacts of heavy metal contaminated soils: A review of literature. *Chemosphere* **2022**, *288*, 132555. [CrossRef] [PubMed]

74. Dhaliwal, S.S.; Sharma, V.; Taneja, P.K.; Shukla, A.K.; Kaur, L.; Verma, G.; Verma, V.; Singh, J. Effect of cadmium and ethylenediamine tetraacetic acid supplementation on cadmium accumulation by roots of *Brassica* species in Cd spiked soil. *Environ. Sci. Pollut. Res.* **2022**, *29*, 6000–6009. [CrossRef]
75. Shinta, Y.C.; Zaman, B.; Sumiyati, S. Citric Acid and EDTA as Chelating Agents in Phytoremediation of Heavy Metal in Polluted Soil: A Review. *IOP Conf. Ser. Earth Environ. Sci.* **2021**, *896*, 012023. [CrossRef]
76. Lee, J.; Sung, K. Effects of chelates on soil microbial properties, plant growth and heavy metal accumulation in plants. *Ecol. Eng.* **2014**, *73*, 386–394. [CrossRef]
77. Singh, J.; Kumar, V.; Kumar, P.; Kumar, P. Kinetics and prediction modeling of heavy metal phytoremediation from glass industry effluent by water hyacinth (*Eichhornia crassipes*). *Int. J. Environ. Sci. Technol.* **2022**, *19*, 5481–5492. [CrossRef]
78. Naaz, M.; Dutta, A.; Kumari, S.; Farooqui, S. Bioaccumulation, phytoremediation and kinetics of uptake of heavy metals (Copper and Zinc) by Eichhornia crassipes. *Res. Rev. J. Ecol.* **2013**, *2*, 2278.
79. Ingole, N.W.; Bhole, A.G. Removal of heavy metals from aqueous solution by water hyacinth (*Eichhornia crassipes*). *J. Water Supply Res. Technol. AQUA* **2003**, *52*, 119–128. [CrossRef]
80. Go, J.L.C.; Madrazo, C.F.; Orbecido, A.H.; de Castro, M.E.G.; Deocaris, C.C.; Belo, L.P. Analysis of the copper removal kinetics of the Philippine giant bamboo (*Dendrocalamus asper*) in hydroponics. *Heliyon* **2021**, *7*, e06208. [CrossRef]
81. Chua, J.; Banua, J.M.; Arcilla, I.; Orbecido, A.; de Castro, M.E.; Ledesma, N.; Deocaris, C.; Madrazo, C.; Belo, L. Phytoremediation potential and copper uptake kinetics of Philippine bamboo species in copper contaminated substrate. *Heliyon* **2019**, *5*, e02440. [CrossRef]
82. Mondal, N.K.; Nayek, P. Hexavalent chromium accumulation kinetics and physiological responses exhibited by *Eichhornia* sp. and *Pistia* sp. *Int. J. Environ. Sci. Technol.* **2020**, *17*, 1397–1410. [CrossRef]
83. Cleide Barbieri de, S.; Gabriel Rodrigues, S. Phytoremediation of effluents contaminated with heavy metals by floating aquatic macrophytes species. In *Biotechnology and Bioengineering*; Eduardo Jacob, L., Leila Queiroz, Z., Eds.; IntechOpen: Rijeka, Croatia, 2019; Chapter 10.
84. Kamalu, C.I.O.; Okere, P.C.; Egbufor, U.C.; Nwandikom, G.I.; Obijiaku, J.C.; Asomugha, C.C. Modeling and optimization of phytoremediation kinetics of metals in soil by a plant hyperacumulator. *Am. J. Eng. Res.* **2017**, *6*, 196–207.

sustainability

MDPI

Review

Black Soldier Fly Larvae (*Hermetia illucens*) for Biodiesel and/or Animal Feed as a Solution for Waste-Food-Energy Nexus: Bibliometric Analysis

Dave Mangindaan [1,2,*], Emil Robert Kaburuan [3] and Bayu Meindrawan [4]

1 Profesional Engineer Program Department, Faculty of Engineering, Bina Nusantara University, Jakarta 11480, Indonesia
2 Waste-Food-Environmental Nexus Research Interest Group, Bina Nusantara University, Jakarta 11480, Indonesia
3 Informatics Engineering Department, Faculty of Computer Science, Mercu Buana University, Jakarta 11650, Indonesia
4 Food Technology Department, Faculty of Agriculture, Sultan Ageng Tirtayasa University, Serang 42124, Indonesia
* Correspondence: dave.mangindaan@binus.ac.id

Abstract: In this research, an emerging study of the utilization of black soldier fly (BSF, *Hermetia illucens*) larvae for the preparation of biodiesel (and organic waste treatment) and the generation of alternative feed for improved food production was mapped bibliometrically from the Scopus database. BSF is a promising biological agent for tackling the waste-food-energy (WFE) nexus, which is a problematic vicious cycle that may threaten Earth's sustainability, hence its emergence. With its short life cycle, ability to consume organic waste equal to its own weight on a daily basis, and ability for conversion to larvae with a high protein and lipid content, BSF larvae is the perfect choice as a one-step solution of the WFE nexus. To further perfect the research of BSF for the WFE nexus, this bibliometric analysis, and the citation evolution profile, were carried out with the objectives of characterizing the progress of publications in the last 10 years (2011–2022) in order to determine future research directions in this field, identify the top publications for wider reach to the public, and identify productive authors and leading countries to visualize opportunities for future collaborations.

Keywords: black soldier fly; biodiesel; bibliometric analysis; waste-food-energy nexus; publications

Citation: Mangindaan, D.; Kaburuan, E.R.; Meindrawan, B. Black Soldier Fly Larvae (*Hermetia illucens*) for Biodiesel and/or Animal Feed as a Solution for Waste-Food-Energy Nexus: Bibliometric Analysis. *Sustainability* **2022**, *14*, 13993. https://doi.org/10.3390/su142113993

Academic Editors: Steven Lim, Shuit Siew Hoong, Santi Chuetor and Pang Yean Ling

Received: 22 September 2022
Accepted: 25 October 2022
Published: 27 October 2022

1. Introduction

Earth is facing a simultaneous crisis of water (wastewater), food, and energy, called the waste-food-energy (WFE) nexus. The population is projected to reach 9 billion people in 2050 [1]. However, this growth is not balanced with the replenishment of natural resources in terms of quantity and quality, especially with the limited supply of land and reduction in fossil fuels. Therefore, there is an urgent need to alleviate this energy issue before it jeopardizes other aspects in the near future. Attempts to obtain alternative sources of renewable energies such as hydro [2,3], wind [4,5], geothermal [6,7], and solar [8,9] have been carried out. In addition, another renewable energy that has received global attention is biodiesel [10–13].

Biodiesel is a product of the transesterification reaction between triglyceride and alcohol (commonly methanol) to produce fatty esters (biodiesel) and glycerol [11]. Biodiesel is an alternative fuel that can be employed in commercial diesel engines without further modification of the engine [14], and does not need to be mixed with fossil fuel. Therefore, the application of renewable energy can be carried out without sacrificing the consumer's convenience.

Initially, biodiesel was produced from edible oils such as palm oil or soybean oil [10,15]. However, due to the conflict of interest with the human food supply, biodiesel has started to be prepared from non-edible plants such as *Jatropha curcas* [16], Chinese tallow [17], and microalgae [18]. However, those non-edible plants possess limitations such as a long time to harvest (several months) and high water content, which definitely reduces the calorific content of these plants [15]. Thus, an alternative is needed to resolve this issue.

Black soldier fly (BSF, *Hermetia illucens*), as shown in Figure 1, is an insect that can consume nutrition from low-cost materials, such as municipal wastes (household, restaurant, food industries, etc.) and manure (chicken, cow, sheep) [15,19]. The larvae of BSF reproduce in a short time (10–20 days), have high protein and lipid contents, and low moisture, therefore possessing a high calorific value [20]. Generally, the larvae of this insect are directly fed to farmed animals without additional processing. However, with the spirit of generating value-added products, BSF larvae are further processed to obtain biodiesel as one of the products [21–24]. Biodiesel from BSF larvae is considered a simultaneous solution for some problems such as the energy crisis and waste processing (without the need for an excessive area for waste storage, and a reduction in the volume of waste that must go to the landfill).

Figure 1. BSFs inside the rearing chamber.

In addition to the aforementioned energy-related advantages, BSF larvae also represent an additional solution for the improvement of the quantity and quality of animal-based food products when implemented as animal feed. This paper bibiliometrically maps the important role of BSF larvae in the fields of organic waste treatment, food (or feed), and alternative energy, or for the waste-food-energy (WFE) nexus.

2. Materials and Methods

The data for this research were produced from 2011 to 2022 and were collected from the Scopus database (Elsevier) on October 2022 using the Scopus search query in Table 1. The first query was "black soldier fly" and "biodiesel", resulting in 488 entries, and the second query was the use of BSF's scientific name "*Hermetia illucens*" and "biodiesel" in order to obtain complete data from the current publications. The two sets of publication data were combined and duplicates removed to obtain a set of 535 publication data, including their keywords.

A preliminary qualitative analysis of the keywords of this set of publication data obtained from https://www.wordclouds.com/ (accessed on 14 October 2022) is shown in Figure 2. The dominant keywords in this research can be seen as "black soldier fly", "*Hermetia illucens*", "larvae", and their components, along with "waste", "bioconversion", "feed", "food', "lipid", "protein", etc. This quick illustration shows that the utilization of BSF larvae is on the right track for solving the waste-food-energy (WFE) nexus. However, the obtained set of 535 publication data must be further analyzed to obtain more in-depth results displaying the metrics, qualities, research trend, and prominent authors in the field of the preparation of biodiesel from BSF. Furthermore, to acquire a wider understanding of the recent research and determine the research trend, the obtained set was further processed to obtain the visualization of similarities (VOS) using VOSViewer software, a powerful software for bibliometric networks visualization (Version 1.6.17, developed by [25], http://www.voswiever.com (accessed on 24 August 2021)).

Table 1. Scopus search query used to obtain the publication data of papers with the keywords of black soldier fly (and/or *Hermetia illucens*) and biodiesel.

Scopus Search Query	Resulting Entries
(TITLE-ABS-KEY (black AND soldier AND fly)) AND (biodiesel)	488
(TITLE-ABS-KEY (hermetia AND illucens)) AND (biodiesel)	431
Combined search result, with duplicated items removed	535

Figure 2. Word cloud extracted from https://www.wordclouds.com/ (accessed on 14 October 2022).

3. Results and Discussion

3.1. Characteristic of Publications (2011–2022)

It was found that the publication of biodiesel from BSF was pioneered in 2011 [15,26], but there was limited development until 2016 (a total of 37 publications). However, between 2017 and 2022, the number increased rapidly (498 publications) or by more than one degree of magnitude under the same span of duration (Table 2). It is believed that this boost was catalyzed by the Sustainable Development Goals established by United Nations General Assembly in 2015 (UN SDGs https://sdgs.un.org (accessed on 1 November 2021)), with goals such as #1 No Poverty, #6 Clean Water and Sanitation, and #7 Affordable and Clean Energy supporting the development of BSF for biodiesel (and other products) to tackle the water-food-energy (WFE) nexus.

Table 2. Number of publications of papers with the keywords of black soldier fly (and/or *Hermetia illucens*) and biodiesel (2011–2022).

Year	Number of Publications
2011	2
2012	5
2013	4
2014	4
2015	13
2016	9
2017	23
2018	50
2019	71
2020	104
2021	119
2022	131

3.2. Top 20 Journals

In addition to the launching of the UN SDGs in 2015, the BSF research to date has been amplified by the establishment of the *Journal of Insects for Food and Feed* (*JIFF*, 2021 impact factor 5.099) by Wageningen Academic Publishers Netherlands. It is clearly shown in Table 3 that this journal leads the top 20 publications supporting the development of the utilization of BSF larvae (although it mainly includes development for food, feed, and their nutritional properties rather than biodiesel fuel), with the highest number of 33 publications. Trailing *JIFF* with 22 publications is the *Journal of Cleaner Production* and *Waste Management*, both from Elsevier (displaying a high 2021 impact factor of 11.072, and 8.816, respectively). It is worth noting that 8 of the top 20 journals in Table 3 are from Elsevier, signifying the role of Elsevier in the development of BSF to solve the WFE nexus, contributing a total of 103 high-impact articles (with an average 2021 impact factor of 8.133) from a wide range of areas such as energy, waste treatment, environmental, feed, etc. Moreover, a considerably new open-access publisher, Multidisciplinary Digital Publishing Institute (MDPI), also collectively published more than *JIFF*, with a cumulative total of 35 articles. In addition, there are also some publishers with publications $\leq$ 10 articles with impact factors ranging from 3.313 to 8.431 (namely Elsevier, Frontiers Media S.A, Public Library of Science (PLoS), Springer Nature, MDPI, and Nature Publishing Group) or without an impact factor (due to the publication of conference papers only, such as the Institute of Physics (IOP) and American Institute of Physics (AIP)).

Table 3. Top 20 journals (with the keywords of black soldier fly (and/or *Hermetia illucens*) and biodiesel) with the highest number of articles.

No.	Journal Name	Publisher	2021 Impact Factor	Total Articles
1.	*Journal of Insects as Food and Feed*	Wageningen Academic Publishers	5.099	33
2.	*Journal of Cleaner Production*	Elsevier	11.072	22
3	*Waste Management*	Elsevier	8.816	22
4.	*Animals*	Multidisciplinary Digital Publishing Institute (MDPI)	3.231	17
5.	*Insects*	MDPI	3.139	15
6.	*IOP Conference Series: Earth and Environmental Science*	Institute of Physics (IOP)	-	15
7.	*Journal of Environmental Management*	Elsevier	8.910	15
8.	*Science of the Total Environment*	Elsevier	10.753	14
9.	*Sustainability*	MDPI	3.889	12
10.	*Frontiers in Microbiology*	Frontiers Media S.A.	6.064	10
11.	*Renewable Energy*	Elsevier	8.634	10
12.	*Aquaculture*	Elsevier	5.135	9
13.	*PLoS ONE*	Public Library of Science (PLoS)	3.752	9
14.	*Waste and Biomass Valorization*	Springer Nature	3.449	9
15.	*AIP Conference Proceedings*	American Institute of Physics (AIP)	-	8
16.	*Processes*	MDPI	3.352	8
17.	*Scientific Reports*	Nature Publishing Group	4.996	8
18.	*Environmental Research*	Elsevier	8.431	6
19.	*Environmental Science and Pollution Research*	Springer Nature	5.190	6
20.	*Animal Feed Science and Technology*	Elsevier	3.313	5

3.3. Top 20 Articles

Among the publishers engaging with the issue of the WFE nexus, there are individual articles that have had a high impact, as demonstrated by their 100 citations or more, as tabulated in Table 4. Please note that there are two articles in 20th position due to the identical number of citations (146 citations). This list is championed by a review of the use of insects as animal feed [27], cited 789 times. This is of high interest since high-quality feed means enhanced production of meat as a protein source (poultry, beef, and fish meat). This is a such a pioneering and fundamental paper that the four articles below it only barely reach half its number of citations (300–400 citations), with all of them discussing the future applications of BSF for food and/or feed as part of solving the WFE nexus [28–31]. In sixth position, there is a paper focused on fly larvae for organic waste treatment (although it is not focused exclusively on the preparation of biodiesel) [32].

In the next 7–20 papers, there are 5 papers that specifically highlight biodiesel production [21,33,34], including the classic trailblazer papers in 2011 that paved the way for the development of BSF for biodiesel [15,26], with around 180–200 citations. The rest of the papers are focused on waste management [35–37] and/or alternative feed or food [38–42]. In addition, the average number of citations in Table 4 is 258.9, with average 2021 impact factor of 6.537.

The top 20 articles in Table 4 were obtained from the search terms in Table 1, which were limited to "black soldier fly", "biodiesel", and "*Hermetia illucens*". These search terms were chosen to reflect the exploration of the research of alternative renewable energy from organic waste (waste-energy nexus) and as the priority in this report. On the other hand, the food-related nexuses (mainly the waste-food nexus, followed by the food-energy nexus) were less prioritized in this study. The analyses of the food-related nexuses were derived from the data collected using the methods defined in Table 1 (for the waste-energy nexus). This limitation was used because of the open debate about the acceptance of BSF larvae for human consumption [31,43], especially regarding its implication to the halal status of meat products [44]. Furthermore, the nexus(es) addressed by the top 20 articles investigated in this study are classified in Table 4, column 6.

Table 4. Top 20 articles (with the keywords of black soldier fly (and/or *Hermetia illucens*) and biodiesel) with the highest number of citations.

No.	Title	Journal Name	Citations	2021 Impact Factor	Type(s) of WFE Nexus	Year	Burst Begins	Burst Ends	Peak	Ref.
1.	State-of-the-art on use of insects as animal feed	*Animal Feed Science and Technology*	789	3.313	Food (or feed)	2014	2015	2022	2021	[27]
2.	Feed conversion, survival and development, and composition of four insect species on diets composed of food by-products	*PLoS ONE*	403	3.752	Food (or feed)	2015	2016	2022	2021	[29]
3.	Nutritional composition of black soldier fly (*Hermetia illucens*) prepupae reared on different organic waste substrates	*Journal of the Science of Food and Agriculture*	395	4.125	Food (or feed)	2017	2017	2022	2021	[30]
4.	Review on the use of insects in the diet of farmed fish: Past and future	*Animal Feed Science and Technology*	385	3.313	Food (or feed)	2015	2015	2022	2021	[28]
5.	Review of black soldier fly (*Hermetia illucens*) as animal feed and human food	*Foods*	312	5.561	Food (or feed)	2017	2019	2022	2021	[31]
6.	The use of fly larvae for organic waste treatment	*Waste Management*	280	8.816	Waste	2015	2016	2022	2022	[32]
7.	Nutritional value of the black soldier fly (*Hermetia illucens* L.) and its suitability as animal feed-a review	*Journal of Insects as Food and Feed*	241	5.099	Food (or feed)	2017	2018	2022	2021	[45]
8.	Environmental impact of food waste bioconversion by insects: Application of Life Cycle Assessment to process using *Hermetia illucens*	*Journal of Cleaner Production*	230	11.072	Waste	2017	2017	2022	2021	[46]
9.	Nutritional value of two insect larval meals (*Tenebrio molitor* and *Hermetia illucens*) for broiler chickens: Apparent nutrient digestibility, apparent ileal amino acid digestibility and apparent metabolizable energy	*Animal Feed Science and Technology*	216	3.313	Food (or feed)	2015	2015	2022	2021	[38]
10.	Bioconversion of organic wastes into biodiesel and animal feed via insect farming	*Renewable Energy*	215	8.634	Waste, energy	2016	2017	2022	2021	[21]
11.	Bioconversion of dairy manure by black soldier fly (Diptera: Stratiomyidae) for biodiesel and sugar production	*Waste Management*	214	8.816	Waste, energy, food	2011	2013	2022	2020	[26]
12.	Evaluation of the suitability of a partially defatted black soldier fly (*Hermetia illucens* L.) larvae meal as ingredient for rainbow trout (*Oncorhynchus mykiss* Walbaum) diets	*Journal of Animal Science and Biotechnology*	213	5.032	Food (or feed)	2017	2018	2022	2021	[39]

Table 4. *Cont.*

No.	Title	Journal Name	Citations	2021 Impact Factor	Type(s) of WFE Nexus	Year	Burst Begins	Burst Ends	Peak	Ref.
13.	Effect of rearing substrate on growth performance, waste reduction efficiency and chemical composition of black soldier fly (*Hermetia illucens*) larvae	*Journal of the Science of Food and Agriculture*	209	4.125	Waste	2018	2019	2022	2022	[35]
14.	Effects of feedstock on larval development and process efficiency in waste treatment with black soldier fly (*Hermetia illucens*)	*Journal of Cleaner Production*	206	11.072	Waste	2019	2019	2022	2022	[36]
15.	Sustainability of insect use for feed and food: Life Cycle Assessment perspective	*Journal of Cleaner Production*	184	11.072	Food (or feed)	2016	2017	2022	2020	[40]
16.	From organic waste to biodiesel: Black soldier fly, *Hermetia illucens*, makes it feasible	*Fuel*	180	6.609	Waste, energy	2011	2012	2022	2021	[15]
17.	Decomposition of biowaste macronutrients, microbes, and chemicals in black soldier fly larval treatment: A review	*Waste Management*	165	8.816	Waste	2018	2019	2022	2021	[37]
18.	Double the biodiesel yield: Rearing black soldier fly larvae, *Hermetia illucens*, on solid residual fraction of restaurant waste after grease extraction for biodiesel production	*Renewable Energy*	160	8.634	Waste, Energy	2012	2015	2022	2020	[33]
19.	Nutritional value of a partially defatted and a highly defatted black soldier fly larvae (*Hermetia illucens* L.) meal for broiler chickens: Apparent nutrient digestibility, apparent metabolizable energy and apparent ileal amino acid digestibility	*Journal of Animal Science and Biotechnology*	147	5.032	Food (or feed)	2017	2018	2022	2021	[41]
= 20a.	Partial or total replacement of soybean oil by black soldier fly larvae (*Hermetia illucens* L.) fat in broiler diets: Effect on growth performances, feed-choice, blood traits, carcass characteristics and meat quality	*Italian Journal of Animal Science*	146	2.217	Food (or feed)	2017	2017	2022	2021	[42]
= 20b.	Biodiesel production from rice straw and restaurant waste employing black soldier fly assisted by microbes	*Energy*	146	8.857	Waste, energy	2012	2015	2022	2020	[34]

In addition to the citation growth of the aforementioned articles in Table 4, the citation burst profiles of these articles were also analyzed (Table 4, column 8–9) and are visualized in Figure 3. The citation burst analysis is useful for assessing the evolution of a topic over time in order to detect emerging research [44,47]. For the papers with the keywords of BSF and/or *Hermetia illucens* and biodiesel, there is a unique observation in that some focused on the nexus of food (or feed), as shown in Figure 3a,b. Both figures show the citation evolution in the nexus of food (or feed), but Figure 3a shows those with >300 total citations while Figure 3b displays those with 140–300 citations.

It can be observed that although the reports have the keywords of BSF and biodiesel, they were also of interest for food (or feed) applications and were popular and mostly cited from 2015–2016 (except for #1, from 2014). The citations of the papers in Figure 3a,b increased up to 2022 but with some signs of slowing down, where the citations peaked in 2021. This diminishing trend might have resulted from the shift from the nexus of food to the nexus of waste, as shown in Figure 3c. The citation burst for the nexus of waste showed variation (between 2015 and 2019 for five papers), and the burst has not finished (up to

2022), with only two papers showing a reduced number of citations (#8 and #17 peaked in 2021).

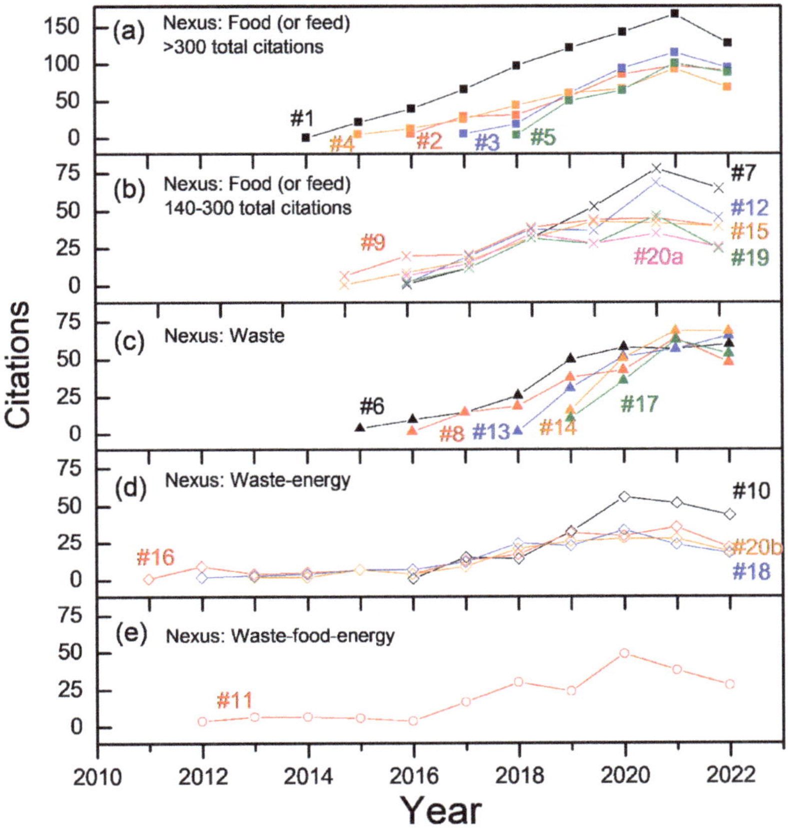

Figure 3. Citation burst of the top 20 articles from Table 4 (with the keywords of black soldier fly (and/or *Hermetia illucens*) and biodiesel) from 2011–2022. (**a**) Nexus: food (or feed), total citations >300, (**b**) nexus: food (or feed), 140–300 citations, (**c**) nexus: waste, (**d**) nexus: waste-energy, and (**e**) nexus: waste-food-energy.

For the papers about BSF in the energy-related nexuses (waste-energy in Figure 3d or waste-food-energy in Figure 3e), the trend is different from that shown in Figure 3a–c. Papers that focused on the energy-related nexus started early (between 2011 and 2013), suggesting the research priority of alternatives to renewable energy during this period. Although the total number of citations is moderate for the papers shown in Figure 3d,e, their longevity must be appreciated. Their citation burst has continued to 2022 while the peak of their citation burst occurred in 2020, which is comparable to those shown in Figure 3a–c (mostly peaked in 2021). Therefore, based on Figure 3, it is suggested that the research directions in terms of BSF for food, feed, waste treatment, biodiesel as renewable energy, etc. are still considered exciting and on the right track for the near future. Based on this finding, the biodiesel production processes from BSF larvae are covered in Table 5 in order to provide the state-of-the-art of this particular research direction.

Table 5. Selected biodiesel production processes from BSF larvae.

No.	Pretreatment	Reactant	BSF Feed	Mixing Ratio	Reaction	Condition	Biodiesel Yield	Unit of Yield	Ref.
1.	None	Dried BSF larvae	Wheat bran	Dried BSFL/methanol/hexane = 1/4/2 *w/v/v*	Acid -catalyzed reaction	H_2SO_4 60% at 120 °C for 90 min	28.4	wt.% per dried BSF larvae	[48]
2.	Solvent extraction with petroleum ether for 6 h	BSF larvae oil	Soya residue	Oil/methanol = 1/8, 1/10, 1/12, 1/14, 1/16 molar ratio	Acid -catalyzed reaction followed with alkaline -catalyzed reaction	H_2SO_4 1% in methanol, 45 °C, 60 min, followed with NaOH (0.5–1.5%), 45–65 °C, 20–40 min	35–90	wt.% per BSF larvae oil	[49]
3.	Solvent extraction with petroleum ether for 24 h	BSF larvae extract	Fermented coconut endosperm waste	Extract/methanol = 1/8 *w/w*,	Acid -catalyzed reaction followed with alkaline -catalyzed reaction	H_2SO_4 1% in methanol, 75 °C, 200 rpm, 1 h, followed with KOH 0.8% in methanol, 65 °C, 200 rpm, 30 min.	35–40	wt.% per powdered BSF larvae	[50]
4.	Solvent extraction with petroleum ether for 24 h	BSF larvae extract	Fermented coconut endosperm waste (0.0–2.5% mixed-bacteria, 0–28 days fermentation	Extract/methanol = 1/8 *w/w*,	Acid -catalyzed reaction followed with alkaline -catalyzed reaction	HCl 1% in methanol, 75 °C, 1 h, followed with KOH 1% in methanol, 65 °C, 30 min.	35–38.5	wt.% per powdered BSF larvae	[51]
5.	Solvent extraction with petroleum ether for 6 h, followed with addition of 1% (*v/v*) concentrated H_3PO_4 (85%)	BSF larvae oil	Restaurant kitchen waste	Oil/methanol = 1/10 molar ratio	Acid -catalyzed reaction followed with alkaline -catalyzed reaction	H_2SO_4 1% in methanol, 50 °C, 41 min, followed with NaOH 1.1% in methanol, 62 °C, 61 min	97	wt.% per BSF larvae oil	[52]
6.	Solvent extraction with petroleum ether for 6 h	BSF larvae oil	Restaurant kitchen waste	Oil/methanol = 1/6–1/14 molar ratio	Acid -catalyzed reaction followed with alkaline -catalyzed reaction	H_2SO_4 1% in methanol, followed with NaOH (0.5–1.5%), 35–65 °C, 40–60 min	24–95	wt.% per BSF larvae oil	[53]
7	Solvent extraction with petroleum ether for 6 h, followed with addition of 1% (*v/v*) concentrated H_3PO_4 (85%)	BSF larvae oil	Pig manure	Grease/methanol = 1/8 *w/w* (for acid -catalyzed reaction), 1/6 (for alkaline-catalyzed reaction)	Acid -catalyzed reaction followed with alkaline -catalyzed reaction	H_2SO_4 1% in methanol, 75 °C, 60 min, followed with NaOH 0.8% in methanol, 65 °C, 30 min	94.91	Wt.% per BSF larvae grease	[54]
8.	Solvent extraction with petroleum ether for 48 h (twice)	BSF larvae extract	Restaurant food waste	Extract/methanol = 1/8 for acid -catalyzed reaction, 1/6 for alkaline -catalyzed reaction	Acid -catalyzed reaction followed with alkaline -catalyzed reaction	H_2SO_4 1% at 75 °C for 1 h, followed with NaOH 0.8% at 65 °C for 30 min	36.3	wt.% per dried BSF larvae	[33]
9.	Solvent extraction with petroleum ether for 16 h	BSF larvae grease	Fresh manure	Grease/methanol = 1/8,	Acid -catalyzed reaction followed with alkaline -catalyzed reaction	H_2SO_4 1% in methanol, 73 °C, 2 h, followed with NaOH 0.8% in methanol, 65 °C, 30 min	96.34	wt.% per BSF larvae grease	[26]
10.	Solvent extraction with petroleum ether for 48 h	BSF larvae extract	Pig manure	Extract/methanol = 1/8 for acid -catalyzed reaction, 1/6 for alkaline -catalyzed reaction	Acid -catalyzed reaction followed with alkaline-catalyzed reaction	H_2SO_4 1% at 75 °C for 1 h, followed with NaOH 0.8% at 65 °C for 30 min	27.9	wt.% per dried BSF larvae	[15]
11.	Solvent extraction with petroleum ether for 12 h	Milled dried BSF larvae	Solid digestate of chicken manure and rapeseed straw	200 larval/150 g digestate	Alkaline transmethylation	Fatty residue dissolved in hexane, KOH 5% in methanol, mixed for 5 min.	14.36	g per kg waste	[55]
12.	Solvent extraction with hexane for 24 h	BSF larvae extract	Food waste	Extract/methanol = 1/8 *w/w*	Alkaline-catalyzed reaction	KOH 5% at 65 °C for 8 h	33.9	wt.% per dried BSF larvae	[56]
13.	Solvent extraction with n-hexane for 48 h	BSF larvae fat	Wheat bran	Fat/methyl acetate = 1/14.64 molar ratio	Enzymatic reaction	Novozym 435 (4% concentration, loaded at 17.58% shaken at 40 °C, 12 h.	96.97	wt.% per BSF	[57]

Table 5. *Cont.*

No.	Pretreatment	Reactant	BSF Feed	Mixing Ratio	Reaction	Condition	Biodiesel Yield	Unit of Yield	Ref.
14.	Solvent extraction with n-hexane for 48 h	BSF larvae fat	Wheat bran	Fat/methanol = 1/6.33 molar ratio	Enzymatic reaction	Novozym 435 (4% concentration, loaded at 20%, shaken at 26 °C, 9.48 h.	96.18	wt.% per BSF fat	[58]
15.	None	Dried BSF larvae	Food waste	Dried BSFL/methanol = 1/10 *w/w*	Non-catalytic reaction	SiO_2 at 390 °C for 1 min	34.7	wt.% per dried BSF larvae	[56]
16.	Solvent extraction with hexane for 24 h	BSF larvae extract	Food waste	Extract/methanol = 1/20 *w/v*	Non-catalytic reaction	SiO_2 at 390 °C for 1 min	34	wt.% per dried BSF larvae	[56]
17.	Solvent extraction with n-hexane for 48 h	Powdered BSF larvae	Wheat bran	Methanol/BSF powder = 4:1 to 10:1 mL/g	Switchable -solvent -catalyzed (using polarity switchable solvent, DBU (1,8-diazabicyclo [5.4.0]undec-7-ene))	DBU/biomass = 8:1 to 20:1 mL/g, 90–120 °C, 30–120 min.	96.2%	actual biodiesel produced per theoretical biodiesel produced	[59]

The selected processes for the production of biodiesel from BSF larvae shown in Table 5 are dominated by the two-step esterification reaction [48–56], i.e., acid-catalyzed reaction (dilute H_2SO_4 or HCl, typically 1% in methanol), followed by alkaline-catalyzed (NaOH or KOH, 0.8–5.0% in methanol) reaction, as pioneered by [15,26,33]. In addition, the grease, fat, or oil from BSF larvae are mostly pretreated using solvent extraction (with petroleum ether or n-hexane) for 6–48 h to extract the lipid content for a further transesterification process to produce biodiesel. The BSF larvae were fed various feed such as wheat bran, restaurant waste, coconut endosperm waste, soya residue, and manure (pig, chicken, human) to achieve larval growth. It is therefore interesting to explore the research on renewable energy from organic waste or underutilized carbon sources to lower the carbon footprint via the utilization of BSF larvae.

Interestingly, there are several novel approaches to the production of biodiesel from BSF larvae in addition to the common acid- and alkaline-catalyzed two-step reactions. A green approach using enzymes represents a breakthrough, where the use of acid and alkaline in methanol is eliminated [57], or a tandem with methyl acetate is used as an alternative to methanol [58]. The elimination of catalyst (acid, alkaline, enzymes) is also quite a radical approach, as shown by [56] by their non-enzymatic process involving SiO_2. However, there is still a drawback where a high temperature of 390 °C is required. In addition, the use of a unique solvent with a switchable polarity (DBU, 1,8-diazabicyclo [5.4.0]undec-7-ene) is also a novelty [59], with relatively mild conditions. However, the price and availability of this kind of solvent must be properly assessed.

3.4. Attributes of the Top 20 Articles

3.4.1. Analysis of Keywords

After the specific investigation of the top publications and top articles with their respective metrics, it is also important obtain a broader vision of the set of 535 publication data by visualizing them in a web of entangled keywords as shown in Figure 4. It was found that there are 342 significant items (that occurred more than 5 times) classified into 6 clusters, with 20,380 links and a total link strength of 54,160. Each cluster is differentiated, as shown in Table 6, with a different color and group to summarize the co-occurrence and connectedness of each keyword represented by the 20,380 links. The top 10 keywords included in Figure 4 are shown in Table 7.

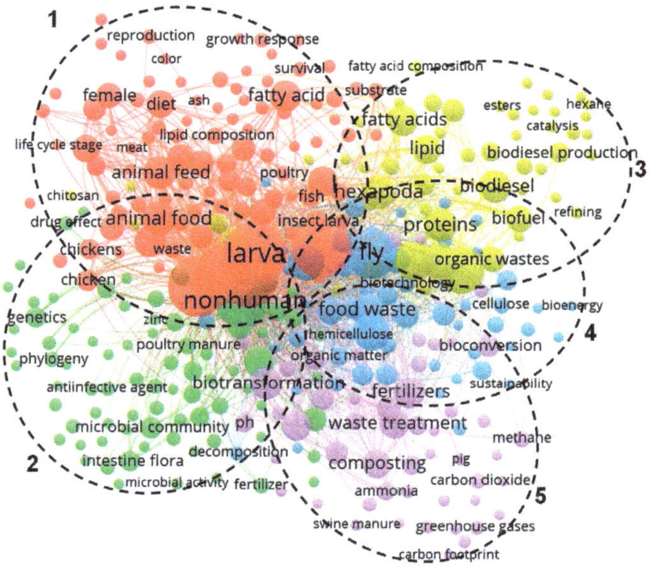

Figure 4. Visualization of the keywords from the 535 articles (with the keywords of black soldier fly (and/or *Hermetia illucens*) and biodiesel). Items = 342, clusters = 5, links = 20,380, total link strength = 54,160. The classification of the clusters can be observed in Table 6.

Table 6. Classification of the clusters in Figure 4.

Cluster	Color	Description
1	Red	Animal feed preparation and experiments
2	Green	Entomology, microbiology, biochemistry
3	Yellow	BSF and biodiesel, biofuels
4	Blue	Agricultural waste management
5	Violet	Municipal waste management

Table 7. Top 10 keywords in each cluster in Figure 4.

Cluster 1	Cluster 2	Cluster 3	Cluster 4	Cluster 5
Hermetia illucens	nonhuman	fly	black soldier fly larvae	biotransformation
larva	maggot	simuliidae	biomass	manure
animals	biodegradation	waste management	black soldier fly	fertilizers
diptera	bioremediation	food waste	hexapoda	waste treatment
animal	heavy metal	organic waste	fatty acids	composting
animal food	microbial community	waste disposal	insect	nitrogen
fatty acid	human	animalia	biodiesel	nutrients
protein	intestine flora	fermentation	proteins	pH
animal experiment	livestock	refuse disposal	lipid	manures
animal feed	rearing	growth rate	biofuel	moisture

3.4.2. Analysis of a Single Keyword ("Biodiesel")

Biodiesel, of particular interest in this manuscript, can be further traced in the VOSViewer software, with the other terms associated with biodiesel shown in Figure 5. This illustrates that one can click a specific circle with ease in VOSViewer software in order to instantaneously observe the interconnection with other keywords. In Table 8, the details of the top 20 keywords that are linked with biodiesel, along with the weight and cluster, are shown. Figure 5 and Table 8 demonstrate that BSF as an insect and its larvae are directly correlated with biodiesel or biofuel ingredients (fatty acid(s), lipid, obtained via the transesterification process) for a greener world with advanced waste management of food waste and for feedstocks to synergistically solve the waste-food-energy nexus.

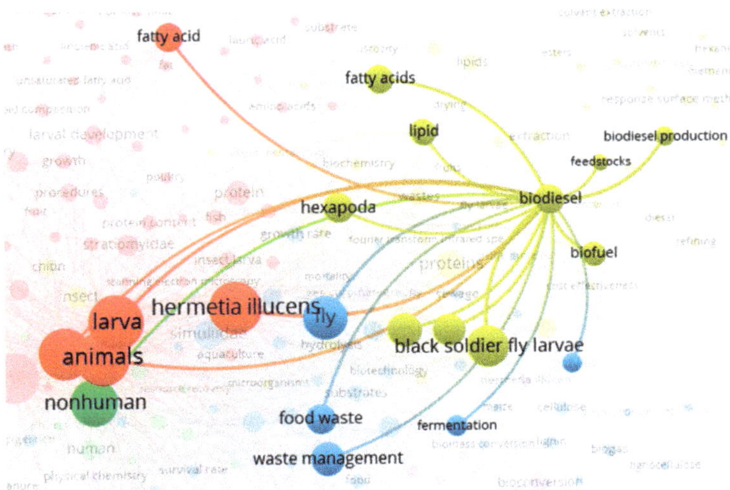

Figure 5. Biodiesel as a specific keyword of interest, along with terms associated with it (from various clusters). Item = biodiesel, links = 218, total link strength = 729.

Table 8. Top 20 terms associated with biodiesel obtained from Figure 5.

No.	Terms Associated with Biodiesel	Weight	Cluster
1.	*Hermetia illucens*	25	4
2.	fly	21	3
3.	black soldier fly larvae	21	4
4.	biofuel	20	3
5.	larva	20	3
6.	biodiesel production	17	4
7.	black soldier fly	16	3
8.	biomass	14	4
9.	nonhuman	14	2
10.	fatty acids	13	3
11.	hexapoda	12	5
12.	food waste	11	4
13.	lipid	9	4
14.	waste management	9	1
15.	fermentation	8	1
16.	feedstocks	8	2
17.	fatty acid	8	3
18.	biofuels	8	5
19.	transesterification	8	5
20.	organic wastes	7	4

3.5. Authors and Countries

In addition to the analysis of the publication data, the data of the authors were also analyzed in order to identify their impact and productivity regarding the research of biodiesel from BSF as one of the solutions of the WFE nexus. The analysis of the authors is shown in Figure 6. Out of 1818 authors, 108 authors have at least 5 publications, and only 94 of them are connected with each other. The authors are classified in 6 distinct clusters, connected with 424 links. The result of Figure 6 was extracted to obtain the top 10 authors, as shown in Table 8.

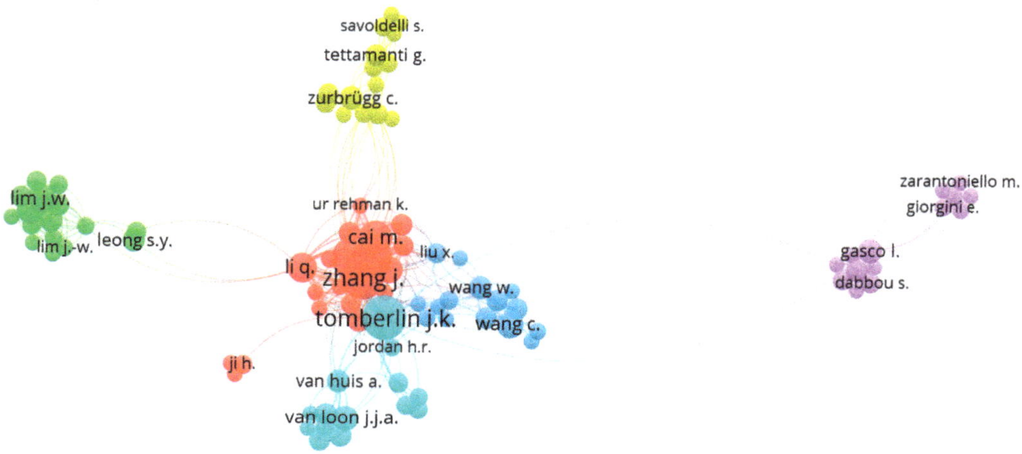

Figure 6. Map of authors with more than five articles (with the keywords of black soldier fly (and/or Hermetia illucens) and biodiesel) and their connectivity with each other. Items = 94, clusters = 6, links = 424, total link strength = 1601.

In Table 9, at least 15 co-authored publications are considered in the top 10 list. In addition, J.K. Tomberlin of Texas A&M University, United States of America, leads the pack with 39 articles, and a total of 7878 Scopus citations. However, in terms of citations, J.J.A. van Loon has around twice the number of J.K. Tomberlin's citations of 14,095, and the highest h-index. However, collectively, Huazhong Agricultural University, China dominates the list, with 5 personnel with more than 3900 citations on average. It is also worth noting that the first two papers on biodiesel from BSF in 2011 are authored by them. In addition to the majority of authors being from China and the Netherlands, there is also a scientist from Malaysia, J.W. Lim, that is on the list with 16 co-authored publications and more than 4500 citations.

Countries that are influential regarding research on the utilization of BSF for biodiesel and other problems in the WFE nexus are displayed in Figure 7, with the top 10 countries shown in Table 10. It can clearly be observed that China and the United States, as global academic powerhouses, lead the list and demonstrate strong cooperation as indicated by the thick link. In the same cluster with China and the United States is also Italy with 22 links of cooperation with Asia, Europe, and Africa.

There is, however, a unique observation in that non-tropical European countries in Table 10 (located 40°–60° North, such as Italy, the Netherlands, Germany, Belgium, and the United Kingdom) are very enthusiastic regarding research on BSF, although BSF research is mainly distributed in 45° North to 40° South (according to the map of Swiss Federal Institute of Aquatic Science and Technology [19]) or geographically distributed in North America, some southern parts of South America, South Africa, and Pacific Asia (Japan, China, Taiwan, Indonesia, Australia) [60].

Table 9. Top 10 authors with the most co-authored articles with the keywords of black soldier fly (and/or *Hermetia illucens*) and biodiesel.

No.	Author Name	Affiliation	Citations	h-Index	Number of Co-Authored Publications
1.	Tomberlin, Jeffrey Keith	Texas A&M University, College Station, the United States of America	7878	48	39
2.	Zhang, Jibin	Huazhong Agricultural University, Wuhan, China	3198	32	37
3.	Yu, Ziniu	Huazhong Agricultural University, Wuhan, China	9479	52	35
4.	Zheng, Longyu	Huazhong Agricultural University, Wuhan, China	2795	30	31
5.	Cai, Minmin.	Huazhong Agricultural University, Wuhan, China	1961	26	23
6.	Li, Qing	Huazhong Agricultural University, Wuhan, China	2075	23	17
7.	Lim, J.-W.	Universiti Teknologi Petronas, Malaysia	4562	39	16
8.	van Loon, J.J.A.	Wageningen University & Research, Wageningen, the Netherlands	14,095	66	15
9.	Li, Wu.	Hubei Polytechnic University, Huangshi, China	1017	13	15
10.	Wang, Cunwen	Wuhan Institute of Technology, Wuhan, China	3912	31	15

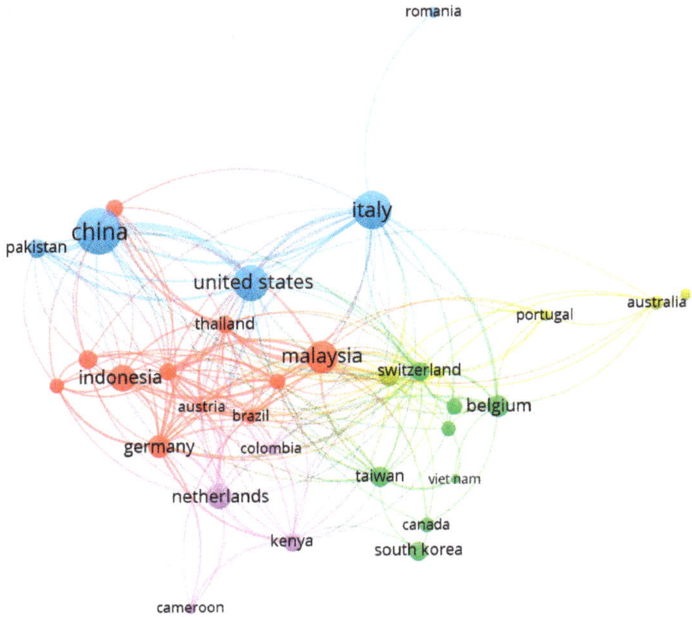

Figure 7. Co-authorship between countries in research on BSF for biodiesel and other applications. Clusters = 5, links = 212, total link strength = 432.

Table 10. Publication productivity of the top 10 countries.

No.	Country	Documents
1.	China	114
2.	Italy	80
3.	The United States	65
4.	Malaysia	55
5.	Indonesia	36
6.	The Netherlands	35
7.	Germany	28
8.	Belgium	25
9.	The United Kingdom	21
10.	Taiwan	21

Based on the co-authorship analysis in Figure 7, the strong desire to study and research alternative energy and food sources of the aforementioned European countries is fulfilled by cooperation with Asian countries (Malaysia, Indonesia, Taiwan, Thailand, and especially China). In addition to this, related to the geographical location, the link strength as a function of the continents is shown in Table 11. It is clearly shown that Europe leads with the highest link strength of 170.5, followed by Asia in second place with a link strength of 131.5, and others (Africa, Americas, Australia) with a combined link strength of 130 (similar to that of Asia). Based on Table 11, it can be suggested that Asia needs to catch up with Europe regarding research on BSF for biodiesel, food waste processing, alternative feed, and other applications related to the waste-food-energy nexus, especially because Asia's geographical position is favorable for BSF.

Table 11. Strength of the connection and research collaboration in different continents related to BSF for biodiesel and other applications.

Continent	Link Strength
Africa	66
America (North, South, Central, Canada)	58.5
Asia (East, South, Southeast)	131.5
Australia (Australia, New Zealand)	5.5
Europe	170.5
Total	432

4. Conclusions

Research from the last 10 years (2011–2022) on black soldier fly (BSF, *Hermetia illucens*) larvae as biodiesel and other promising applications (organic waste treatment, food and feed source) was bibliographically analyzed. The publication data captured by a trusted research database, Scopus, was retrieved, filtered, and sorted to reveal the characteristics of publications in the last decade, which has experienced a substantial increase since 2017. In addition, the top 20 journals and top 20 articles were also identified based on numerous citations, demonstrating their high impact on the research in this field. Based on the citation burst analysis, it is suggested that the research topics of BSF larvae for food, feed, organic waste treatment, and biodiesel as a renewable energy are still sustainable research directions for the foreseeable future.

The current publication data has a vast array of keywords, which were analyzed and distinguished using VOSViewer software to form five clusters of research. The research on biodiesel is, in fact, related to five distinct clusters, revealing the multifaceted approach to BSF larvae as a biological agent to tackle the global issue of the water-food-energy nexus. Therefore, research on BSF is on the right track towards the avenues of (1) waste treatment that is highly useful for (2) creating alternative feed to generate a sustainable food source, and the (3) production of green bioenergy. Furthermore, prominent authors and influential countries regarding research on BSF for biodiesel (also waste treatment and alternative feed or food) were also identified. Global BSF research on biodiesel and other applications in the waste-food-energy nexus was also mapped in this study, which is currently led by European countries. However, Asian countries are encouraged to reach the research performance of European countries in the future, particularly for tropical Asian countries for which BSF is comfortable with.

Author Contributions: Conceptualization, D.M., E.R.K., B.M.; methodology, D.M.; software, D.M.; formal analysis, D.M.; writing—original draft preparation, D.M., E.R.K., B.M.; writing—review and editing, D.M.; visualization, D.M.; project administration, E.R.K.; funding acquisition, D.M. and E.R.K. All authors have read and agreed to the published version of the manuscript.

Funding: This work was supported by the Directorate of Resources, Directorate General of Higher Education, Ministry of Education, Culture, Research, and Technology, Republic of Indonesia, as a part of Penelitian Dasar Unggulan Perguruan Tinggi (PDUPT) to Bina Nusantara University with contract number: 155/E5/PG.02.00.PT/2022; 410/LL3/AK.04/2022; 126/VR.RTT/VI/2022.

Acknowledgments: The authors would like to thank Directorate of Resources, Directorate General of Higher Education, Ministry of Education, Culture, Research, and Technology, Republic of Indonesia for the support.

Conflicts of Interest: The authors declare no conflict of interest.

References

1. Zhang, Y.-H.P. Next generation biorefineries will solve the food, biofuels, and environmental trilemma in the energy–food–water nexus. *Energy Sci. Eng.* **2013**, *1*, 27–41. [CrossRef]
2. Stocks, M.; Stocks, R.; Lu, B.; Cheng, C.; Blakers, A. Global Atlas of Closed-Loop Pumped Hydro Energy Storage. *Joule* **2021**, *5*, 270–284. [CrossRef]
3. Saklani, U.; Shresth, P.P.; Mukherji, A.; Scott, C.A. Hydro-energy cooperation in South Asia: Prospects for transboundary energy and water security. *Environ. Sci. Policy* **2020**, *114*, 22–34. [CrossRef]
4. Jung, C.; Schindler, D. Introducing a new approach for wind energy potential assessment under climate change at the wind turbine scale. *Energy Convers. Manag.* **2020**, *225*, 113425. [CrossRef]
5. Rezaeiha, A.; Montazeri, H.; Blocken, B. A framework for preliminary large-scale urban wind energy potential assessment: Roof-mounted wind turbines. *Energy Convers. Manag.* **2020**, *214*, 112770. [CrossRef]
6. Longa, F.D.; Nogueira, L.P.; Limberger, J.; van Wees, J.-D.; van der Zwaan, B. Scenarios for geothermal energy deployment in Europe. *Energy* **2020**, *206*, 118060. [CrossRef]
7. Wang, Y.; Liu, Y.; Dou, J.; Li, M.; Zeng, M. Geothermal energy in China: Status, challenges, and policy recommendations. *Util. Policy* **2020**, *64*, 101020. [CrossRef]
8. Gorjian, S.; Sharon, H.; Ebadi, H.; Kant, K.; Scavo, F.B.; Tina, G.M. Recent technical advancements, economics and environmental impacts of floating photovoltaic solar energy conversion systems. *J. Clean. Prod.* **2021**, *278*, 124285. [CrossRef]
9. Zhang, Y.; Ren, J.; Pu, Y.; Wang, P. Solar energy potential assessment: A framework to integrate geographic, technological, and economic indices for a potential analysis. *Renew. Energy* **2020**, *149*, 577–586. [CrossRef]
10. Van Gerpen, J. Biodiesel processing and production. *Fuel Process. Technol.* **2005**, *86*, 1097–1107. [CrossRef]
11. Knothe, G.; Krahl, J.; Van Gerpen, J. *The Biodiesel Handbook*; AOCS Press: Urbana, IL, USA, 2010.
12. Demirbas, A. Importance of biodiesel as transportation fuel. *Energy Policy* **2007**, *35*, 4661–4670. [CrossRef]
13. Singh, D.; Sharma, D.; Soni, S.L.; Sharma, S.; Sharma, P.K.; Jhalani, A. A review on feedstocks, production processes, and yield for different generations of biodiesel. *Fuel* **2020**, *262*, 116553. [CrossRef]
14. Wang, W.G.; Lyons, D.W.; Clark, N.N.; Gautam, M.; Norton, P.M. Emissions from Nine Heavy Trucks Fueled by Diesel and Biodiesel Blend without Engine Modification. *Environ. Sci. Technol.* **2000**, *34*, 933–939. [CrossRef]
15. Li, Q.; Zheng, L.; Cai, H.; Garza, E.; Yu, Z.; Zhou, S. From organic waste to biodiesel: Black soldier fly, *Hermetia illucens*, makes it feasible. *Fuel* **2011**, *90*, 1545–1548. [CrossRef]
16. Lu, H.; Liu, Y.; Zhou, H.; Yang, Y.; Chen, M.; Liang, B. Production of biodiesel from *Jatropha curcas* L. oil. *Comput. Chem. Eng.* **2009**, *33*, 1091–1096. [CrossRef]

17. Gao, Y.-Y.; Chen, W.-W.; Lei, H.; Liu, Y.; Lin, X.; Ruan, R. Optimization of transesterification conditions for the production of fatty acid methyl ester (FAME) from Chinese tallow kernel oil with surfactant-coated lipase. *Biomass Bioenergy* **2009**, *33*, 277–282. [CrossRef]
18. Chisti, Y. Biodiesel from microalgae. *Biotechnol. Adv.* **2007**, *25*, 294–306. [CrossRef]
19. Dortmans, B.M.A.; Diener, S.; Verstappen, B.M.; Zurbrügg, C. *Black Soldier Fly Biowaste Processing-A Step-by-Step Guide*; Eawag: Swiss Federal Institute of Aquatic Science and Technology: Dübendorf, Switzerland, 2017.
20. Kamarulzaman, M.K.; Abdullah, A.; Mamat, R. Combustion, performances, and emissions characteristics of *Hermetia illucens* larvae oil in a direct injection compression ignition engine. *Energy Sources A* **2019**, *41*, 1483–1496. [CrossRef]
21. Surendra, K.C.; Olivier, R.; Tomberlin, J.K.; Jha, R.; Khanal, S.K. Bioconversion of organic wastes into biodiesel and animal feed via insect farming. *Renew. Energy* **2016**, *98*, 197–202. [CrossRef]
22. Wang, C.; Qian, L.; Wang, W.; Wang, T.; Deng, Z.; Yang, F.; Xiong, J.; Feng, W. Exploring the potential of lipids from black soldier fly: New paradigm for biodiesel production (I). *Renew. Energy* **2017**, *111*, 749–756. [CrossRef]
23. Feng, W.; Qian, L.; Wang, W.; Wang, T.; Deng, Z.; Yang, F.; Xiong, J.; Wang, C. Exploring the potential of lipids from black soldier fly: New paradigm for biodiesel production (II)—Extraction kinetics and thermodynamic. *Renew. Energy* **2018**, *119*, 12–18. [CrossRef]
24. Su, C.H.; Nguyen, H.C.; Bui, T.L.; Huang, D.L. Enzyme-assisted extraction of insect fat for biodiesel production. *J. Clean. Prod.* **2019**, *223*, 436–444. [CrossRef]
25. Van Eck, N.J.; Waltman, L. Software survey: VOSviewer, a computer program for bibliometric mapping. *Scientometrics* **2010**, *84*, 523–538. [CrossRef] [PubMed]
26. Li, Q.; Zheng, L.; Qiu, N.; Cai, H.; Tomberlin, J.K.; Yu, Z. Bioconversion of dairy manure by black soldier fly (Diptera: Stratiomyidae) for biodiesel and sugar production. *Waste Manag.* **2011**, *31*, 1316–1320. [CrossRef]
27. Makkar, H.P.S.; Tran, G.; Heuzé, V.; Ankers, P. State-of-the-art on use of insects as animal feed. *Anim. Feed Sci. Technol.* **2014**, *197*, 1–33. [CrossRef]
28. Henry, M.; Gasco, L.; Piccolo, G.; Fountoulaki, E. Review on the use of insects in the diet of farmed fish: Past and future. *Anim. Feed Sci. Technol.* **2015**, *203*, 1–22. [CrossRef]
29. Oonincx, D.G.A.B.; Van Broekhoven, S.; Van Huis, A.; Van Loon, J.J.A. Feed conversion, survival and development, and composition of four insect species on diets composed of food by-products. *PLoS ONE* **2015**, *10*, e0144601. [CrossRef]
30. Spranghers, T.; Ottoboni, M.; Klootwijk, C.; Ovyn, A.; Deboosere, S.; De Meulenaer, B.; Michiels, J.; Eeckhout, M.; De Clercq, P.; De Smet, S. Nutritional composition of black soldier fly (*Hermetia illucens*) prepupae reared on different organic waste substrates. *J. Sci. Food Agric.* **2017**, *97*, 2594–2600. [CrossRef]
31. Wang, Y.S.; Shelomi, M. Review of black soldier fly (*Hermetia illucens*) as animal feed and human food. *Foods* **2017**, *6*, 91. [CrossRef]
32. Čičková, H.; Newton, G.L.; Lacy, R.C.; Kozánek, M. The use of fly larvae for organic waste treatment. *Waste Manag.* **2015**, *35*, 68–80. [CrossRef]
33. Zheng, L.; Li, Q.; Zhang, J.; Yu, Z. Double the biodiesel yield: Rearing black soldier fly larvae, *Hermetia illucens*, on solid residual fraction of restaurant waste after grease extraction for biodiesel production. *Renew. Energy* **2012**, *41*, 75–79. [CrossRef]
34. Zheng, L.; Hou, Y.; Li, W.; Yang, S.; Li, Q.; Yu, Z. Biodiesel production from rice straw and restaurant waste employing black soldier fly assisted by microbes. *Energy* **2012**, *47*, 225–229. [CrossRef]
35. Meneguz, M.; Schiavone, A.; Gai, F.; Dama, A.; Lussiana, C.; Renna, M.; Gasco, L. Effect of rearing substrate on growth performance, waste reduction efficiency and chemical composition of black soldier fly (*Hermetia illucens*) larvae. *J. Sci. Food Agric.* **2018**, *98*, 5776–5784. [CrossRef] [PubMed]
36. Lalander, C.; Diener, S.; Zurbrügg, C.; Vinnerås, B. Effects of feedstock on larval development and process efficiency in waste treatment with black soldier fly (*Hermetia illucens*). *J. Clean. Prod.* **2019**, *208*, 211–219. [CrossRef]
37. Gold, M.; Tomberlin, J.K.; Diener, S.; Zurbrügg, C.; Mathys, A. Decomposition of biowaste macronutrients, microbes, and chemicals in black soldier fly larval treatment: A review. *Waste Manag.* **2018**, *82*, 302–318. [CrossRef]
38. De Marco, M.; Martínez, S.; Hernandez, F.; Madrid, J.; Gai, F.; Rotolo, L.; Belforti, M.; Bergero, D.; Katz, H.; Dabbou, S.; et al. Nutritional value of two insect larval meals (*Tenebrio molitor* and *Hermetia illucens*) for broiler chickens: Apparent nutrient digestibility, apparent ileal amino acid digestibility and apparent metabolizable energy. *Anim. Feed Sci. Technol.* **2015**, *209*, 211–218. [CrossRef]
39. Renna, M.; Schiavone, A.; Gai, F.; Dabbou, S.; Lussiana, C.; Malfatto, V.; Prearo, M.; Capucchio, M.T.; Biasato, I.; Biasibetti, E.; et al. Evaluation of the suitability of a partially defatted black soldier fly (*Hermetia illucens* L.) larvae meal as ingredient for rainbow trout (*Oncorhynchus mykiss* Walbaum) diets. *J. Anim. Sci. Biotechnol.* **2017**, *8*, 1–13. [CrossRef] [PubMed]
40. Smetana, S.; Palanisamy, M.; Mathys, A.; Heinz, V. Sustainability of insect use for feed and food: Life Cycle Assessment perspective. *J. Clean. Prod.* **2016**, *137*, 741–751. [CrossRef]
41. Schiavone, A.; De Marco, M.; Martínez, S.; Dabbou, S.; Renna, M.; Madrid, J.; Hernandez, F.; Rotolo, L.; Costa, P.; Gai, F.; et al. Nutritional value of a partially defatted and a highly defatted black soldier fly larvae (*Hermetia illucens* L.) meal for broiler chickens: Apparent nutrient digestibility, apparent metabolizable energy and apparent ileal amino acid digestibility. *J. Anim. Sci. Biotechnol.* **2017**, *8*, 1–9. [CrossRef] [PubMed]

42. Schiavone, A.; Cullere, M.; De Marco, M.; Meneguz, M.; Biasato, I.; Bergagna, S.; Dezzutto, D.; Gai, F.; Dabbou, S.; Gasco, L.; et al. Partial or total replacement of soybean oil by black soldier fly larvae (*Hermetia illucens* L.) fat in broiler diets: Effect on growth performances, feed-choice, blood traits, carcass characteristics and meat quality. *Ital. J. Anim. Sci.* **2017**, *16*, 93–100. [CrossRef]
43. Higa, J.E.; Ruby, M.B.; Rozin, P. Americans' acceptance of black soldier fly larvae as food for themselves, their dogs, and farmed animals. *Food Qual. Prefer.* **2021**, *90*, 104119. [CrossRef]
44. Li, W.; Dong, H.; Yu, H.; Wang, D.; Yu, H. Global characteristics and trends of research on ceramic membranes from 1998 to 2016: Based on bibliometric analysis combined with information visualization analysis. *Ceram. Int.* **2018**, *44*, 6926–6934. [CrossRef]
45. Barragan-Fonseca, K.B.; Dicke, M.; van Loon, J.J.A. Nutritional value of the black soldier fly (*Hermetia illucens* L.) and its suitability as animal feed-a review. *J. Insects Food Feed* **2017**, *3*, 105–120. [CrossRef]
46. Salomone, R.; Saija, G.; Mondello, G.; Giannetto, A.; Fasulo, S.; Savastano, D. Environmental impact of food waste bioconversion by insects: Application of Life Cycle Assessment to process using *Hermetia illucens*. *J. Clean. Prod.* **2017**, *140*, 890–905. [CrossRef]
47. Liang, K.; Li, W.; Wen, J.; Ai, W.; Wang, J. Research characteristics and trends of power sector carbon emissions: A bibliometric analysis from various perspectives. *Environ. Sci. Pollut. Res.* **2022**, *in press*. [CrossRef] [PubMed]
48. Nguyen, H.C.; Liang, S.H.; Li, S.Y.; Su, C.H.; Chien, C.C.; Chen, Y.J.; Huong, D.T.M. Direct transesterification of black soldier fly larvae (*Hermetia illucens*) for biodiesel production. *J. Taiwan Inst. Chem. Eng.* **2018**, *85*, 165–169. [CrossRef]
49. Kamari, A.; Ishak, S.; Hussin, M.I.A.M.; Wong, S.T.S.; Jumadi, J.; Yahaya, N.M. Optimisation and characterisation studies of biodiesel production from black soldier fly larvae fed by soya residue. *IOP Conf. Ser. Mater. Sci. Eng.* **2020**, *980*, 012057. [CrossRef]
50. Wong, C.Y.; Lim, J.W.; Chong, F.K.; Lam, M.K.; Uemura, Y.; Tan, W.N.; Bashir, M.J.K.; Lam, S.M.; Sin, J.C.; Lam, S.S. Valorization of exo-microbial fermented coconut endosperm waste by black soldier fly larvae for simultaneous biodiesel and protein productions. *Environ. Res.* **2020**, *185*, 109458. [CrossRef]
51. Wong, C.Y.; Rosli, S.S.; Uemura, Y.; Ho, Y.C.; Leejeerajumnean, A.; Kiatkittipong, W.; Cheng, C.K.; Lam, M.K.; Lim, J.W. Potential protein and biodiesel sources from black soldier fly larvae: Insights of larval harvesting instar and fermented feeding medium. *Energies* **2019**, *12*, 1570. [CrossRef]
52. Ishak, S.; Kamari, A. Biodiesel from black soldier fly larvae grown on restaurant kitchen waste. *Environ. Chem. Lett.* **2019**, *17*, 1143–1150. [CrossRef]
53. Ishak, S.; Kamari, A.; Yusoff, S.N.M.; Halim, A.L.A. Optimisation of biodiesel production of Black Soldier Fly larvae rearing on restaurant kitchen waste. *J. Phys. Conf. Ser.* **2018**, *1097*, 012052. [CrossRef]
54. Li, W.; Li, Q.; Zheng, L.; Wang, Y.; Zhang, J.; Yu, Z.; Zhang, Y. Potential biodiesel and biogas production from corncob by anaerobic fermentation and black soldier fly. *Bioresour. Technol.* **2015**, *194*, 276–282. [CrossRef] [PubMed]
55. Elsayed, M.; Ran, Y.; Ai, P.; Azab, M.; Mansour, A.; Jin, K.; Zhang, Y.; Abomohra, A.E.F. Innovative integrated approach of biofuel production from agricultural wastes by anaerobic digestion and black soldier fly larvae. *J. Clean. Prod.* **2020**, *263*, 121495. [CrossRef]
56. Jung, S.; Jung, J.M.; Tsang, Y.F.; Bhatnagar, A.; Chen, W.H.; Lin, K.Y.A.; Kwon, E.E. Biodiesel production from black soldier fly larvae derived from food waste by non-catalytic transesterification. *Energy* **2022**, *238*, 121700. [CrossRef]
57. Nguyen, H.C.; Liang, S.-H.; Chen, S.-S.; Su, C.-H.; Lin, J.-H.; Chien, C.-C. Enzymatic production of biodiesel from insect fat using methyl acetate as an acyl acceptor: Optimization by using response surface methodology. *Energy Conv. Manag.* **2018**, *158*, 168–175. [CrossRef]
58. Nguyen, H.C.; Liang, S.-H.; Doan, T.T.; Su, C.-H.; Yang, P.-C. Lipase-catalyzed synthesis of biodiesel from black soldier fly (*Hermetica illucens*): Optimization by using response surface methodology. *Energy Convers. Manag.* **2017**, *145*, 335–342. [CrossRef]
59. Nguyen, H.C.; Nguyen, M.L.; Liang, S.H.; Su, C.H.; Wang, F.M. Switchable Solvent-Catalyzed Direct Transesterification of Insect Biomass for Biodiesel Production. *Bioenergy Res.* **2020**, *13*, 563–570. [CrossRef]
60. Singh, A.; Kumari, K. An inclusive approach for organic waste treatment and valorisation using Black Soldier Fly larvae: A review. *J. Environ. Manag.* **2019**, *251*, 109569. [CrossRef] [PubMed]

sustainability

MDPI

Article

Valorization of Oil Palm Empty Fruit Bunch for Cellulose Fibers: A Reinforcement Material in Polyvinyl Alcohol Biocomposites for Its Application as Detergent Capsules

Jia Ying Tan [1], Wah Yen Tey [2,3,4], Joongjai Panpranot [5], Steven Lim [6,*] and Kiat Moon Lee [1,4,*]

1 Department of Chemical & Petroleum Engineering, Faculty of Engineering, Technology and Built Environment, UCSI University, Kuala Lumpur 56000, Malaysia
2 Department of Mechanical Engineering, Faculty of Engineering, Technology and Built Environment, UCSI University, Kuala Lumpur 56000, Malaysia
3 Department of Mechanical Precision Engineering, Malaysia-Japan International Institute of Technology, Universiti Teknologi Malaysia, Kuala Lumpur 54100, Malaysia
4 UCSI-Cheras Low Carbon Innovation Hub Research Consortium, UCSI University, Kuala Lumpur 56000, Malaysia
5 Department of Chemical Engineering, Faculty of Engineering, Chulalongkorn University, Bangkok 10330, Thailand
6 Department of Chemical Engineering, Lee Kong Chian Faculty of Engineering and Science, Universiti Tunku Abdul Rahman, Kajang 43000, Malaysia
* Correspondence: stevenlim@utar.edu.my (S.L.); leekm@ucsiuniversity.edu.my (K.M.L.)

Abstract: Cellulose fibers isolated from oil palm empty fruit bunches (OPEFB) have been studied as a potential reinforcement for polyvinyl alcohol (PVA) biocomposite. Analysis of variance (ANOVA) showed that all three parameters—hydrolysis temperature, time and acid concentration, as well as their interactions—significantly affected the yield of cellulose. Moving Least Squares (MLS) and Multivariable Power Least Squares (MPLS) models demonstrated good fitness. The model also proved that acid concentration was the dominant parameter, supported by the Fourier transform infrared spectroscopy (FTIR) analysis. Hydrolysis using 54% acid at 35 °C and 15 min achieved the highest cellulose yield of 80.72%. Cellulose-reinforced PVA biocomposite films demonstrated better mechanical strength, elongation at break, moisture barrier properties, thermal stability and poorer light transmission rate compared to neat PVA due to the high aspect ratio, crystallinity and good compatibility of cellulose fibers. These findings suggested the potential of cellulose fibers-reinforced PVA biocomposite film as water-soluble detergent capsules.

Keywords: oil palm empty fruit bunch; polyvinyl alcohol; cellulose fiber; biocomposite; Moving Least Square; Multivariate Power Least Square

Citation: Tan, J.Y.; Tey, W.Y.; Panpranot, J.; Lim, S.; Lee, K.M. Valorization of Oil Palm Empty Fruit Bunch for Cellulose Fibers: A Reinforcement Material in Polyvinyl Alcohol Biocomposites for Its Application as Detergent Capsules. *Sustainability* **2022**, *14*, 11446. https://doi.org/10.3390/su141811446

Academic Editor: Elio Dinuccio

Received: 13 June 2022
Accepted: 29 July 2022
Published: 13 September 2022

1. Introduction

Over the years, the demand for packaging materials has been rising rapidly due to the growing population worldwide and the demand for more convenient products with extended shelf life. Plastic materials offer outstanding qualities such as lightweight, hygienic, high strength, versatile, durable and flexible, making them the first choice for most packaging applications. Nevertheless, plastics are generally non-biodegradable and usually take thousands of years to decay. The associated environmental issues such as plastic pollution [1], increased emissions of greenhouse gasses [2] and global warming [3] have encouraged the development of "greener" or environmental-friendly materials to replace conventional petroleum-based plastics. These materials are usually biodegradable. In addition, they are extracted from renewable biomass sources such as crops and lignocellulosic biomass, promoting sustainability.

Polyvinyl alcohol (PVA) is a colorless polymer that is water-soluble and highly biodegradable. It is widely used in industries with good commercial value, given several advantages, such as good film formation, strong conglutination, high thermal stability and flexibility [4]. In addition, PVA-based films or composites possess high moisture absorption properties due to hydrophilic hydroxyl groups. Another unique feature of PVA is its low water barrier properties. These properties favor its application in the laundry detergent industry. In recent years, detergent capsules have gained interest in the laundry and home care sector [5]. The modern innovation of detergent capsules concentrates and packs the active ingredients of liquid detergent in a single unit dose. The capsule will dissolve upon contact with water and its content will release. Despite the favorable characteristics mentioned earlier, PVA films experienced low elongation at break, poor decomposition temperature and low glass transition temperature [6]. These make their usage as eco-friendly packaging materials very challenging. Much research proposed incorporating other polymers or fillers into the PVA matrix to improve its mechanical properties [7–9].

Cellulose fiber-reinforced biocomposites have attracted significant attention due to their appealing properties such as low density, non-toxic, low cost, non-abrasive and low resistance to biodegradability [10]. There are many applications of cellulose fibers in this context. To name a few, cellulose fibers can be oxidized for the preparation of poly(methyl methacrylate) (PMMA) nanocomposites [11], poly(2-acrylamido-2-methylpropanesulfonic acid/poly(acrylic acid (PAMPS/PAA) hydrogels [12], polygluronic acid [13] and bioadsorbent for paraquat [14]. In addition, various works have proven that adding cellulose fibers into PVA could provide a significant reinforcing effect to tackle the poor mechanical properties of the PVA film [15–18]. Cellulose can be extracted from agricultural wastes such as rice husk, corn stover, palm oil bunches, etc. [19–21]. In Malaysia, about 3.31 million hectares of land are under oil palm cultivation [22], and the amount of biomass wastes generated is an average of 231.5 kg dry weight/year [23]. Oil palm biomass wastes consist of the oil palm empty fruit bunches (OPEFB), oil palm trunk (OPT) and oil palm fronds (OPF). These wastes are of low commercial value and are usually landfilled. Therefore, incorporating cellulose fibers from agricultural waste into the PVA matrix to form cellulose fibers-reinforced PVA biocomposite film as water-soluble detergent capsules could promote sustainability.

In the present work, cellulose was isolated from the OPEFB and incorporated with PVA to form cellulose-reinforced PVA biocomposite film. OPEFB was selected as the raw material because of its high cellulose content, i.e., as high as 40% to 65% reported [24–26]. In order to maximize the recovery of cellulose fibers from OPEFB, it is vital to investigate the effect of different process parameters on the efficiency of cellulose fiber isolation during the hydrolysis reaction. The Moving Least Squares (MLS) and Multivariable Power Least Squares (MPLS) methods were applied in combination as the numerical tools for regression modeling to study the process parameters correlations in cellulose isolation from the biomass. The MLS method, proposed by Lancaster and Salkauskas [27], can produce a continuous equation in a successive series of arbitrary functions. Meanwhile, the MPLS method [28] could correlate the multivariate data as a power function. The MPLS method applied in this work aims to investigate the significance level of manipulated variables and provide a regression model to represent the hydrolysis reaction. The MLS method has been widely applied in finite element analysis and image processing, and the method has potential to be used for parameter study. Meanwhile, the MPLS method was recently proposed in our previous work [28] as an alternative for design of an experiment tool. In this article, we combined both methods to form the novel MLS–MPLS method, which is new yet robust in parameter study. Last but not least, the effect of different cellulose loadings in the PVA matrix was evaluated by comparing the physical, mechanical, moisture barrier and thermal properties of the biocomposite films formed.

2. Materials and Methods

2.1. Materials

OPEFB was donated by Kwantas Oil Sdn Bhd, Malaysia. The biomass was washed thoroughly with distilled water and dried in an oven (Carbolite AX 120, Verder Scientific, Derbyshire, UK) at 60 °C for 24 h. The dried OPEFB underwent size reduction and was screened through a mesh size of 0.5 mm. The sample was stored in an air-tight container before being used. Sodium hydroxide (99% purity), acetic acid glacial (99.8% purity), sodium chlorite (80% purity), sulphuric acid (95–98% purity) and polyvinyl alcohol (>95% purity) were the commercial source, and they were used without further purification.

2.2. Alkaline and Bleaching Treatment of OPEFB

Alkaline pretreatment of OPEFB was conducted with 4% (*w*/*v*) sodium hydroxide solution at a solid-to-liquid ratio of 1:30. The reaction mixture was heated at 70 °C for 3 h under continuous stirring, as suggested by Zailuddin and Husseinsyah [29]. This pretreatment process was repeated three times to ensure high lignin degradation to facilitate the extraction of cellulosic fibers from the OPEFB. In each pretreatment, the pretreated fibers were filtered and washed with distilled water to achieve pH 7 to ensure complete removal of residual alkali, followed by drying in the oven at 90 °C overnight. The resulting dried fibers were subjected to bleaching treatment by mixing the fibers with a solution of equal parts of acetic acid buffer (2.7 g sodium hydroxide and 7.5 mL acetic acid in 100 mL of distilled water) and aqueous sodium chlorite (1.7%, *w*/*v*), at a solid-to-liquid ratio of 1:20. The bleaching treatment was performed at 80 °C for 4 h under continuous stirring. The bleached fibers were filtered, washed and dried in the oven at 90 °C overnight.

2.3. Isolation of Cellulose through Acid Hydrolysis

Acid hydrolysis was performed by mixing 1 g of bleached OPEFB fibers with 10 mL sulphuric acid. The reaction was performed at three different acid concentrations: 54%, 60% and 64%, and temperatures: 35 °C, 45 °C and 55 °C, for 60 min with sampling at 15 min intervals. Hydrolysis was performed under constant stirring at 85 rpm. Soon after the hydrolysis reaction, 20 mL of cold distilled water was added to stop the reaction. The cellulose suspension was immediately subjected to the washing process under centrifugation (Universal 320R, Hettich, UK) at 8000 rpm for 15 min. The washing process was repeated until the supernatant showed pH 7. Lastly, the drying process was carried out in the oven at 40 °C overnight. The weight of extracted cellulose was recorded, and the yield was computed using Equation (1):

$$\text{Yield of cellulose (\%)} = w_c / w_{\text{treatedOPEFB}} \times 100 \quad (1)$$

where w_c and $w_{\text{treatedOPEFB}}$ represent the weight of dried cellulose after acid hydrolysis (g) and the weight of dried, bleached OPEFB (g), respectively. All experiments were conducted in duplicate.

2.4. Preparation of Cellulose-Reinforced PVA Biocomposite Films

Biocomposite films containing PVA and cellulose fibers were formed through solution casting method. PVA pellets were soaked in distilled water (10%, *w*/*v*) for 1 h to enhance their solubility, followed by heating the mixture at 97 °C for 30 min under continuous stirring. Cellulose fibers of different loadings (1, 2.5 and 5%, *w*/*w*) were added to the PVA solution, and the mixtures were stirred for 1 h. The mixed polymers solution of 5 g was cast on a Petri dish and dried at 40 °C for 24 h. A thin film of biocomposite was formed once the polymer solution was completely dried. A control sample (known as neat PVA) was formed by repeating the experiment without adding cellulose fibers.

2.5. Moving Least Squares (MLS) Method

MLS method is a method to construct a continuous function from a set of scattered data, as shown in Equation (2):

$$y^h = \sum_{j=1}^{m} P_j(\boldsymbol{x})\mu_j = \mathbf{P}^{\mathrm{T}}\boldsymbol{\mu} \tag{2}$$

where y^h is the predicted response, P is the user-defined polynomial interpolants, μ is their associating coefficients, $\boldsymbol{x} = [x_1\ x_2\ \ldots\ x_m]$ is the manipulated factors, while m is the number of manipulated variables. The discrete norm can be determined using Equation (3):

$$J = \sum_{i=1}^{n} W_i R^2 = \sum_{i=1}^{n} W_i \left[\left(\sum_{i=1}^{m} (P_j(x)\mu_j) \right)_i - y_i \right]^2 \tag{3}$$

where n is the number of variables within the specific set of manipulated variables, while y represents the actual response. W is the weight function, which can be defined using a quartic spline function as shown in Equation (4):

$$W_i = 1 - 6{r_i}^2 + 8{r_i}^3 - 3{r_i}^4 \tag{4}$$

where

$$r = \frac{|x - \mathbf{x}|}{\|\boldsymbol{x}\|} \tag{5}$$

In the MLS method, the coefficients $\boldsymbol{\mu}$ can be determined by taking the minimization of the discrete norm such that:

$$\frac{\partial J}{\partial \alpha} = 0 \rightarrow \mathbf{A}\boldsymbol{\mu} = \mathbf{B} \rightarrow \boldsymbol{\mu} = \mathbf{A}^{-1}\mathbf{B} \tag{6}$$

where

$$\mathbf{A} = \sum_{i=1}^{n} W_i P_i(x_i) {P_i}^T(x_i) \tag{7}$$

$$\mathbf{B} = W_i P_i(x_i) \tag{8}$$

2.6. Multivariate Power Least Squares (MPLS) Method

MPLS method was developed through modification of MLS model and presented as [28]:

$$y^h = a{x_1}^{b_1}{x_2}^{b_2}\cdots{x_m}^{b_m} \tag{9}$$

where a is the MPLS coefficient while $\mathbf{b} = [b_1\ b_2\ \ldots\ b_\mathrm{m}]$ is the index for $\boldsymbol{x}$. The discrete norm can be obtained as:

$$J = \sum_{i=1}^{n} \left(\mathrm{In}(y_i) - \mathrm{In}\left({y_i}^h\right) \right)^2 \tag{10}$$

Upon minimizing Equation (10), the value of a and $\mathbf{b}$ can be determined by solving Equation (11) and the matrix in Equation (12), respectively.

$$a = \exp\left[\frac{1}{n} \left(\left(\sum_{i=1}^{n} \mathrm{In}(y_i) \right) - \sum_{j=1}^{m} \left(b_j \sum_{i=1}^{n} \sum_{j=1}^{m} \mathrm{In}(x_{j,i}) \right) \right) \right] \tag{11}$$

$$\begin{pmatrix} \gamma_1 & -\lambda_{12} & \cdots & -\lambda_{1m} \\ -\lambda_{21} & \gamma_2 & \cdots & -\lambda_{2m} \\ \vdots & \vdots & \ddots & \vdots \\ -\lambda_{m1} & -\lambda_{m2} & \cdots & \gamma_m \end{pmatrix} \begin{pmatrix} b_1 \\ b_2 \\ \vdots \\ b_m \end{pmatrix} = \begin{pmatrix} \xi_1 \\ \xi_2 \\ \vdots \\ \xi_m \end{pmatrix} \tag{12}$$

where

$$\gamma_j = n\left\{\sum_{i=1}^{n}\sum_{j=1}^{m}[\text{In}(x_{j,i})]^2\right\} - \left\{\sum_{i=1}^{n}\sum_{j=1}^{m}[\text{In}(x_{j,i})]\right\}^2 \tag{13}$$

$$\xi_j = n\left[\sum_{i=1}^{n}\sum_{j=1}^{m}\text{In}(x_{j,i})\text{In}(y_j)\right] - \left[\sum_{i=1}^{n}\sum_{j=1}^{m}\text{In}(x_{j,i})\right]\left[\sum_{i=1}^{n}\text{In}(y_j)\right] \tag{14}$$

$$\lambda_{j,k} = \left[\sum_{i=1}^{n}\sum_{j=1}^{m}\text{In}(x_{j,i})\right]\left[\sum_{i=1}^{n}\sum_{j=1}^{m}\text{In}(x_{k,i})\right] - n\left[\sum_{i=1}^{n}\sum_{j=1}^{m}\text{In}(x_{j,i})\text{In}(x_{k,i})\right] \tag{15}$$

The subscript k represents the value of the second subscript of the term λ.

A normalized x, X as expressed in Equation (16) can be used to obtain the normalized MPLS equation:

$$X = C + \frac{x - x_{\min}}{x_{\max} - x_{\min}} \tag{16}$$

where C is a positive integer, and in this paper, $C = 1$ is applied. The analysis of the significance level can be conducted by comparing the magnitude of index **b**. The larger the magnitude of **b**, the more significant the manipulated variable is. However, the significance level analysis is only valid if the coefficient of determination R^2 for Equation (9) exceeds 0.5 [28]. The coefficient of determination, R^2 can be defined mathematically as in Equation (17):

$$R^2 = \frac{\sum_{i=1}^{n}\left(y^h{}_i - \overline{y}\right)^2}{\sum_{i=1}^{n}(y_i - \overline{y})^2} \tag{17}$$

where $\overline{y}$ is the average value for y.

2.7. Scanning Electron Microscopy (SEM) Analysis

Surface morphology of raw OPEFB, bleached-OPEFB and isolated cellulose fibers was observed using a scanning electron microscope (Hitachi VP-SEM SU1510, Hitachi, Japan), operating at an accelerating voltage of 15 kV. Before the examination, the specimens were sputter-coated with gold using Hitachi E1010 Ion Sputter (Japan) to prevent electrostatic charging and poor resolution. Few images were taken at different magnifications, and the sizes of fibers were also measured for comparison purposes.

2.8. Fourier Transform Infrared Spectroscopy (FTIR) Analysis

Fourier transform infrared spectroscopy (FTIR) analysis was conducted using FTIR Spectrometer (PerkinElmer Spectrum 400, PerkinElmer, USA) to identify the variation in the chemical compositions in the OPEFB fibres after each treatment. About 0.5 mg to 1.0 mg sample was mixed with potassium bromide (KBr), and the mixture was pressed to form a disk. The sample was scanned with the spectra transmittance region between a wavenumber 4000 cm^{-1} to 500 cm^{-1} at resolution of 4 cm^{-1}.

2.9. Transparency Analysis

Transparency of the biocomposite films was measured with the aid of Secomam UviLine spectrophotometer 9400 (France). The biocomposite film was cut into a rectangular shape of 5 cm × 2.5 cm (length × width) and placed at the cuvette position. The light absorbance of the film was measured for wavelength ranging from 190 nm to 400 nm. Transparency value of the film was calculated using Equation (18) [30]:

$$\text{Transpancy value} = A_{400/}x \tag{18}$$

where A_{400} is the light absorbance value at 400 nm wavelength, and x is the thickness of the film in mm.

2.10. Mechanical Properties

Mechanical properties for the biocomposite films were carried out according to the ASTM D882 standard using a tensile machine (Zwicki Z5.0TH, Zwick Roell, Germany). Each sample was cut into a rectangular shape of 15 cm × 2.5 cm (length × width). A stress-strain curve was obtained to study the mechanical performance of biocomposite films in terms of tensile strength, percentage of elongation at break, and Young's Modulus. The values of the tensile properties are the averages of five measurements.

2.11. Water Vapor Transmission Rate (WVTR) Test

Water vapor transmission rate (WVTR) test was performed based on the standard testing method specified under ASTM E96. Before the test, the biocomposite films were placed in a drying cabinet for 5 days to remove the film's moisture. A plastic cup was used as an impermeable dish, some silica gels were placed inside the cup, and the cup's mouth was firmly sealed with biocomposite film. The whole assembly was weighed before placing in a controlled environment (desiccator) along with a plastic cup of distilled water. The assembly was taken from the desiccator and weighted every 24 h. The experiment was conducted for 5 consecutive days. Subsequently, the WVTR was calculated using Equation (19):

$$\mathrm{WVTR}\ ((\mathrm{g/day})/\mathrm{m}^2) = G/(t \times S) \tag{19}$$

where G is the weight gain of the biocomposite film (g), t is the testing duration (day), and S is the exposed surface area of the film (m^2).

2.12. Thermagravimetric Analysis (TGA)

Thermagravimetric analysis (TGA) was carried out using the thermal analysis instrument (Q50, TA Instrument, USA). It was carried out under nitrogen atmosphere with a heating rate of 20 °C/min. The weight of the film was maintained, and heating scans were conducted in a temperature range from 30 °C to 600 °C.

3. Results and Discussion

3.1. Cellulose Isolation via Acid Hydrolysis

The compact arrangement of cellulose, hemicellulose and lignin within the OPEFB makes the biomass recalcitrance and limits the accessibility to its cellulose component. To isolate the high quality of cellulose from the biomass, pretreatment of OPEFB is essential. Alkaline pretreatment was applied to break the linkage between the lignin-carbohydrate complex to increase the accessibility of carbohydrates such as cellulose. Further treatment of the OPEFB with sodium chlorite increased the cellulose percentage in the OPEFB by removing the lignin and hemicellulose components. Subsequent acid hydrolysis removed the amorphous regions of the cellulose and retained its crystalline regions [31], resulting in agglomerated cellulose in micrometer size. The dissolution of cellulose in acid hydrolysis was affected by its process parameters. The following sections discuss their individual and interaction effects and significance level in detail.

3.1.1. Effect of Hydrolysis Temperature and Time on the Yield of Cellulose

The effect of acid hydrolysis temperature and time on the yield of cellulose was presented in Table 1. At the temperature of 35 °C, the highest cellulose yield of 80.72% was obtained at the hydrolysis condition of 35 °C and 15 min. With an increment of 10 °C in hydrolysis temperature, the cellulose yield decreased to 78.18% and subsequently to 68.42% at 55 °C, at the same hydrolysis time and acid concentration. This observation implies that temperature favored the hydrolysis reaction, and as a result, cellulose isolation suffered due to extensive depolymerization. Similarly, the hydrolysis trend was reported by Zhang and his co-authors [32]. This phenomenon is because a higher hydrolysis temperature encourages the breakage of glycosidic bonds by protons (H^+) within the cellulose structure. On top of that, high temperature not just hydrolyzed the disordered amorphous domains of cellulose but also has the potential to hydrolyze the highly ordered crystalline domains of cellulose [33].

Table 1. Average yield of cellulose at various hydrolysis temperatures and times.

Acid Concentration (%)	Temperature (°C)	Time (min)	Yield of Cellulose (%)
54	35	15	80.72 ± 0.31
		30	79.09 ± 0.56
		45	78.36 ± 0.02
		60	75.63 ± 0.18
	45	15	78.18 ± 0.59
		30	72.64 ± 0.20
		45	73.14 ± 0.24
		60	71.93 ± 1.11
	55	15	68.42 ± 1.58
		30	50.29 ± 1.91
		45	58.41 ± 0.55
		60	50.51 ± 0.92

For all the investigated hydrolysis temperatures, isolation of cellulose increased gradually with the reduction of hydrolysis time. The crystalline regions of cellulose were converted to amorphous cellulose during the acid hydrolysis. Therefore, the amount of cellulose isolated reduced with hydrolysis time. An analysis of variance (ANOVA) study was conducted to examine the significance level of each parameter and its interaction on the yield of cellulose isolated. ANOVA for the yield of cellulose based on hydrolysis temperature and time is presented in Supplementary Data (Table S1). F_0 for hydrolysis temperature, time and temperature–time interaction was compared to $F_{critical}$ using α value of 0.05 as the significance level. All F_0 values were greater than their corresponding $F_{critical}$ values. Therefore, it can be concluded that the main effect of hydrolysis temperature and time were significant, as well as the interaction between the hydrolysis temperature and time.

3.1.2. Effect of Acid Concentration and Time on the Yield of Cellulose

Table 2 shows the effect of acid concentration and hydrolysis time on the yield of cellulose. Among the acid concentrations investigated, the lowest acid concentration of 54% yielded the highest amount of cellulose. Relatively lower cellulose yields were obtained at higher acid concentrations at all reaction times. As a result, the lowest cellulose yield of 13.10% was obtained for acid hydrolysis with 64% acid concentration for 60 min. The low yield of cellulose was suspected due to the extreme dissolution of the cellulose caused by high acid concentration. Concentrated acid has the potential to cleave the hydrogen bonds holding cellulose in the crystalline state [34], thus converting it to its monomer (sugars) and resulting in lower cellulose isolation. A similar report was found in Hamelinck et al. [35], which claimed that complete cellulose hydrolysis to sugar monomers at low temperatures was possible when a long reaction time and a highly concentrated acid were applied. The amorphous cellulose was degraded to undesired products of sugar monomers at a high acid concentration [36].

From Table 2, a similar trend of cellulose yield can be observed as shown in the effect of temperature discussed earlier, where the amount of cellulose isolated progressively decreased with hydrolysis time regardless of the acid concentration used. The effect of hydrolysis time was more pronounced at a higher acid concentration of 64%. The cellulose yield was drastically reduced from 60.36% to 13.10% when acid hydrolysis was performed at 15 min and 60 min, respectively. The longer the hydrolysis time, the more cellulose was converted into simple glucose [37], resulting in less cellulose isolated from the OPEFB. ANOVA on the yield of cellulose isolated based on acid concentration and hydrolysis time is available in Supplementary Data (Table S2). Similarly, all F_0 values were greater than their corresponding $F_{critical}$ values, suggesting that the acid concentration, time and their interaction significantly affected cellulose yield.

Table 2. Average yield of cellulose at various acid concentrations and hydrolysis times.

Temperature (°C)	Acid Concentration (%)	Time (min)	Yield of Cellulose (%)
35	54	15	80.72 ± 0.31
		30	79.09 ± 0.56
		45	78.36 ± 0.02
		60	75.63 ± 0.18
	60	15	69.37 ± 0.28
		30	68.30 ± 0.51
		45	63.37 ± 1.94
		60	60.38 ± 2.28
	64	15	60.36 ± 0.69
		30	34.19 ± 1.12
		45	16.10 ± 1.57
		60	13.10 ± 0.39

3.2. Process Modeling Using MLS and MPLS Methods

Mathematical modeling was constructed via least squares methods, particularly the MLS and MPLS methods. The models obtained for these two regression methods are presented in Equations (20) and (21), respectively.

$$y^h = -1650.7524 + 68.1638x_1 - 15.7560x_2 + 4.8021x_3 - 0.7042{x_1}^2 \\ -0.0632{x_2}^2 + 0.0075{x_3}^2 + 0.3879x_1x_2 - 0.0118x_2x_3 - 0.0914x_1x_3 \quad (20)$$

$$y^h = 1.1345{x_1}^{-1.4582}{x_2}^{-0.6287}{x_3}^{-0.6153} \quad (21)$$

where y^h, x_1, x_2, and x_3 denote yield of cellulose (%), acid concentration (%), hydrolysis temperature (°C) and time (min), respectively. The coefficient of determination R^2 for the MLS and MPLS models were 0.9725 and 0.7693, respectively. Since the R^2 obtained were larger than 0.5 indicated the goodness of the models in representing the reaction. The fitness of the models can also be visualized graphically using predicted value versus actual value plot, as shown in Figure 1. The closeness of the data to the diagonal lines signified the excellent representation of the regression models to the actual situation.

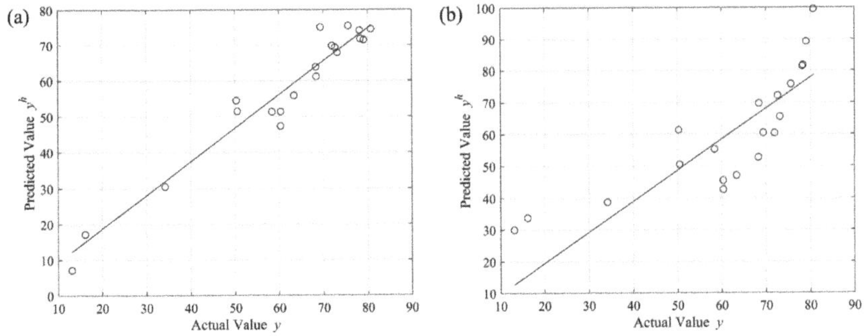

Figure 1. Predicted value versus actual value plot for the yield of cellulose using (**a**) MLS method and (**b**) MPLS method.

The significance of process parameters can be evaluated through the indices of the factors in the MPLS model. The higher the index of the factor, the greater the significance level of the parameter towards the response. Therefore, the most dominant parameter affecting the yield of cellulose was acid concentration (x_1), followed by hydrolysis temperature (x_2) and time (x_3). In acid hydrolysis, H^+ ions tend to attack oxygen atoms in the β-1,4-glycosidic bond within cellulose polymers, resulting in a shorter chain of sugar molecules [38]. By using concentrated acid, the decomposition rate of the cellulose was accelerated as the concentration of H^+ ions increased. Therefore, acid concentration was the most significant parameter affecting the cellulose isolation efficiency from the OPEFB. The MPLS model could also provide information on the interactions between parameters in affecting the cellulose yield. The positive coefficients are expected to cause a positive impact on the yield, while negative coefficients indicate that the parameters would negatively affect the cellulose yield. Based on Equation (21), all the three parameters are negative coefficients, which agreed with the discussion made earlier, whereby the cellulose yield was favored at low acid concentration, low hydrolysis temperature and short hydrolysis time.

3.3. *Morphology of Cellulose Fibers*

The surface morphology of cellulose isolated from the OPEFB through acid hydrolysis was compared with raw and bleached OPEFB. As shown in Figure 2, significant physical differences were observed from the SEM images after each chemical treatment. Initially, the raw OPEFB was coated with irregular deposition of residual waxes, lignin or other inorganic substances (Figure 2a), which was similar to other findings reported earlier [39,40]. After pretreated with an alkaline solution and bleached with acidified sodium chlorite, the surface of the fibers became cleaner and rougher, where more fibers were exposed, as shown in Figure 2b. The clean fibers' surface was due to the removal of unwanted substances such as wax and cuticle through the interaction with sodium ions during the alkaline pretreatment [41]. The elimination of lignin increased the contact area, where the fibers became more exposed during the bleaching treatment.

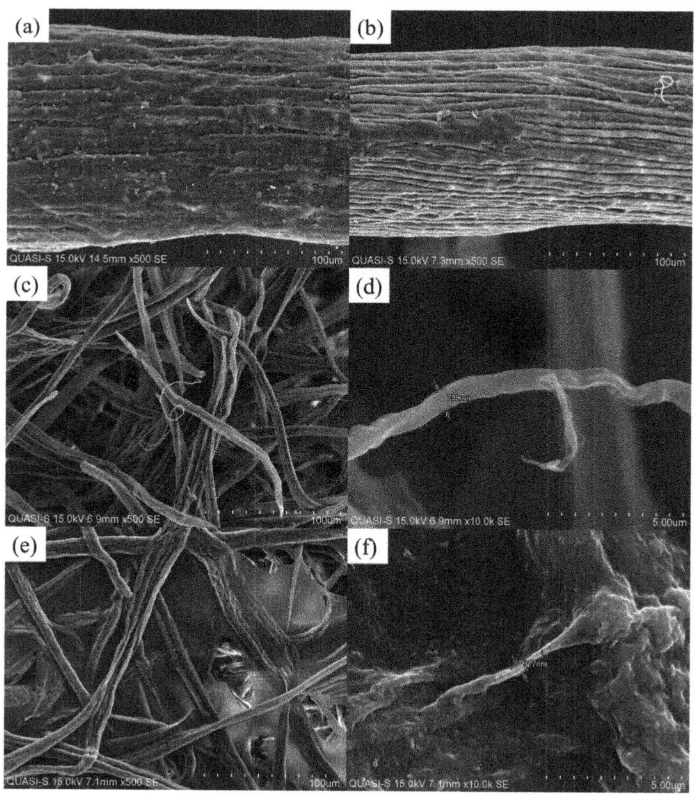

Figure 2. SEM images of (**a**) raw OPEFB fiber; (**b**) bleached OPEFB fiber; (**c**,**d**) isolated cellulose (35 °C, 15 min and 54%); (**e**,**f**) isolated cellulose (35 °C, 15 min and 60%).

Further treatment with sulphuric acid caused the formation of individual long cellulose fibrils, which were segregated from the thick fiber bundles (Figure 2c,e). The SEM images have revealed that highly purified cellulose was successfully isolated from the OPEFB with the removal of lignin and hemicellulose. A similar observation was reported by Chimentão and his co-authors [42]. It can be seen that the bundle sizes of fibers were gradually reduced after every treatment. Eventually, micro-size cellulose fibers were obtained. As depicted in Figure 2d, the fiber diameter of 739 nm was observed for cellulose isolated under the condition of 35 °C, 15 min and 54% acid concentration. At the higher acid concentration of 60% under the same hydrolysis temperature and time, the fibers obtained were of smaller diameter of approximately 317 nm, as shown in Figure 2f. This observation further confirmed that a stronger acid was more effective in cellulose hydrolysis and produced cellulose fibers of smaller size, but at the same time suffered from a lower cellulose yield due to extensive depolymerization and product degradation.

3.4. Structural Analysis of Cellulose Fibers

FTIR spectroscopy is a useful tool to monitor the change of functional groups in samples. FTIR spectra of raw OPEFB fibers, bleached OPEFB fibers and cellulose fibers are illustrated in Figure 3a. From the figure, few peaks (1735 cm^{-1} and 1228 cm^{-1}) in the raw OPEFB fibers sample were not found in the bleached OPEFB fibers and cellulose fibers spectra. Peak at 1735 cm^{-1} is associated with the presence of the acetyl and uronic ester groups of hemicelluloses or the ester carbonyl groups of lignin [43,44], while peak at 1228 cm^{-1} is attributed to the syringyl ring unit and C–O stretching in lignin and xylan [45]. The disappearance of these peaks in the spectra of bleached OPEFB fiber and cellulose fiber verified the removal of hemicelluloses and lignin during the bleaching and hydrolysis of OPEFB fibers.

FTIR spectra of the treated and untreated fibers showed a broad absorption band at 3340 and 2900 cm^{-1}. These peaks correspond to the OH stretching vibration in cellulose and C–H stretching vibration of CH_2 and CH_3, respectively [44,46]. Peak at 3340 cm^{-1} can also be attributed to the hydroxyl groups in cellulose [45]. Peak at 1035 cm^{-1} appeared in all samples attributed to the C–O–C pyranose ring (anti-symmetric in phase ring) stretching vibration [47]. From the figure, sharper peaks were observed after each chemical treatment. This was due to the removal of non-cellulosic components in the OPEFB fiber during the treatment processes.

Figure 3b–d shows the FTIR spectra of cellulose fibers with variations in hydrolysis temperature, time and acid concentration, respectively. All samples displayed similar spectra, suggesting no changes in their chemical composition or new bonds formation during the acid hydrolysis. Same vibration bands at 3340, 2900, 1635, 1427, 1373, 1035, and 894 cm^{-1} were found in all cellulose fibers. Peak at 1635 cm^{-1} originates from the bending vibrations of the OH groups of cellulose [48,49]. Peaks at 1427 cm^{-1} and 1373 cm^{-1} reflect the symmetric bending of CH_2 and the bending vibrations of the C–H and C–O groups of the aromatic rings in cellulose, respectively [50,51]. Peak at 894 cm^{-1} relates to the characteristic of cellulose with β-glycoside bonds of glucose ring [52]. By comparing the FTIR spectra of cellulose fibers at all investigated parameters, the difference in peaks intensity was more pronounced with variation of acid concentration. This finding was consistent with the earlier discussion, whereby acid concentration was the most dominant parameter for isolating cellulose fibers through acid hydrolysis.

3.5. Light Transparency of Cellulose-Reinforced PVA Biocomposite Films

Different formulations of cellulose-reinforced PVA biocomposite films were investigated to examine the suitability of the biocomposite films as packaging material for detergent capsules. Figure 4a shows the physical appearance of all formulations of PVA films. The blue background can be seen through for all the films, showing that these films were transparent. For better evaluation, the films' transparency values were calculated, and the results were tabulated in Table 3. Neat PVA film exhibited excellent transparency compared to cellulose-reinforced biocomposite films, whereby the transparency value of the biocomposite films gradually decreased with the increase in cellulose loading. This is due to the presence of opaque particles of cellulose fibers that blocked the light passage through the films.

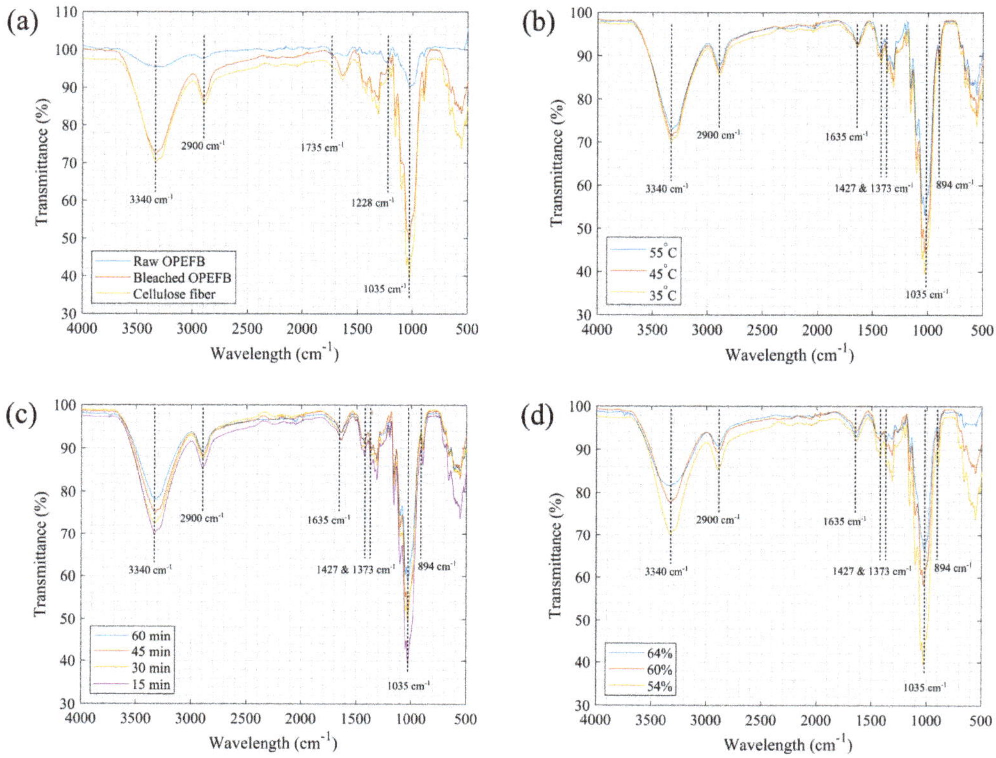

Figure 3. FTIR spectra of (**a**) raw OPEFB, bleached OPEFB and cellulose fibers; (**b**) cellulose fibers at different hydrolysis temperatures; (**c**) cellulose fibers at different hydrolysis times; and (**d**) cellulose fibers at different acid concentrations.

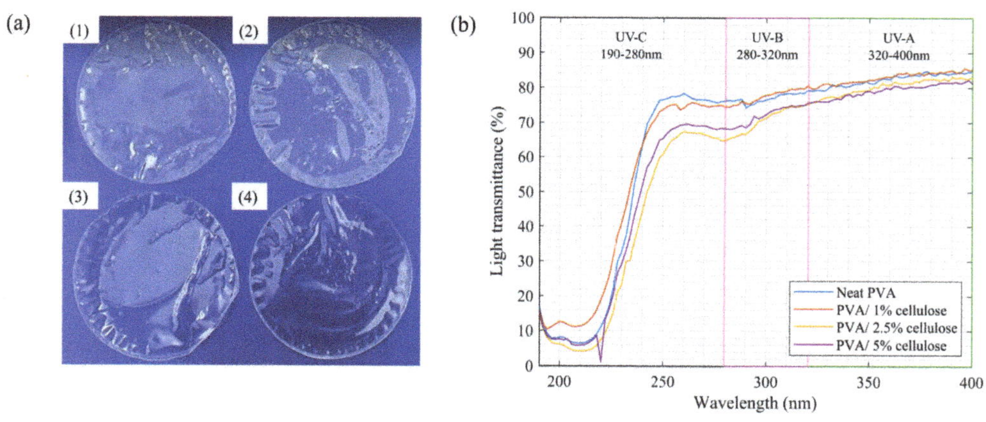

Figure 4. (**a**) Physical appearance of (1) neat PVA; (2) PVA/1% cellulose; (3) PVA/2.5% cellulose; and (4) PVA/5% cellulose. (**b**) UV-vis spectra analysis of neat PVA and biocomposite films in the UV wavelength range.

Table 3. Transparency value of neat PVA film and biocomposite films at different cellulose loadings.

Sample	Transparency Value
Neat PVA	19.4
PVA/1% cellulose	16.1
PVA/2.5% cellulose	14.8
PVA/5% cellulose	14.7

Film with good UV-barrier properties reported longer shelf life of packed detergent products due to the delay of active ingredients degradation, which will diminish the quality of the product. For example, propane-1,2-diol, which is a common liquid laundry detergent ingredient, is known to discolor upon exposure to sunlight [53] and to form a colored complex with iron (III) [54]. UV-vis spectrophotometer was used to evaluate the light transmittance of the fabricated films. Spectrum study was conducted by measuring the light transmittance throughout the wavelength of 190 to 400 nm. The chosen wavelength range covers three UV light radiations, including UV-A (320–400 nm), UV-B (280–320 nm) and UV-C (190–280 nm).

Figure 4b illustrates the percentage of light transmittance for all the biocomposite films. Neat PVA was included as the control sample. Overall, light transmittances of biocomposite films were close to light transmittance of neat PVA film. A decrease in light transmittance can be observed with the increase in cellulose fibers loading. At the wavelength of 400 nm, neat PVA film was transparent with light transmittance of 84.9%. The biocomposite film with 1% cellulose fibers showed a slight decrease in light transmittance, resulting in 84.5% of light transmittance. The light transmittance continued leveling off to 83.0% and 81.3% for the biocomposite films containing 2.5% and 5% cellulose fibers, respectively. This result indicated that the presence of cellulose fibers in the biocomposite films had blocked the passage of light, which eventually reduced the percentage of light transmittance of the films. Another reason was light reflection caused by the non-uniformity dispersion of the cellulose fibers within the biocomposite films, which reduced the light transmittance through the cellulose-reinforced biocomposite films.

Naduparambath and his co-authors [15] reported similar observations, in which a lower light transmittance was observed at higher microcrystalline cellulose (MCC) loading in PVA/MCC biocomposite films. Likewise, Ching and his co-authors [55] observed that the percentage of light transmittance reduced for nanocellulose-reinforced PVA films when the amount of nanocellulose increased. It was due to the presence of agglomerated fibers in the biocomposites. In short, the cellulose-reinforced PVA biocomposite film is preferred as it is more resistant to light transmission, which could provide a better protective layer to the detergent capsule from UV light emitted from sunlight.

3.6. *Mechanical Properties of Cellulose-Reinforced PVA Biocomposite Films*

The packaging material should exhibit adequate mechanical strength to prevent and reduce the chances of detergent leaking and product contamination from broken detergent capsules within the same storage container. Inadequate mechanical strength may also lead to premature bursting of detergent capsules during packing, transportation and storage. Therefore, it is essential to evaluate the mechanical properties of cellulose-reinforced PVA biocomposite films by analyzing their tensile strength, elongation at break and Young's modulus.

Table 4 tabulates the tensile properties of different formulations of cellulose-reinforced PVA biocomposite films. The results showed a significant improvement in tensile strength when cellulose fibers were reinforced in the PVA matrix. The highest tensile strength of 36.09 MPa was obtained for the film that consisted of 2.5% cellulose fibers. The increase in tensile performance was due to the formation of hydrogen bonds between the cellulose fibers and the PVA, which contributed to the dispersion of fibers in the polymer network [56]. However, biocomposite film's tensile strength was slightly reduced when its cellulose loading increased to 5%. The same finding was obtained by Li and his co-authors, where the 5% and 10% addition of cellulose nanofibers from pea hull waste into carboxymethyl cellulose film had a tensile strength of 32.95 MPa and 25.02 MPa, respectively. The decreased tensile strength may be due to the imperfection in fibers dispersion at high cellulose loading, which is also supported by the works reported by Asser et al. [57] and Oyeoka et al. [58]. The study has a higher tensile strength compared to the literature. For example, biocomposite produced from Areca nut husk cellulose fibers, PVA and chitosan film had a tensile strength of 9.46 MPa [59], and water hyacinth cellulose fibers, PVA-gelatin films had tensile strength ranges from 7.91 MPa to 13.83 MPa [58].

Table 4. Tensile strength, elongation at break and Young's modulus of neat PVA film and biocomposite films at different cellulose loadings.

Sample	Tensile Strength (MPa)	Elongation at Break (%)	Young's Modulus (MPa)
Neat PVA	28.65	97.18	3.05
PVA/1% cellulose	34.60	125.55	2.77
PVA/2.5% cellulose	36.09	119.96	2.96
PVA/5% cellulose	31.63	92.19	3.63

Similarly, elongation at break had greatly enhanced with the increment in fibers loading up to 2.5%. Among all, PVA/1% cellulose possessed the highest elongation at break at 125.55%. This observation might be due to the optimized compatibility between the cellulose and the PVA, which made the biocomposite film a highly toughened and flexible material. With the increase in cellulose loading to 5%, elongation at break was reduced due to particle saturation, which resulted in phase segregation. Heterogeneity of the sample affected its ability to be elongated. Similarly, this work has a better elongation at break than the literature. Biocomposite films produced from Areca nut husk cellulose fibers-PVA-chitosan film and water hyacinth cellulose fibers-PVA-gelatin films had an elongation at break of 19.35% [59], and 45.8% to 81.2% [58].

On the other hand, an increase in Young's modulus with cellulose loading indicated that stress had been successfully transferred from the PVA matrix to the cellulose fibers due to effective dispersion and adhesion of cellulose fibers in the biocomposite film. This observation is due to the formation of a large interfacial area when many cellulose fibers of relatively smaller size were used as fillers [60]. The stress-strain behavior can also be illustrated using a stress-strain plot, as shown in Figure 5. It shows that PVA/5% cellulose samples had the highest stress-strain properties compared to neat PVA and biocomposite films containing 1% and 2.5% cellulose fibers.

In this study, the size of cellulose fibers used to reinforce the PVA was not consistent. Different fiber lengths have different effects on mechanical properties. For example, longer fibers could contribute to better tensile strength than shorter fibers [61]. This may be one reason for having results that were not aligned for all the three mechanical properties evaluated. Nevertheless, the best composition was PVA reinforced with 2.5% cellulose fibers because it had the best mechanical properties, except for a slight reduction in film stiffness. On top of that, tensile strength is the most crucial property among all the mechanical properties evaluated as it is used to measure the premature bursting of detergent capsules.

3.7. Moisture Barrier Properties of Cellulose-Reinforced PVA Biocomposite Films

The moisture barrier of biocomposite film is another critical property to be considered in packaging materials. This is because water vapor is one of the main permeates present in the environment, which could quickly diffuse through the packaging film—as a result, affecting the product quality and shelf life of content. The film should exhibit good moisture barrier properties as a detergent packaging material. By reducing the chances of water vapor being captured by the biocomposite film, dilution of detergent concentration and film solubility alteration could be minimized. WVTR is a standard indicator to measure the rate of water vapor permeates through the composite film at a specific thickness, temperature and relative humidity. Figure 6 shows the WVTR of all the samples over 4 days. It was observed that all the biocomposite films fabricated possessed better moisture barrier properties than neat PVA. Hydrophilic groups in the PVA increased their interactions with water molecules, leading to higher WVTR. On the other hand, the cellulose fibers in the biocomposite films served as impermeable physical barriers in the PVA matrix. These fibers prevented the water molecules from traveling and passing through the film by forming tortuous pathways, as shown in Figure 7. As a result, improvement in moisture barrier performance was observed in all the cellulose-reinforced PVA biocomposite films.

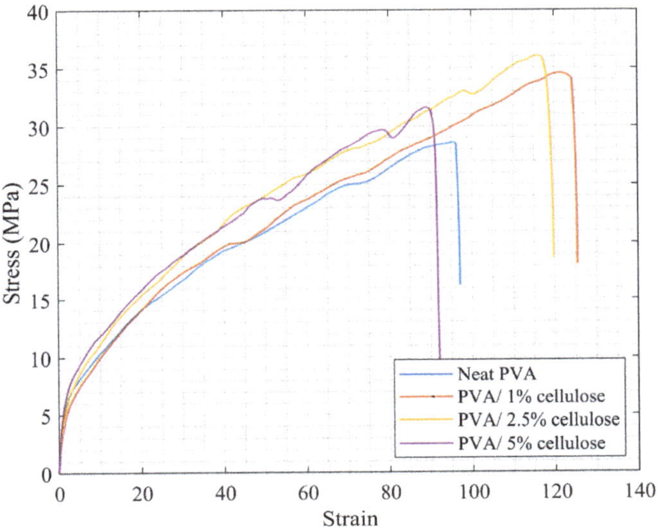

Figure 5. Stress-strain plot of neat PVA and biocomposite films at different cellulose loadings.

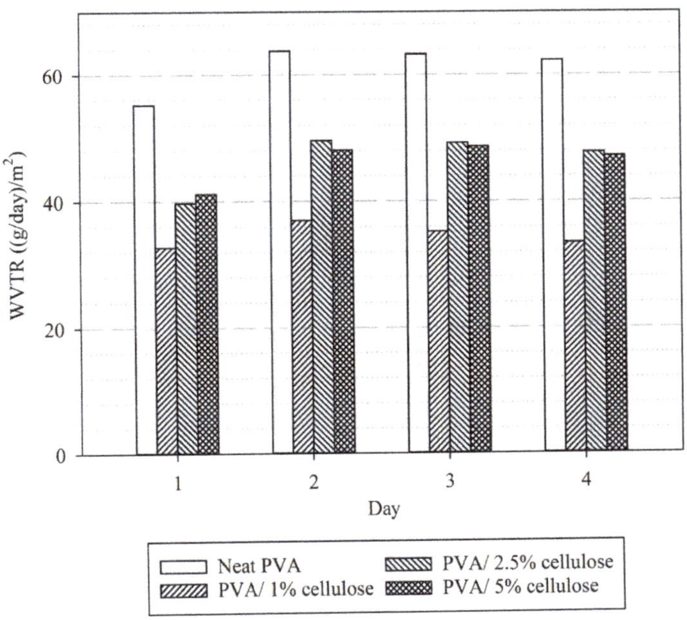

Figure 6. Water vapor transmission rate (WVTR) of neat PVA and biocomposite films at different cellulose loadings.

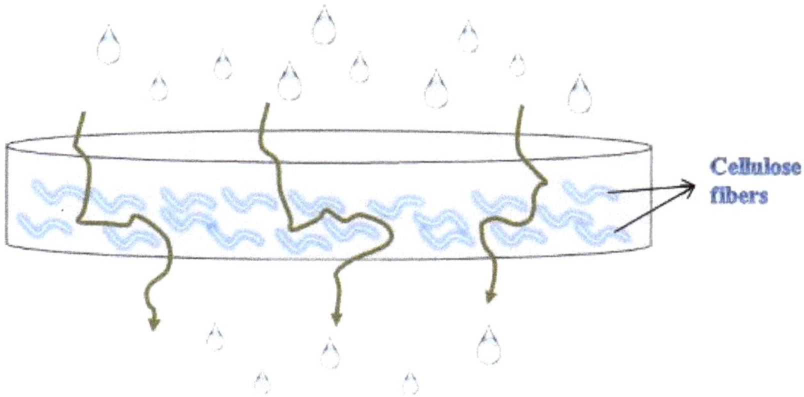

Figure 7. Schematic illustration of the tortuous pathway for water vapor in cellulose-reinforced PVA biocomposite film.

Among all the biocomposite films fabricated, the film with 1% cellulose fibers loading exhibited the best moisture barrier properties. The observation was due to fibers agglomeration at higher cellulose loading, which was ineffective in creating tortuous pathways for water vapor to pass through the film. As a result, the WVTR was significantly reduced with increased cellulose fibers loading in the biocomposite film. Similar results were reported by Shankar et al. [62] and Yang et al. [63]. Ali and his co-authors [6] also reported an increasing trend of WVTR with MCC loading in the PVA matrix. In another work by Patil et al. [64], a negative relationship was also obtained between the WVTR and cellulose fibers loading. The observation was due to the hydrogen bond interactions between the MCC and the PVA, which increased the hydrophilic groups in the biocomposite and, subsequently, the WVTR. These findings signified the importance of incorporating cellulose fibers into PVA in a suitable ratio so that favorable characteristics of biocomposite film could be formed to suit its application.

3.8. Thermal Stability of Cellulose-Reinforced PVA Biocomposite Films

TGA was used to evaluate the thermal performance of neat PVA and different formulations of cellulose-reinforced PVA biocomposite films. TGA thermogram in Figure 8a shows the percentage of weight loss with respect to the temperature for a heating profile of 30 °C to 600 °C. The figure shows that all the films exhibited three main weight loss regions. The first region appeared in a temperature range of 80 °C to 140 °C with weight loss of approximately 7%. It corresponded to the evaporation of the adsorped moisture in the films [58,65]. The significant weight loss of approximately 60% happened in the second region of 230 °C to 370 °C. The weight loss was due to the structural depolymerization, dehydration and decomposition of the biocomposite films, as confirmed by Salh and Raswl [66] and Oyeoka et al. [58]. The third region happened at temperatures above 370 °C. The weight lost observed could be due to the cleavage backbone of PVA composite films or the combustion of carbon materials [67,68]. Constant weight of 4% to 6% was observed at temperatures of 500 °C onwards.

As seen in Figure 8a, the incorporation of cellulose fibers in the PVA film had a negligible effect on the thermal profile of the PVA film. Cellulose-reinforced PVA film with 1% cellulose loadings exhibited the highest thermal stability. The observation will be more significant by presenting the thermogram in differential thermagravimetry (DTG), which plots the rate of material weight changes upon heating against temperature, as shown in Figure 8b. It can be seen that the curve for 1% cellulose-reinforced biocomposite film was moved to a higher temperature compared to the neat PVA. The increase in thermal stability in cellulose-reinforced PVA film could probably be due to the strong hydrogen bonding between the hydroxyl groups of cellulose and the PVA matrix, as reported by Lee and his co-authors [69]. Another plausible reason could be the insulation effect and mass transport barrier exhibited by the cellulose fibers, which inhibited the generation of volatile products during decomposition [70].

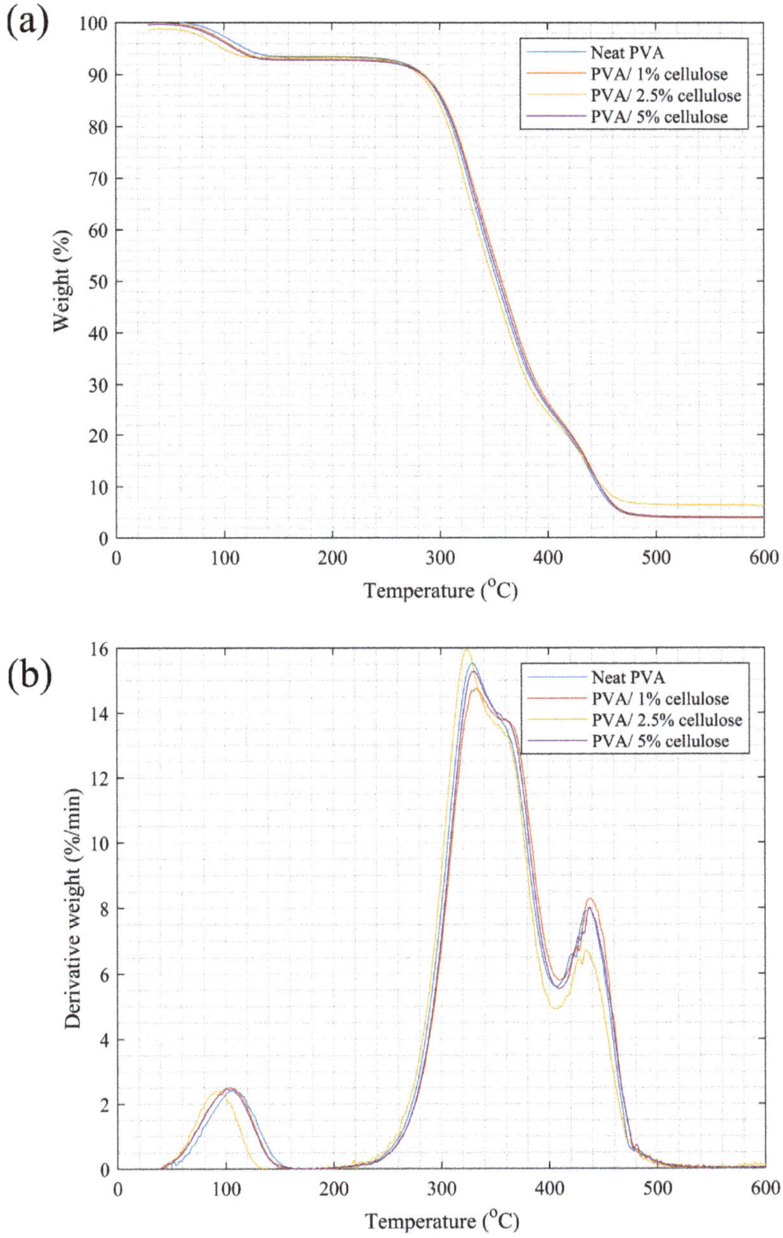

Figure 8. (**a**) TGA thermograms, and (**b**) DTG thermograms; of neat PVA and biocomposite films at different cellulose loadings.

4. Conclusions

Cellulose fibers were successfully isolated from the OPEFB through sulphuric acid hydrolysis. The reaction was greatly affected by acid concentration, hydrolysis temperature and time in descending order. An apparent interaction effect was also observed between hydrolysis temperature and time as well as acid concentration and hydrolysis time. The cellulose yield was favored at low acid concentration, low hydrolysis temperature and short hydrolysis time, supported by both experiments and mathematical modeling. A maximum cellulose yield of 80.72% was attained at hydrolysis using 54% acid concentration at 35 °C for 15 min. The well-dispersed cellulose fibers showed good reinforcement effects in the PVA matrix. Tensile strength and elongation at break were greatly enhanced. In addition, the biocomposite films exhibited reduced light transmission rate, good absorption of UV rays and thermal stability, as well as reduced water vapor transmission rate. These properties are essential, especially in the detergent packing industry, to prevent the breakage of packaging material and product deterioration. Among all the biocomposite formulations investigated, PVA/2.5% cellulose was the most promising biocomposite film as it exhibited good mechanical and moisture barrier properties and blocked UV rays. The results indicated that cellulose-reinforced PVA biocomposite film is a potential material for biodegradable detergent capsules packaging.

Supplementary Materials: The following supporting information can be downloaded at: https://www.mdpi.com/article/10.3390/su141811446/s1, Table S1: ANOVA for the yield of cellulose based on hydrolysis temperature and time; Table S2: ANOVA for the yield of cellulose based on acid concentration and hydrolysis time.

Author Contributions: Conceptualization, K.M.L.; methodology, J.Y.T. and W.Y.T.; formal analysis, K.M.L., J.Y.T. and W.Y.T.; software, W.Y.T.; investigation, J.Y.T.; writing—original draft, J.Y.T.; writing—review and editing, K.M.L., W.Y.T., J.P. and S.L.; supervision, K.M.L.; funding acquisition, K.M.L. All authors have read and agreed to the published version of the manuscript.

Funding: This work was supported by the UCSI University Research Excellence & Innovation Grant (Grant number: REIG-FETBE-2020/038).

Institutional Review Board Statement: Not applicable.

Informed Consent Statement: Not applicable.

Data Availability Statement: All the data are available inside this paper and in the Supplementary Materials.

Acknowledgments: This work was supported by the UCSI University Research Excellence & Innovation Grant (Grant number: REIG-FETBE-2020/038). The authors would like to thank Kwantas Oil Sdn. Bhd., Sabah, Malaysia, for donating the OPEFB.

Conflicts of Interest: The authors declare no conflict of interest. The funders had no role in the design of the study; in the collection, analyses, or interpretation of data; in the writing of the manuscript, or in the decision to publish the results.

References

1. Chen, H.L.; Nath, T.K.; Chong, S.; Foo, V.; Gibbins, C.; Lechner, A.M. The plastic waste problem in Malaysia: Management, recycling and disposal of local and global plastic waste. *SN Appl. Sci.* **2021**, *3*, 437. [CrossRef]
2. Latif, S.N.A.; Chiong, M.S.; Rajoo, S.; Takada, A.; Chun, Y.-Y.; Tahara, K.; Ikegami, Y. The Trend and Status of Energy Resources and Greenhouse Gas Emissions in the Malaysia Power Generation Mix. *Energies* **2021**, *14*, 2200. [CrossRef]
3. Tai, N.L.; Ghasemlou, M.; Adhikari, R.; Adhikari, B. Starch-based isocyanate- and non-isocyanate polyurethane hybrids: A review on synthesis, performance and biodegradation. *Carbohydr. Polym.* **2021**, *265*, 118029. [CrossRef] [PubMed]
4. Alias, N.F.; Ismail, H.; Wahab, M.K.A.; Ragunathan, S.; Ardhyananta, H.; Ting, S.S. Development of New Material Based on Polyvinyl Alcohol/Palm Kernel Shell Powder Biocomposites. *Adv. Environ. Stud.* **2018**, *2*, 98–107. [CrossRef]
5. Byrne, D.; Boeije, G.; Croft, I.; Hüttmann, G.; Luijkx, G.; Meier, F.; Parulekar, Y.; Stijntjes, G. Biodegradability of Polyvinyl Alcohol Based Film Used for Liquid Detergent Capsules. *Tenside Surfactants Deterg.* **2021**, *58*, 88–96. [CrossRef]
6. Ali, N.A.; Huseen, S.I.; Jaffer, H.I. Barrier, Mechanical and Thermal of Polyvinyl Alchol/Microcrystalline Cellouse Composites in Packaging Application. *Am. J. Phys. Appl.* **2017**, *5*, 46–51. [CrossRef]

7. Abdulkhani, A.; Hojati Marvast, E.; Ashori, A.; Hamzeh, Y.; Karimi, A.N. Preparation of cellulose/polyvinyl alcohol biocomposite films using 1-n-butyl-3-methylimidazolium chloride. *Int. J. Biol. Macromol.* **2013**, *62*, 379–386. [CrossRef] [PubMed]
8. Rao, J.K.; Raizada, A.; Ganguly, D.; Mankad, M.M.; Satyanarayana, S.V.; Madhu, G.M. Enhanced mechanical properties of polyvinyl alcohol composite films containing copper oxide nanoparticles as filler. *Polym. Bull.* **2015**, *72*, 2033–2047. [CrossRef]
9. Chang, J.-H. Comparative Analysis of Properties of PVA Composites with Various Nanofillers: Pristine Clay, Organoclay, and Functionalized Graphene. *Nanomaterials* **2019**, *9*, 323. [CrossRef] [PubMed]
10. Tajeddin, B. Cellulose-Based Polymers for Packaging Applications. In *Lignocellulosic Polymer Composites*; Wiley: Hoboken, NJ, USA, 2014; pp. 477–498, ISBN 9781118773949.
11. Tu, C.-W.; Tsai, F.-C.; Chang, C.-J.; Yang, C.-H.; Kuo, S.-W.; Zhang, J.; Chen, T.; Huang, C.-F. Surface-Initiated Initiators for Continuous Activator Regeneration (SI ICAR) ATRP of MMA from 2,2,6,6–tetramethylpiperidine–1–oxy (TEMPO) Oxidized Cellulose Nanofibers for the Preparations of PMMA Nanocomposites. *Polymers* **2019**, *11*, 1631. [CrossRef]
12. Tu, C.-W.; Tsai, F.-C.; Chen, J.-K.; Wang, H.-P.; Lee, R.-H.; Zhang, J.; Chen, T.; Wang, C.-C.; Huang, C.-F. Preparations of Tough and Conductive PAMPS/PAA Double Network Hydrogels Containing Cellulose Nanofibers and Polypyrroles. *Polymers* **2020**, *12*, 2835. [CrossRef] [PubMed]
13. Isogai, A.; Yumiko, K. Preparation of Polyuronic Acid from Cellulose by TEMPO-Mediated Oxidation. *Cellulose* **1998**, *5*, 153–164. [CrossRef]
14. Huang, C.-F.; Tu, C.-W.; Lee, R.-H.; Yang, C.-H.; Hung, W.-C.; Lin, K.-Y.A. Study of various diameter and functionality of TEMPO-oxidized cellulose nanofibers on paraquat adsorptions. *Polym. Degrad. Stab.* **2019**, *161*, 206–212. [CrossRef]
15. Naduparambath, S.; Sreejith, M.P.; Jinitha, T.V.; Shaniba, V.; Aparna, K.B.; Purushothaman, E. Development of green composites of poly (vinyl alcohol) reinforced with microcrystalline cellulose derived from sago seed shells. *Polym. Compos.* **2018**, *39*, 3033–3039. [CrossRef]
16. Santi, R.; Cigada, A.; Del Curto, B.; Farè, S. Modulable properties of PVA/cellulose fiber composites. *J. Appl. Biomater. Funct. Mater.* **2019**, *17*. [CrossRef] [PubMed]
17. Liu, D.; Sun, X.; Tian, H.; Maiti, S.; Ma, Z. Effects of cellulose nanofibrils on the structure and properties on PVA nanocomposites. *Cellulose* **2013**, *20*, 2981–2989. [CrossRef]
18. Lani, N.S.; Ngadi, N.; Johari, A.; Jusoh, M. Isolation, Characterization and Application of Nanocellulose from Oil Palm Empty Fruit Bunch (EFB) Fiber as Nanocomposites. *Nanomater. Environ.* **2014**, *2014*, 702538.
19. Andalia, R.; Rahmi, R.; Julinawati, J.; Helwati, H. Isolation and characterization of cellulose from rice husk waste and sawdust with chemical method. *J. Nat.* **2020**, *20*, 6–9. [CrossRef]
20. Yang, S.; Zhang, Y.; Yue, W.; Wang, W.; Wang, Y.-Y.; Yuan, T.-Q.; Sun, R.-C. Valorization of lignin and cellulose in acid-steam-exploded corn stover by a moderate alkaline ethanol post-treatment based on an integrated biorefinery concept. *Biotechnol. Biofuels* **2016**, *9*, 238. [CrossRef] [PubMed]
21. Chin, D.W.K.; Lim, S.; Pang, Y.L.; Lim, C.H.; Shuit, S.H.; Lee, K.M.; Chong, C.T. Effects of Organic Solvents on the Organosolv Pretreatment of Degraded Empty Fruit Bunch for Fractionation and Lignin Removal. *Sustainability* **2021**, *13*, 6757. [CrossRef]
22. Teoh, C.H. Land Use and the Oil Palm Industry in Malaysia. *Abridged Report Produced for the WWF Forest Information System Database*. 2000. Available online: http://d2ouvy59p0dg6k.cloudfront.net/downloads/oplanduseabridged.pdf (accessed on 12 June 2022).
23. Abdul Khalil, H.P.S.; Jawaid, M.; Hassan, A.; Paridah, M.T.; Zaido, A. Oil Palm Biomass Fibres and Recent Advancement in Oil Palm Biomass Fibres Based Hybrid Biocomposites. In *Composites and Their Applications*; Hu, N., Ed.; Intech: Rijeka, Croatia, 2012. [CrossRef]
24. Hassan, A.; Salema, A.A.; Ani, F.N.; Bakar, A.A. A review on oil palm empty fruit bunch fiber-reinforced polymer composite materials. *Polym. Compos.* **2010**, *31*, 2079–2101. [CrossRef]
25. Rahmi, L.; Julinawati, S. Shabrina Preparation of chitosan composite film reinforced with cellulose isolated from oil palm empty fruit bunch and application in cadmium ions removal from aqueous solutions. *Carbohydr. Polym.* **2017**, *170*, 226–233. [CrossRef] [PubMed]
26. Lee, K.M.; Zanil, M.F.; Chan, K.K.; Chin, Z.P.; Liu, Y.C.; Lim, S. Synergistic ultrasound-assisted organosolv pretreatment of oil palm empty fruit bunches for enhanced enzymatic saccharification: An optimization study using artificial neural networks. *Biomass-Bioenergy* **2020**, *139*, 105621. [CrossRef]
27. Lancaster, P.; Salkauskas, K. Surfaces Generated by Moving Least Squares Methods. *Math. Comput.* **1981**, *37*, 141–158. [CrossRef]
28. Tey, W.Y.; Lee, K.M.; Asako, Y.; Tan, L.K.; Arai, N. Multivariable power least squares method: Complementary tool for Response Surface Methodology. *Ain Shams Eng. J.* **2020**, *11*, 161–169. [CrossRef]
29. Zailuddin, N.L.I.; Husseinsyah, S. Tensile Properties and Morphology of Oil Palm Empty Fruit Bunch Regenerated Cellulose Biocomposite Films. *Procedia Chem.* **2016**, *19*, 366–372. [CrossRef]
30. Tee, Y.B.; Wong, J.; Tan, M.C.; Talib, R.A. Development of Edible Film from Flaxseed Mucilage. *BioResources* **2016**, *11*, 10286–10295. [CrossRef]
31. Lee, H.V.; Hamid, S.B.A.; Zain, S.K. Conversion of Lignocellulosic Biomass to Nanocellulose: Structure and Chemical Process. *Sci. World J.* **2014**, *2014*, 631013. [CrossRef]
32. Zhang, Y.; Xu, Y.; Yue, X.; Dai, L.; Ni, Y. Isolation and Characterization of Microcrystalline Cellulose from Bamboo Pulp through Extremely Low Acid Hydrolysis. *J. Wood Chem. Technol.* **2019**, *39*, 242–254. [CrossRef]

33. Zhao, T.; Chen, Z.; Lin, X.; Ren, Z.; Li, B.; Zhang, Y. Preparation and characterization of microcrystalline cellulose (MCC) from tea waste. *Carbohydr. Polym.* **2018**, *184*, 164–170. [CrossRef]
34. Steinbach, D.; Kruse, A.; Sauer, J. Pretreatment technologies of lignocellulosic biomass in water in view of furfural and 5-hydroxymethylfurfural production- A review. *Biomass-Convers. Biorefinery* **2017**, *7*, 247–274. [CrossRef]
35. Hamelinck, C.N.; van Hooijdonk, G.; Faaij, A.P.C. Ethanol from lignocellulosic biomass: Techno-economic performance in short-, middle- and long-term. *Biomass-Bioenergy* **2005**, *28*, 384–410. [CrossRef]
36. Beck-Candanedo, S.; Roman, M.; Gray, D.G. Effect of Reaction Conditions on the Properties and Behavior of Wood Cellulose Nanocrystal Suspensions. *Biomacromolecules* **2005**, *6*, 1048–1054. [CrossRef] [PubMed]
37. Roni, K.A.; Hastarina, M.; Herawati, N. Effect of Time and Concentration of Sulfuric Acid on Yield Bioethanol Produced In Making Bioethanol from Peat Soil. *J. Physics Conf. Ser.* **2019**, *1167*, 012056. [CrossRef]
38. Xiang, Q.; Lee, Y.Y.; Pettersson, P.O.; Torget, R.W. Heterogeneous Aspects of Acid Hydrolysis of α-Cellulose. *Appl. Biochem. Biotechnol-Part A Enzym. Eng. Biotechnol.* **2003**, *107*, 505–514. [CrossRef]
39. Sai, Y.W.; Lee, K.M. Enhanced cellulase accessibility using acid-based deep eutectic solvent in pretreatment of empty fruit bunches. *Cellulose* **2019**, *26*, 9517–9528. [CrossRef]
40. Lee, K.M.; Hong, J.Y.; Tey, W.Y. Combination of ultrasonication and deep eutectic solvent in pretreatment of lignocellulosic biomass for enhanced enzymatic saccharification. *Cellulose* **2021**, *28*, 1513–1526. [CrossRef]
41. Norul Izani, M.A.; Paridah, M.T.; Anwar, U.M.K.; Mohd Nor, M.Y.; H'ng, P.S. Effects of fiber treatment on morphology, tensile and thermogravimetric analysis of oil palm empty fruit bunches fibers. *Compos. Part. B Eng.* **2013**, *45*, 1251–1257. [CrossRef]
42. Chimentão, R.J.; Lorente, E.; Gispert-Guirado, F.; Medina, F.; López, F. Hydrolysis of dilute acid-pretreated cellulose under mild hydrothermal conditions. *Carbohydr. Polym.* **2014**, *111*, 116–124. [CrossRef] [PubMed]
43. Goh, K.Y.; Ching, Y.C.; Chuah, C.H.; Abdullah, L.C.; Liou, N.-S. Individualization of microfibrillated celluloses from oil palm empty fruit bunch: Comparative studies between acid hydrolysis and ammonium persulfate oxidation. *Cellulose* **2016**, *23*, 379–390. [CrossRef]
44. Sain, M.; Panthapulakkal, S. Bioprocess preparation of wheat straw fibers and their characterization. *Ind. Crop. Prod.* **2006**, *23*, 1–8. [CrossRef]
45. Lai, D.S.; Osman, A.F.; Adnan, S.A.; Ibrahim, I.; Alrashdi, A.A.; Salimi, M.N.A.; Ul-Hamid, A. On the Use of OPEFB-Derived Microcrystalline Cellulose and Nano-Bentonite for Development of Thermoplastic Starch Hybrid Bio-Composites with Improved Performance. *Polymers* **2021**, *13*, 897. [CrossRef] [PubMed]
46. Ilharco, L.M.; Garcia, A.R.; Lopes Da Silva, J.; Vieira Ferreira, L.F. Infrared Approach to the Study of Adsorption on Cellulose: Influence of Cellulose Crystallinity on the Adsorption of Benzophenone. *Langmuir* **1997**, *13*, 4126–4132. [CrossRef]
47. Lei, W.; Fang, C.; Zhou, X.; Yin, Q.; Pan, S.; Yang, R.; Liu, D.; Ouyang, Y. Cellulose nanocrystals obtained from office waste paper and their potential application in PET packing materials. *Carbohydr. Polym.* **2018**, *181*, 376–385. [CrossRef] [PubMed]
48. Tang, Y.; Yang, S.; Zhang, N.; Zhang, J. Preparation and characterization of nanocrystalline cellulose via low-intensity ultrasonic-assisted sulfuric acid hydrolysis. *Cellulose* **2014**, *21*, 335–346. [CrossRef]
49. Ngwabebhoh, F.A.; Erdem, A.; Yildiz, U. A design optimization study on synthesized nanocrystalline cellulose, evaluation and surface modification as a potential biomaterial for prospective biomedical applications. *Int. J. Biol. Macromol.* **2018**, *114*, 536–546. [CrossRef] [PubMed]
50. Fackler, K.; Stevanic, J.S.; Ters, T.; Hinterstoisser, B.; Schwanninger, M.; Salmén, L. FT-IR imaging microscopy to localise and characterise simultaneous and selective white-rot decay within spruce wood cells. *Holzforschung* **2011**, *65*, 411–420. [CrossRef]
51. Xu, F.; Yu, J.; Tesso, T.; Dowell, F.; Wang, D. Qualitative and quantitative analysis of lignocellulosic biomass using infrared techniques: A mini-review. *Appl. Energy* **2013**, *104*, 801–809. [CrossRef]
52. Hemmati, F.; Jafari, S.M.; Kashaninejad, M.; Barani Motlagh, M. Synthesis and characterization of cellulose nanocrystals derived from walnut shell agricultural residues. *Int. J. Biol. Macromol.* **2018**, *120*, 1216–1224. [CrossRef] [PubMed]
53. Missler, S.R.; Vredeveld, D.J.; Westrate, E.D.; Sliva, P.G.; Brouwer, S.J. Investigation of Color Instability in a Liquid Laundry Detergent. *J. Surfactants Deterg.* **2014**, *17*, 839–847. [CrossRef]
54. Nagy, L.; Burger, K.; Kürti, J.; Mostafa, M.A.; Korecz, L.; Kiricsi, I. Iron (III) complexes of sugar-type ligands. *Inorganica Chim. Acta* **1986**, *124*, 55–59. [CrossRef]
55. Ching, Y.C.; Rahman, A.; Ching, K.Y.; Sukiman, N.L.; Chuah, C.H. Preparation and Characterization of Polyvinyl Alcohol-Based Composite Reinforced with Nanocellulose and Nanosilica. *BioResources* **2015**, *10*, 3364–3377. [CrossRef]
56. Schoeler, M.N.; Scremin, F.R.; de Mendonça, N.F.; Benetti, V.P.; de Jesus, J.A.; de Oliveira Basso, R.L.; Stival Bittencour, P.R. Cellulose Nanofibers from Cassava Agro-Industrial Waste as Reinforcement in PVA Films. *Quim. Nova* **2020**, *43*, 711–717. [CrossRef]
57. Aseer, J.R.; Sankaranarayanasamy, K. Effect of fiber content on tensile retention properties of Cellulose Microfiber Reinforced Polymer Composites for Automobile Application. *IOP Conf. Ser. Mater. Sci. Eng.* **2017**, *272*, 012020. [CrossRef]
58. Oyeoka, H.C.; Ewulonu, C.M.; Nwuzor, I.C.; Obele, C.M.; Nwabanne, J.T. Packaging and degradability properties of polyvinyl alcohol/gelatin nanocomposite films filled water hyacinth cellulose nanocrystals. *J. Bioresour. Bioprod.* **2021**, *6*, 168–185. [CrossRef]
59. Perumal, A.B.; Nambiar, R.B.; Sellamuthu, P.S.; Sadiku, E.R.; Li, X.; He, Y. Extraction of cellulose nanocrystals from areca waste and its application in eco-friendly biocomposite film. *Chemosphere* **2022**, *287*, 132084. [CrossRef]

60. Uma Maheswari, C.; Obi Reddy, K.; Muzenda, E.; Guduri, B.R.; Varada Rajulu, A. Extraction and characterization of cellulose microfibrils from agricultural residue-*Cocos nucifera* L. *Biomass-Bioenergy* **2012**, *46*, 555–563. [CrossRef]
61. Yang, H.-S.; Gardner, D.J. Mechanical Properties of Cellulose Nanofibril-Filled Polypropylene Composites. *Wood Fiber Sci.* **2011**, *43*, 143–152.
62. Shankar, S.; Wang, L.-F.; Rhim, J.-W. Effect of melanin nanoparticles on the mechanical, water vapor barrier, and antioxidant properties of gelatin-based films for food packaging application. *Food Packag. Shelf Life* **2019**, *21*, 100363. [CrossRef]
63. Yang, W.; Qi, G.; Kenny, J.M.; Puglia, D.; Ma, P. Effect of Cellulose Nanocrystals and Lignin Nanoparticles on Mechanical, Antioxidant and Water Vapour Barrier Properties of Glutaraldehyde Crosslinked PVA Films. *Polymers* **2020**, *12*, 1364. [CrossRef] [PubMed]
64. Patil, S.; Bharimalla, A.K.; Nadanathangam, V.; Dhakane-Lad, J.; Mahapatra, A.; Jagajanantha, P.; Saxena, S. Nanocellulose reinforced corn starch-based biocomposite films: Composite optimization, characterization and storage studies. *Food Packag. Shelf Life* **2022**, *33*, 100860. [CrossRef]
65. Kakroodi, A.R.; Cheng, S.; Sain, M.; Asiri, A. Mechanical, Thermal, and Morphological Properties of Nanocomposites Based on Polyvinyl Alcohol and Cellulose Nanofiber from Aloe vera Rind. *J. Nanomater.* **2014**, *2014*, 903498. [CrossRef]
66. Salh, S.H.; Raswl, D.A. Thermal Stability of Polymer Composite Films Based on Polyvinyl Alcohol Doped with Different Fillers. *Open Access J. Phys.* **2018**, *2*, 5–10.
67. Holland, B.J.; Hay, J.N. The thermal degradation of poly (vinyl alcohol). *Polymer* **2001**, *42*, 6775–6783. [CrossRef]
68. Wang, Z.; Ding, Y.; Wang, J. Novel Polyvinyl Alcohol (PVA)/Cellulose Nanocrystal (CNC) Supramolecular Composite Hydrogels: Preparation and Application as Soil Conditioners. *Nanomaterials* **2019**, *9*, 1397. [CrossRef]
69. Lee, S.-Y.; Mohan, D.J.; Kang, I.-A.; Doh, G.-H.; Lee, S.; Han, S.O. Nanocellulose reinforced PVA composite films: Effects of acid treatment and filler loading. *Fibers Polym.* **2009**, *10*, 77–82. [CrossRef]
70. de Carvalho Benini, K.C.C.; Voorwald, H.J.C.; Cioffi, M.O.H.; Rezende, M.C.; Arantes, V. Preparation of nanocellulose from Imperata brasiliensis grass using Taguchi method. *Carbohydr. Polym.* **2018**, *192*, 337–346. [CrossRef] [PubMed]

sustainability

MDPI

Article

Ganodiesel: A New Biodiesel Feedstock from Biomass of the Mushroom *Ganoderma lucidum*

Besek Mariam Mohamad Jahis [1,2], Zul Ilham [3,4], Sugenendran Supramani [1,5], Mohamad Nor Azzimi Sohedein [1], Mohamad Faizal Ibrahim [2,*], Suraini Abd-Aziz [2], Neil Rowan [6] and Wan Abd Al Qadr Imad Wan-Mohtar [1,3,6,*]

1 Functional Omics and Bioprocess Development Laboratory, Institute of Biological Sciences, Faculty of Science, Universiti Malaya, Kuala Lumpur 50603, Malaysia
2 Department of Bioprocess Technology, Faculty of Biotechnology and Biomolecular Sciences, Universiti Putra Malaysia, Serdang 43400, Malaysia
3 Bioresources and Bioprocessing Research Group, Institute of Biological Sciences, Faculty of Science, Universiti Malaya, Kuala Lumpur 50603, Malaysia
4 Department of Biological and Environmental Engineering, College of Agriculture and Life Sciences, Cornell University, Ithaca, NY 14850, USA
5 School of Bioscience, Faculty of Medicine, Bioscience and Nursing, MAHSA University, Jenjarom 42610, Malaysia
6 Bioscience Research Institute, Technical University Shannon Midlands Midwest, N37 HD68 Athlone, Ireland
* Correspondence: faizal_ibrahim@upm.edu.my (M.F.I.); qadyr@um.edu.my (W.A.A.Q.I.W.-M.); Tel.: +603-97691936 (M.F.I.); +603-79674128 (W.A.A.Q.I.W.-M.)

check for updates

Citation: Mohamad Jahis, B.M.; Ilham, Z.; Supramani, S.; Sohedein, M.N.A.; Ibrahim, M.F.; Abd-Aziz, S.; Rowan, N.; Wan-Mohtar, W.A.A.Q.I. Ganodiesel: A New Biodiesel Feedstock from Biomass of the Mushroom *Ganoderma lucidum*. *Sustainability* **2022**, *14*, 10764. https://doi.org/10.3390/su141710764

Academic Editors: Steven Lim, Shuit Siew Hoong, Santi Chuetor and Pang Yean Ling

Received: 26 July 2022
Accepted: 25 August 2022
Published: 29 August 2022

Abstract: There is a pressing demand for new sustainable eco-friendly approaches to producing green energy worldwide. This study represents the novel production of biodiesel feedstock from the medicinal mushroom *Ganoderma lucidum* QRS 5120 using state-of-the-art biotechnology tools. Response surface methodology (RSM) was used to enhance *G. lucidum* production in a repeated-batch fermentation strategy. By referring to the broth replacement ratio (BRR) and broth replacement time point (BRTP), RSM that was formulated using a central composite design (CCD) resulted in a significant model for all tested variables, which are exopolysaccharide (EPS), endopolysaccharide (ENS) and biomass, with BRR (%) of 60, 75 and 90, and BRTP (days) of 11, 13 and 15. The model was validated using the optimised conditions, and the results showed 4.21 g/L of EPS (BRR 77.46% and BRTP 12 days), 2.44 g/L of ENS (BRR 60% and BRTP 12.85 days), and 34.32 g/L of biomass (BRR 89.52% and BRTP 10.96 days) were produced. Biomass produced from the *G. lucidum* was subsequently used as feedstock for biodiesel production. Approximately 20.36% of lipid was successfully extracted from the dried *G. lucidum* biomass via a solvent extraction and subsequently converted to Ganodiesel through a transesterification process. The Ganodiesel produced fulfilled most of the international standards, i.e., US (ASTM D6751-08) and EU (EN 14214). Overall, this study demonstrates the optimised *G. lucidum* production and its lipid production as a new biodiesel feedstock.

Keywords: biodiesel; bioreactor biomass; *Ganoderma lucidum*; mushroom cultivation; repeated-batch fermentation

1. Introduction

The consumption of food, water, energy and electricity has risen dramatically due to the rapid population growth linked with the commensurate need for improvements in the lifestyle around the world [1,2]. The global population is estimated to reach 10 billion people by 2050, which will place increasing pressure on energy and food production [3,4]. Currently, fossil fuels dominate the world's demands for energy and the economy, especially in the transportation sector. A high consumption of fuel is greatly affecting the environment [2,5]. Moreover, fossil fuels are non-renewable, with day-on-day depletion. Therefore, bio-based fuels such as biodiesel have been proposed as an alternative to fossil fuels as they are sustainable, cost-effective, free of toxic chemicals (such as sulphur) and have greater lubricity (for automobiles) [6].

Although it is greener to utilise biodiesel than fossil fuel, the production of biodiesel also faces technical challenges [7]. For instance, it relies upon high yields of materials that may affect market demand, i.e., oil from crops [6,8]. Consequently, it has been estimated that using crops such as sunflower seed or rapeseed requires large-scale plantations and long durations to achieve existing biodiesel goals. For these reasons, it is important to find new sources of raw materials (the feedstocks) that can reduce the production price of biodiesel without competing with food security [9]. Recently, researchers tend to focus on biodiesel production from various feedstocks including plant-based oils, animal fats, waste oils, algal oils or even biomass [8]. However, there is an apparent gap in evidence-based research on the smart use of fungi-based feedstock to meet this pressing opportunity. Fungi are fast-growing and have been used for decades for biotech applications, thus, the production of biodiesel from fungal species could serve as a potential resource for new energy [10,11].

Ganoderma lucidum has been extensively used in traditional Asian medicine for more than 2000 years [12,13]. This mushroom is edible and has many health benefits such as for the treatment of various diseases, most commonly cancer [12–16]. *G. lucidum* undergoes four stages of its life cycle; (1) spores, (2) spore germination, (3) mycelium and (4) fruiting body. *G. lucidum* can easily be cultivated from its mycelium in a short period with the help of biotechnological practices and a suitable set of nutrients. A longer cultivation duration is required if growing *G. lucidum* using its fruiting body or spores [14,17]. Based on a recent study, *G. lucidum* mycelium can be cultivated using submerged fermentation (SmF), which takes 3 to 6 months to be fully completed [12,18,19]. A high biomass yield could be obtained after 10 days of the fermentation process, which can reduce the time needed to produce by-products and prevent the risk of contamination. Moreover, the active ingredients of *G. lucidum* are usually extracted for medicinal purposes. Apart from medicinal benefits, *Ganoderma lucidum* is also used as a reliable source of biomass and polysaccharides [11,20]. There is a growing awareness of the transformative potential of medicinal mushrooms, including the use of *G. lucidum*, in supporting and enabling green enterprises where its disruptive innovation potential will be actualised and accelerated through the forging of collaborative partnerships internationally between subject-matter experts in academia and industries, using connected innovation hubs [3,4].

In order to evaluate biomass production, a response surface methodology (RSM) can be applied using the broth replacement ratio (BRR) and broth replacement time point (BRTP) as the key parameters. By looking at the interaction and correlation between the main parameters, RSM is used to find the optimum conditions for the cultivation of polysaccharides and biomass production [21]. On the other hand, repeated-batch fermentation (RBF) is conducted, and it is a modification of the fermentation technique. This technique involves extracting a portion of the medium and, without changing the existing culture, thus replacing a fresh medium for continuous growth [22,23]. RBF offers many benefits, such as the formation of a long-term supply and a reduced workload, resulting in great savings of time and manpower [24–26].

Therefore, the extraction of lipids from *G. lucidum* for biodiesel production requires knowledge of the appropriate extraction techniques and factors that can affect the lipid yield. There are three extraction methods, namely physical, mechanical and chemical. These include Soxhlet extraction (SXE), solvent extraction (SVE), supercritical fluid extraction, aqueous enzyme extraction and pressurised liquid extraction [27]. Lipid extraction consumes around 90% of the energy used in biodiesel production and is costly, making the process difficult [28]. Thus, the implementation of green extraction techniques that can extract lipids with less duration, energy and solvent consumption without losing lipid quality is the key factor for cost-effective and environmentally sustainable processes. Earlier studies have shown that the amount of oil yielded from extraction methods is influenced by the sample size, temperature, solvent volume, solvent types and time [29,30].

The identification of the BRR and BRTP in producing the highest polysaccharide and biomass concentration from the mycelium of *Ganoderma lucidum* QRS 5120 will be the primary aim of this study. Therefore, it focuses on different extraction strategies; SXE, SVE and ultrasonic-assisted extraction (UAE), for the extraction of *G. lucidum* mycelia biomass lipids (GMBLs) cultured through SmF. The yield of the GMBL will be identified and compared to determine the most effective extraction technique for GMBLs. Thus, maximum yield with the shortest fermentation cycle can be achieved to reach an industrial level and meet the global market demand to produce biomass lipids.

2. Materials and Methods

2.1. Microorganism and Medium

G. lucidum strain QRS 5120 was obtained from Functional Omics and Bioprocess Development Laboratory, University Malaya. Medium compositions for both seed culture and repeated-batch fermentation were (in g/L): yeast extract (1), glucose (30), NH_4Cl (4), $MgSO_4$ (0.5), K_2HPO_4 (0.5) and KH_2PO_4 (0.5) [21,23].

2.2. Stock Culture and Inoculum Preparation

Stock culture of *G. lucidum* (Figure 1a) was prepared by cutting its mycelium from the mother plate into a size of 1 × 1 cm and sub-culturing it onto the potato dextrose agar (PDA). The inoculated culture was incubated at 30 °C for 7 days and stored in a 4 °C chiller until further use. The inoculum was prepared by cultivating the first seed for 10 days and then the second seed for 7–15 days with BRTP set at 30 °C and 100 rpm. From a seven-day plate (Figure 1b), three mycelial agar squares were cut into a size of 1 × 1 cm and inoculated in 100 mL of prepared medium. After 10 days, formed mycelium in the first seed culture was collected and homogenised using a sterile Waring blender. Then, 20 mL (20% *v*/*v*) of the first seed culture was transferred to another flask containing 80 mL and incubation was continued for another 7–15 days. The second seed culture was collected and homogenised for the fermentation process.

2.3. Batch Fermentation

Batch fermentation was conducted using an incubator shaker with the conditions set at an initial pH of 4.0, a temperature of 30 °C, an agitation speed of 100 rpm and an aeration rate of 2.0 vvm for regulated dissolved oxygen. To optimise the BRR, three different predetermined BRRs [60%, 75% and 90%] were studied. In addition, three batch fermentation growth curves obtained from previous research [11] were chosen to identify the appropriate BRTP as shown in Figure 1c. There are three main phases in each curve: the logarithmic, transition and stationery, which represent an increasing, highest and stabilising concentration, respectively.

2.4. Optimisation of Medium Compositions Using RSM

The optimisation of medium compositions to produce *G. lucidum* was conducted using RSM. To perform the RSM, a CCD was used to construct the experimental set using Design Expert version 12.0 software by StatEase®, Minneapolis, MN, USA. A BRR of 60% and BRTP of 7 days were set as the lowest variable, while a BRR of 90% and BRTP of 15 days were set as the highest variable. Meanwhile, the α-value was set to 1.0. EPS yield (g/L), ENS yield (g/L) and biomass concentration (g/L) were set as the response parameters of this study. At the end of the process, an analysis of variance (ANOVA) was generated to evaluate the optimisation model and validated using the optimised experimental parameters.

2.5. Repeated-Batch Fermentation Using Optimised Condition

Repeated-batch fermentation (RBF) was carried out until the maximum possible number of cycles to achieve the maximum biomass concentration until the fermentation broth reached the highest viscosity. The process was conducted by refilling the current mixture of medium fermentation with 80 mL of fresh medium for every cycle. As the main purpose of RBF is to eliminate the lag phase, five days of the feeding interval were chosen for the full fermentation cycle.

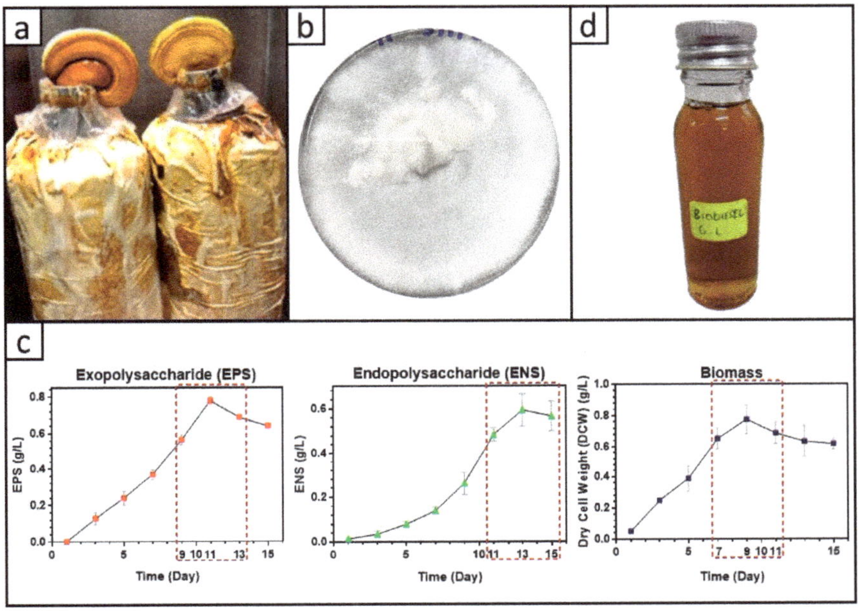

Figure 1. (**a**) Fruiting body of *Ganoderma lucidum* strain QRS 5120 located at the Functional Omics and Bioprocess Development Laboratory, Institute of Biological Sciences, Universiti Malaya; (**b**) Mycelia state of the *G. lucidum* on a PDA plate after 7 days; (**c**) Growth curves discovered from previous studies were chosen as broth replacement time points (BRTPs) (red dashed box) for exopolysaccharide (EPS) 9–11–13, endopolysaccharide (ENS) 11–13–15 and biomass 7–9–11. Red circles = EPS production, Green triangles = ENS production, Purple squares = Biomass production; (**d**) Biodiesel extracted from *G. lucidum* QRS 5120.

2.6. Extraction of Lipid

The extraction of lipids from *G. lucidum* mycelial biomass was conducted using three different extraction techniques, which are Soxhlet extraction (SXE), solvent extraction (SVE) and ultrasonic-assisted extraction (UAE). Prior to the extraction process, mycelium was dried at 60 °C for 3 days in a drying oven. Then, the dried biomass obtained was ground into a fine powder. All three extraction techniques used hexane as a solvent and 5 g of *G. lucidum* mycelia biomass with the heating temperature set at 60 °C. After extraction, the lipids of *G. lucidum* (Figure 1d) were collected by filtering, followed by the evaporation of the lipid mixture using a rotary evaporator to remove the solvent [27].

2.6.1. Soxhlet Extraction

SXE was carried out using Soxhlet equipment with 5 g of Ganoderma biomass powder added in 50, 100 and 150 mL of hexane heated at 60 °C for 3, 6 and 9 h.

2.6.2. Solvent Extraction

SVE was conducted using 500 mL of a sample bottle containing 100, 150 and 200 mL of hexane mixed with 5 g of biomass powder and put on a hot plate at 60 °C with magnetic stirring of 200 rpm for 1, 2 and 3 h.

2.6.3. Ultrasonic-Assisted Extraction

UAE was performed using an ultrasonic water bath equipped with a 500 mL sample bottle containing 5 g of biomass powder mixed with 100, 200 and 300 mL of hexane and put in an ultrasonic water bath at 60 °C for 1, 3 and 5 h.

2.7. Ganodiesel Production through Transesterification

Using modified transesterification techniques [28], extracted lipids were converted to biodiesel. Acid-catalysed transesterification was conducted in accordance with the European Standards (EN 14103) in a 1000 mL round-bottomed glass flask. The reaction product was evaporated in a rotary evaporator at 70 °C for 20 min followed by gravity separation that separated Ganodiesel in the form of fatty acid methyl ester (FAME) from glycerol as a by-product. FAME was later washed with distilled water to neutralise the catalyst, then purified using silica gel and dehydrated with anhydrous sodium sulphate before gas chromatography (GC) analysis. FAME was filtered using a disposable polytetrafluoroethylene (PTFE) filter fitted to a glass syringe. The mixture of 1.5 mL of oil extract and 1 mL of 15% H_2SO_4 in methanol was then poured into a boiling tube. The tube was placed in a heating block at 70 °C for two hours. After heating, 1 mL of distilled water was added to the tube before aggressively shaking the mixture. The solution was left to allow the oil layer to separate. The FAME layer was further diluted with n-hexane before being injected for Gas Chromatography (GC) analysis. The FAME was examined with a GC–MS (Shimadzu, Nakagyo-ku, Kyoto, Japan) equipped with an Agilent Zorbax C18 column (80 Å, 3.5 µm, 1 × 150 mm, Agilent, Santa Clara, CA, USA).

2.8. Analytical Methods

2.8.1. Determination of Extracellular Polysaccharide Yield

EPS precipitation was collected by adding a 4:1 ratio of 95% ethanol (v/v) to the supernatant and kept at 4 °C overnight. EPS precipitate was filtered through filter paper that had been pre-dried and weighted and rinsed with 5 mL of 95% ethanol twice. The weight of EPS was recorded after drying in a food dehydrator to a constant weight (Figure 2a).

2.8.2. Determination of Endopolysaccharide Yield

ENS was isolated from the pellet mycelium after being drained from the extracted fermentation broth. Filtered mycelium was dried using a food dehydrator overnight and weighed to measure the amount of distilled water needed. The extraction of ENS was performed by rinsing dried mycelium with the ratio of 1:20 (w/v) of distilled water and then heating at 121 °C for 30 min. Then, the ENS was further treated by mixing the supernatant with ethanol at the ratio of 4:1 and kept overnight at 4 °C. The formed precipitate was filtered using pre-dried and weighed filter paper, followed by drying to a constant weight in the food dehydrator (Figure 2a) (Supplementary Materials, Table S1).

2.8.3. Determination of Biomass Concentration

The filtration of harvested fermentation broth through a Büchner funnel filter set attached to a water pump produced biomass. The mycelial biomass was washed repeatedly (three times) with distilled water, followed by a drying process in a food dehydrator to a constant weight. The weight of the dried mycelial biomass (Figure 2a) was recorded.

2.8.4. Determination of Lipid Yield

The lipid remained dissolved in the hexane solvent after the extraction process. By using a rotary evaporator, the excess solvent was eliminated by applying the concept of heat and pressure [27,30]. The mass of the recovered lipid after the removal of solvent was weighed. The extracted lipid yield (%) was estimated by using the equation below [30]:

$$Lipid\ yield\ (\%) = mi/ms \times 100 \quad (1)$$

where the coefficient mi (g/L) is the mass of recovered lipid while ms (g/L) is the mass of dried material (5 g of *G. lucidum* mycelial biomass) used for the extraction of SXE, SVE and UAE.

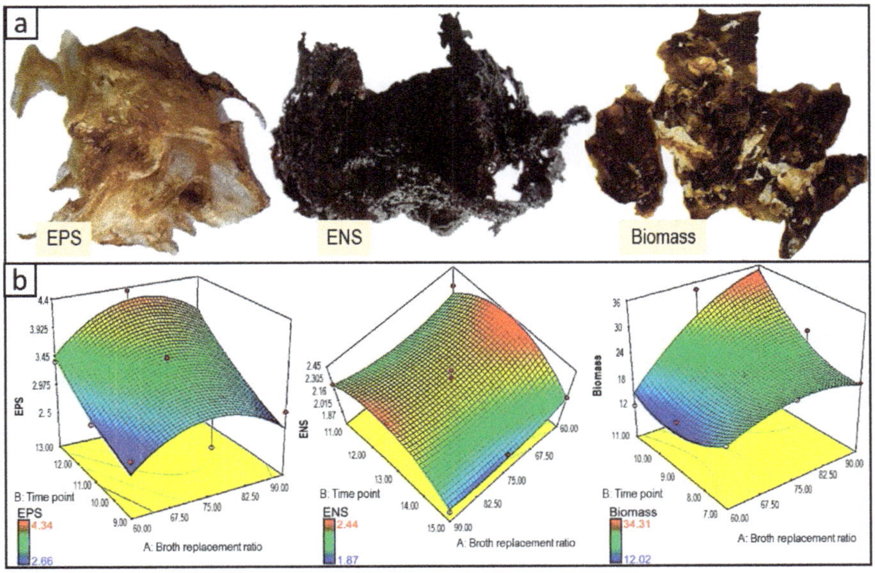

Figure 2. Dried EPS, ENS and biomass extracted from *Ganoderma lucidum* QRS 5120 (**a**), and response surface curve (3D plot) showing the EPS, ENS, biomass production and interaction between broth replacement ratio (BRR) and broth replacement time point (BRTP) from *Ganoderma lucidum* (**b**).

2.9. Kinetic Calculations

By following [31], the kinetic parameters were calculated.

$$EPS/ENS/Biomass\ productivity, P_{EPS}/P_{ENS}/P_X\left(\tfrac{g}{L}day^{-1}\right) = \frac{X_{max}-X_0}{the\ time\ for\ product\ recovery\ at\ certain\ cycle\ in\ repeated{-}batch\ culture} \quad (2)$$

where the X_{max} is the maximum cell concentration achieved at the stationary phase and X_0 is the initial cell concentration at inoculation.

2.10. Statistical Analysis

Triplicates for each sample were analysed and GraphPad Prism Version 9 by Dotmatics (GraphPad Software, San Diego, CA, USA) was used to calculate the mean and standard deviation with displayed error bars. Optimum BRR and BRTP for EPS, ENS and biomass were then assessed through response surface methodology (RSM).

Hence, the equation below demonstrates the influential factors and correlation, which are based on a quadratic second-order model for the responses:

$$Y = b'_0 + \sum_{i=1}^{n} b_i X_i + \sum_{i=1}^{n} b_{ii} X_i^2 + \sum_{i=1}^{n} \sum_{j>1}^{n} b_{ij} X_i X_j \quad (3)$$

where

Y: predicted response
b'_0: constant coefficient
b_i: linear coefficient
b_{ij}: interaction coefficient
b_{ii}: quadratic coefficient
$X_i X_j$: coded values

Meanwhile, statistical analysis for lipid extraction from *Ganoderma lucidum* mycelial biomass was carried out by Minitab version 18 software. Extraction techniques (SXE, SVE, UAE) were conducted in triplicates. Analysis of variance (ANOVA) was used to calculate the mean ± S.D.

2.11. Determination of Biodiesel Properties

The biodiesel characteristics of *Ganoderma lucidum* fatty acid methyl esters (GLFAMEs) were examined by conducting a variety of tests following international guidelines [32,33]. Mini Pour/Cloud Point Tester MPC-102 was used to determine the cloud and pour point at −60 °C to 51 °C, which was subsequently evaluated according to ASTM D6749 and ASTM D2500, respectively. The automatic Kinematic Viscosity Measuring System AKV-201 was used to measure the kinematic viscosity at 40 °C, as mentioned in ASTM D445. Following ASTM D637, the measurement for the cold filter plugging point was conducted using Automated Cold Filter Plugging Point Tester AFP-102. By referring to the ASTM D4530, a test for carbon residue was carried out using Micro Carbon Residue Tester ACR-M3. The ignition point test was performed using Pensky-Martens Closed Cup Automated Flash Point Tester APM-7 by referring to the ASTM D93. Certified instrumentations provided by TANAKA Scientific Ltd., Tokyo, Japan, were used to carry out all the tests. In addition, the stability of the oxygen has also been tested using Rancimat 743 (Metrohm, Herisau, Switzerland) following ASTM 14,112. Additionally, tests for ester content, monoglyceride, diglyceride, triglyceride, total glycerol content, acid number, iodine value and water content were measured in accordance with European Standard Methods [32].

3. Results

3.1. Optimization

The influences of BRR and BRTP on the production of EPS–ENS–biomass from *Ganoderma lucidum* strain QRS 5120 mycelium were analysed using RSM. Tables 1 and 2 show the experimental designs and responses. The tables summarise that thirteen sets of conditions for culture were applied for RSM optimisation. This was defined by CCD and will be used to assess the coefficients using non-linear regression analysis. The determination of the model coefficient was analysed using analysis of variance, and the study reported the significance was $p < 0.05$. Based on Table 1, EPS and ENS concentrations were found to be the highest during day 13 of the BRTP with 4.34 g/L and 2.44 g/L, respectively. A total of 75% BRR at day 11 BRTP was the optimised variable for biomass production with the actual response of 34.31 g/L, as mentioned in Table 2.

Table 1. Experimental design and responses for production of EPS and ENS from *G. lucidum* mycelia. The experiments for the actual responses were conducted under controlled conditions.

Run No.	EPS (g/L)				ENS (g/L)			
	Variables		Responses		Variables		Responses	
	BRR (%)	BRTP (Day)	Actual	Predicted	BRR (%)	BRTP (Day)	Actual	Predicted
1	75	11	3.63	3.56	75	11	2.02	2.10
2	75	11	3.63	3.56	60	13	2.36	2.44
3	75	11	3.63	3.56	75	13	2.29	2.30
4	90	11	2.91	3.15	75	13	2.25	2.30
5	75	13	4.34	4.19	60	11	2.26	2.18
6	60	11	2.80	2.9	75	15	1.98	1.92
7	75	9	2.66	3.15	75	13	2.35	2.30
8	90	9	2.96	2.68	90	15	1.87	1.94
9	60	9	2.75	2.54	75	13	2.43	2.30
10	75	11	3.63	3.56	90	11	2.24	2.23
11	90	13	3.79	3.83	90	13	2.44	2.38
12	60	13	3.37	3.48	60	15	2.11	2.11
13	75	11	3.63	3.56	75	13	2.21	2.30

Table 2. Experimental design and responses for production of biomass from *G. lucidum* mycelia. The experiments for the actual responses were conducted under controlled conditions.

Run No.	Biomass (g/L)			
	Variables		Responses	
	BRR (%)	BRTP (Day)	Actual	Predicted
1	60	11	12.02	14.70
2	75	9	20.39	21.82
3	60	7	18.05	17.71
4	75	9	20.39	18.91
5	75	7	22.44	21.82
6	75	9	20.39	21.82
7	90	9	26.26	27.45
8	60	9	15.57	16.20
9	75	9	20.39	21.82
10	90	11	32.38	34.78
11	90	7	20.73	20.11
12	75	11	34.31	24.74
13	75	9	20.39	21.82

3.1.1. Optimization of EPS Production

Table 3 shows the ANOVA for EPS production and the 3D plot is expressed in Figure 2b. According to the table, the expected determination of the coefficient ($R^2 = 0.8404$) suggests that this model can express 84.04% of the response variability and as $p < 0.05$, thus it proved the significance of the model. The validity of the model is indicated by the value of the modified coefficient determination (Adj. $R^2 = 0.727$), which was adequate in comparison to the expected R^2 value. In the equation below, the model was regressed in terms of actual EPS variables and expressed as a final equation in terms of absolute factors.

$$EPS = (-8.65135) + 0.34713 \times BRR - 0.46309 \times BRTP + 1.75000e^{-3} \times BRR \times BRTP - 2.38774e^{-3} \times BRR^2 + 0.026940 \times BRTP^2 \quad (4)$$

Table 3. Experimental results of the CCD quadratic model using ANOVA for EPS production from *Ganoderma lucidum*.

Source	Sum of Squares	DF	Mean Square	*F* Value	Prob > *F*	
Model	2.560	5	0.510	7.39	0.0103 *	significant
A: BRR	0.091	1	0.091	1.32	0.2889	
B: BRTP	1.630	1	1.630	23.55	0.0018 *	significant
AB	0.011	1	0.011	0.16	0.7019	
A^2	0.800	1	0.800	11.50	0.0116 *	significant
B^2	0.032	1	0.032	0.46	0.5183	
Residual	0.490	7	0.069			
Lack of Fit	0.490	3	0.160			not significant
Pure Error	0	4	0			
Cor Total	3.05	12				
Std. Dev. = 0.26			R^2 = 0.8408	Adeq Precision = 9.232		
Mean = 3.36			Adjusted R^2 = 0.727			

* Significant value.

3.1.2. Optimization of ENS Production

Table 4 shows the ANOVA for the production of ENS, and Figure 2b displays the 3D plot. The model's R-squared was 0.8341, which means it could represent 83% of the response. The *p*-value was 0.0117, showing the model was significant as $p < 0.05$. It clearly shows that more variances can be explained if the number of R^2 is high so that the optimum model can be generated. The adjusted R^2 was 0.7155, indicating that the model was significant and showed consent to the expected R^2 value. If more non-significant variables are added to the model, the difference between the R^2 and the adjusted R^2 will thus be greater. An equation as shown below has been formulated and regressed in terms of the actual factor of ENS, by considering the significant terms from the model:

$$ENS = (-8.46282) - 0.049201 \times BRR + 1.99825 \times BRTP - 1.83333e^{-3} \times BRR \times BRTP + 4.73563e^{-4} \times BRR^2 - 0.073362 \times BRTP^2 \quad (5)$$

Table 4. Experimental results of the CCD quadratic model using ANOVA for ENS production from *Ganoderma lucidum*.

Source	Sum of Squares	DF	Mean Square	*F* Value	Prob > *F*	
Model	0.310	5	0.062	7.04	0.0117 *	significant
A: BRR	5.40×10^{-3}	1	5.40×10^{-3}	0.62	0.4578	
B: BRTP	0.052	1	0.052	5.98	0.0445 *	significant
AB	0.012	1	0.012	1.38	0.2779	
A^2	0.031	1	0.031	3.59	0.1002	
B^2	0.240	1	0.240	27.19	0.0012 *	significant
Residual	0.061	7	8.75×10^{-3}			
Lack of Fit	0.031	3	0.010	1.39	0.3664	not significant
Pure Error	0.03	4	7.48×10^{-3}			
Cor Total	0.37	12				
Std. Dev. = 0.094 Mean = 2.22			R^2 = 0.8341 Adjusted R^2 = 0.7155		Adeq Precision = 8.237	

* Significant value.

3.1.3. Optimization of Biomass Production

Table 5 shows the ANOVA for mycelium biomass production, with a 3D graph in Figure 2b. The predicted coefficient determination demonstrated that this model can explain 83.74% (R^2 = 0.8374) of the response variability. The model ($p < 0.05$) is applicable. The model's validity was suggested by the R^2 value of 0.7213, which was in reasonable agreement with the predicted R^2 value. The model was regressed and expressed as an equation in terms of the actual biomass variables by considering the relevant terms.

$$Biomass = +77.50102 + 1.30728 \times BRR - 27.91759 \times BRTP + 0.14733 \times BRR \times BRTP - 0.015057 \times BRR^2 + 1.01806 \times BRTP^2 \quad (6)$$

Table 5. Experimental results of the CCD quadratic model using ANOVA for biomass production from *Ganoderma lucidum*.

Source	Sum of Squares	DF	Mean Square	*F* Value	Prob > *F*	
Model	375.44	5	75.09	7.21	0.0110 *	significant
A: BRR	189.62	1	189.62	18.21	0.0037 *	significant
B: BRTP	50.98	1	50.98	4.90	0.0626	
AB	78.15	1	78.15	7.50	0.0289 *	significant
A^2	31.70	1	31.70	3.04	0.1246	
B^2	45.80	1	45.80	4.40	0.0742	
Residual	72.89	7	10.41			
Lack of Fit	72.89	3	24.30			not significant
Pure Error	0	4	0			
Cor Total	448.34	12				
Std. Dev. = 3.23 Mean = 21.82			R^2 = 0.8374 Adjusted R^2 = 0.7213		Adeq Precision = 10.332	

* Significant value.

3.1.4. Verification of Optimised Conditions

Verification for a statistical model of the highest EPS–ENS–biomass is shown in Table 6 by applying the optimised variables. Various experiments were performed to validate the strength and precision of the model according to the Equation EPS/ENS/biomass. Under optimised conditions, EPS production gained was 4.21 g/L, ENS production was 2.44 g/L and 34.32 g/L of mycelium biomass was obtained. These correspond to the prediction values of the EPS, ENS and biomass (g/L); 4.19, 2.44 and 34.78, respectively. Therefore, the validity of the model (Equation EPS/ENS/biomass) is verified for EPS, ENS and biomass production. However, due to different broth replacement time points for EPS, ENS and biomass (BRTP: refer to Section 2.3), the maximum production for the combination run was inapplicable and only an individual response for each run was achievable in this study.

Table 6. Verification of model with the optimised variables.

Run	Variables		Responses		
	BRR	BRTP	EPS (g/L)	ENS (g/L)	Biomass (DCW g/L)
EPS	77.46	13.00	4.21	–	–
ENS	60.00	12.85	–	2.44	–
Biomass	89.52	10.96	–	–	34.32

– Not available.

3.2. Repeated-Batch Fermentation Using Validated EPS–ENS–biomass Condition

3.2.1. Effect of BRR

Figure 3 displays the growth curves of EPS–ENS–biomass production in a repeated-batch fermentation. The growth curves included the extracted EPS–ENS–biomass for all six cycles (R1–R6). R0 was derived from the previously defined growth profile (Figure 1d). Using the pre-optimised conditions, 75%, 60%, and 90% were set as BRR for EPS, ENS, and biomass, respectively. The BRR procedure involved the removal and addition of fresh media to the flasks. Samples withdrawn in 120 h (5 days) intervals were analysed for their polysaccharide and biomass content.

3.2.2. Effect of BRTP

A study on the effect of time points was conducted based on the pre-optimised conditions. The broth replacement was carried out on day 13 for EPS and ENS, while biomass broth replacement occurred on day 11. According to Table 7, the production of EPS–ENS–biomass was observed and a total of six batches in a row could be conducted using the RBF process. It demonstrated the capability of *G. lucidum* cells toward this technique.

Table 7. EPS–ENS–biomass productivity from pre-optimised conditions in a repeated-batch fermentation strategy at day 13 (75%), day 13 (60%) and day 11 (90%).

BRTP (Day)	BRR (%)	Kinetics [a]	RBF Cycles [b]						Sum (R1–R6)
			R1	R2	R3	R4	R5	R6	
13	75	P_{EPS} (g/L day^{-1})	0.3467	0.0720	0.0813	0.1413	0.1080	0.0413	0.7906
13	60	P_{ENS} (g/L day^{-1})	0.0087	0.1160	0.0813	0.0927	0.0613	0.0833	0.4433
11	90	P_X (g/L day^{-1})	0.6880	0.7480	0.6600	0.5860	1.0160	1.1200	4.8180

[a] P_{EPS} (g/L day^{-1}) = EPS productivity, P_{ENS} (g/L day^{-1}) = ENS productivity, P_X (g/L day^{-1}) = biomass productivity. [b] Cycles repetition (R1–R6)

3.3. Evaluation of Lipid Yield

The extraction of GMBLs using SXE, SVE and UAE were evaluated. It was observed that all three techniques were able to extract GMBLs with different lipid yields obtained. The amount of lipid yield was not as high as expected. The highest GMBL yield, of 20.36%, was achieved using SVE (3 h/200 mL), followed by 18.8% through SXE (6 h/150 mL) and 7.5% through UAE (5 h/200 mL). Based on research, 1 g of mycelium of *G. lucidum* has a lipid content of approximately 1.67% (Hou et al., 2017). Therefore, Table 8 shows statistical analysis using analysis of variance for all techniques. Two factors have been tested upon the yield of GMBL which were the solvent volume and extraction time as well as the interaction between both independent variables.

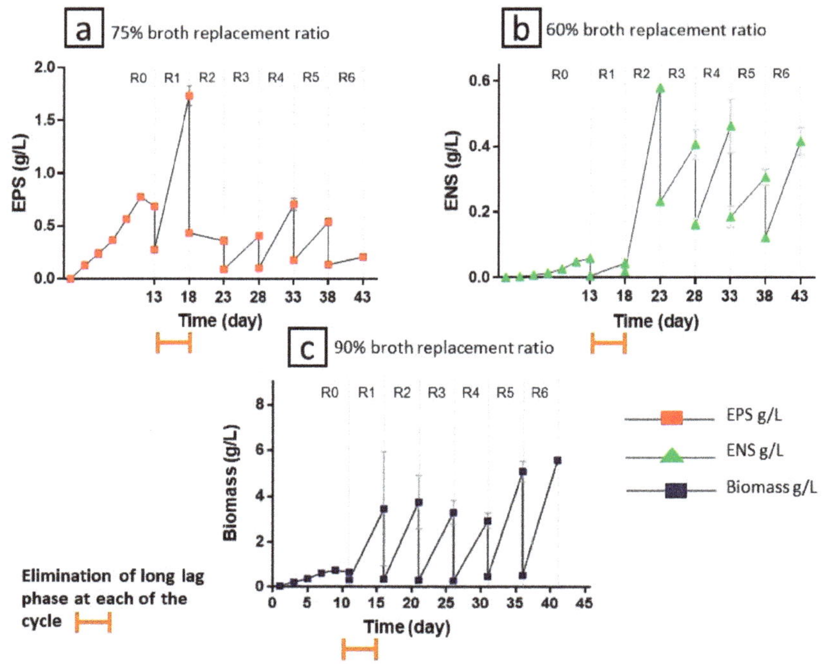

Figure 3. Production of EPS (**a**), ENS (**b**) and biomass (**c**) from the mycelium of *Ganoderma lucidum* QRS 5120 in a repeated-batch fermentation strategy from the pre-optimised conditions (EPS, 75% broth replacement ratio (BRR); ENS, 60% BRR; and biomass, 90% BRR).

Table 8. Experimental results of SXE, SVE and UAE using ANOVA.

Extraction Techniques [a]	Source	DF	Sum of sq	Mean sq	*F*-Value	*p*-Value
(a) SXE	Solvent volume	2	14.560	7.2800	67.55	0.000
	Extraction time	2	139.253	69.6267	646.02	0.000
	Solvent volume Extraction time	4	22.897	5.7242	53.11	0.000
	Error	9	0.970	0.1078		
	Mean = 12.12 Variance = 10.45 Std. Dev. = 3.03		R^2 = 0.9945 Adjusted R^2 = 0.987 Predicted R^2 = 0.9782			
(b) SVE	Solvent volume	2	9.906	4.9532	31.60	0.000
	Extraction time	2	137.224	68.6118	437.73	0.000
	Solvent volume Extraction time	4	43.766	10.9415	69.80	0.000
	Error	9	1.411	0.1567		
	Mean = 13.44 Variance = 11.31 Std. Dev. = 3.36		R^2 = 0.9927 Adjusted R^2 = 0.9261 Predicted R^2 = 0.9707			

Table 8. *Cont.*

Extraction Techniques [a]	Source	DF	Sum of sq	Mean sq	*F*-Value	*p*-Value
(c) UAE	Solvent volume	2	1.0844	0.54222	9.76	0.006
	Extraction time	2	19.1878	9.59389	172.69	0.000
	Solvent volume Extraction time	4	1.4056	0.35139	6.33	0.010
	Error	9	0.5000	0.05556		
	Mean = 9.24 Variance = 5.34 Std. Dev. = 2.31		R^2 = 0.9775 Adjusted R^2 = 0.9574 Predicted R^2 = 0.9098			

[a] SXE, Soxhlet Extraction; SVE, Solvent Extraction; UAE, Ultrasonic-assisted Extraction.

3.4. *Effect of Solvent Volume and Extraction Time*

The volume of hexane and time of extraction affects the yield of the GMBLs. Lipid extraction using SXE recorded a highest yield of 150 mL (6 h) and a lowest yield of 50 mL (3 h), SVE demonstrated a high of 200 mL (3 h) and a low of 100 mL (1 h), and UAE a high of 200 mL (5 h) and a low of 100 mL (1 h). The results obtained clearly show that lipid yield increased with the increase of the solvent volume and extraction time.

4. Discussion

4.1. *Optimization of EPS–ENS–biomass Production*

BRTP (B) and BRR^2 (A^2) of EPS, as mentioned in Table 3, show that both factors influence their production at $p < 0.05$ compared to others (A, B^2 and AB). Figure 2b illustrates the correlation effect between the BRR and BRTP for EPS production. From the figure, an increase in the BRTP leads to an increase in the production of EPS, and BRR at all ranges records the lowest EPS production. According to this, it was shown that the BRTP influenced the production of EPS and there was no correlation between the BRR and BRTP. Maximum production of EPS (red area) was 75% on day 13 with 4.34 g/L of EPS achieved.

According to the ENS model (Table 4), both the BRTP (B) and quadratic terms of BRTP (B^2) show the effect on the production of ENS as $p < 0.05$. However, BRR (A) and other quadratic terms (A^2 and AB) present a negative result. Figure 2b demonstrated the combined effect of BRR and BRTP on the production of ENS. From the figure, the BRTP on day 13 shows the most production of ENS and is least affected by the BRR. By this, it was concluded that the production of ENS was strongly affected by the broth replacement time point (BRTP) and there was no interaction between the BRR and BRTP. In contrast with EPS and biomass, the ENS model showed two red areas which indicate the optimised production of ENS. The highest distribution of red area fell on day 13 (60%) compared with the least red area, also on day 13 (90%). This also might be due to the insignificance of BRR which does not influence ENS production. Therefore, the highest production of ENS (2.44 g/L) obtained was on day 13, with 60% BRR.

Therefore, BRR (A) and a quadratic term (AB) of the biomass model (Table 5) present significant values as both terms record *p*-values of less than 0.05. However, BRTP (B) and other quadratic terms (A^2 and B^2) show a negative effect on the production of biomass. Figure 2b explained the relationship between BRR and BRTP. According to the figure, all ranges of BRTP record the lowest production of biomass. Nevertheless, BRR and BRTP show a linear correlation as the production of biomass increased when their value increased. From this, it can be summarised that there was a relationship between BRR and BRTP. The maximum production of biomass (34.31 DCW g/L) achieved was at 90% on day 11.

However, the total error Sum of Square (SS) and Mean of Square (MS) has indeed been zero for EPS and biomass as shown in Tables 3 and 5, respectively. All these values were very small and insignificant. The explanation for these findings may have been because the model's SS value of regression was much higher than the residual SS value. The null value of pure error, both SS and MS, showed that the lack of fit (LOF) test had no *p*-value and F value. Adequate Precision calculated the ratio of signal to noise and it is preferable if this value is greater than 4. As the model for EPS and biomass had a 9.232 and 10.332 ratio, respectively, the signal was adequate, and this shows that the model is applicable for navigating the design area.

4.2. Repeated-Batch Fermentation Using Validated EPS–ENS–biomass Condition

Effect of Broth Replacement Ratio and Broth Replacement Time Point

As mentioned in Figure 3, fermentation for EPS–ENS–biomass production can be tolerated for up to six cycles (R6) when they show sustained continuous growth of the fungus during the RBF. The repeated-batch cell cycle has demonstrated its ability with repeated use. In this repeated-batch fermentation study, the lag phase has been reduced by sampling at five day intervals. Therefore, it resulted in the continuous production of EPS, ENS and biomass. The amount of EPS–ENS–biomass produced could be similar to or higher than that of batch fermentation (R0). As mentioned by Naritomi et al., productivity from repeated batch culture is higher rather than productivity from the standard batch culture [26]. Pre-optimised BRRs and BRTPs were used during the RBF process, with EPS (75%, day 13), ENS (60%, day 13) and biomass (90%, day 11).

According to Figure 3, the lowest development of EPS, ENS and biomass were observed on the sixth (R6), first (R1) and fourth (R4) days, respectively. The lowest production was due to some fungi attempting to cope with the new environment and conditions as new media were introduced, making some of the fungi enter the phase of death caused by stress. Meanwhile, in the first (R1), second (R2) and sixth (R6) cycles, the highest performance of EPS, ENS and biomass was recorded. This is due to the capability of the fungus to stabilise and preserve its production until the environmental viscosity was high for further growth. As similar cells of fungus (*G. lucidum*) have been used throughout the RBF process, which was in between the first cycle (R1) and the sixth cycle (R6), it was crucial to replace the volume at an optimised time point by changing the media to maintain the level toxicity of the EPS–ENS–biomass concentration. However, towards the end of the cycle for the present study, the EPS–ENS–biomass was still being produced. Therefore, future work can be conducted to identify the highest cycle for their production.

In general, the consistency of EPS and ENS productivity was recorded after the second cycle (R2) with (g/L day^{-1}) 0.0413–0.1413 and 0.0813–0.1160, while biomass was observed along with the six batches (R1-R6) with 0.5860–1.1200 g/L day^{-1}. Overall, total P_{EPS}, P_{ENS} and P_X (g/L day^{-1}) recorded were 0.7906, 0.4433 and 4.8180 with average readings of 0.1318, 0.0739 and 0.8030, respectively, from the six cycles. Furthermore, the highest productivity was reported for EPS (0.3467 g/L day^{-1}) at the first cycle (R1), ENS (0.1160 g/L day^{-1}) at the second cycle (R2) and biomass (1.1200 g/L day^{-1}) at the sixth cycle (R6).

4.3. Evaluation of Lipid Yield

Table 8 shows the fitness models analysed which were 0.9945 for SXE, 0.9927 for SVE and 0.9775 for UAE. These indicate that the lipid yield was strongly affected by the volume of solvent and time for extraction. The table displays that there were significant effects and interactions between the factors for all three extraction techniques (p-value < 0.05). A few factors have been suggested by previous findings as affecting the yield of lipids, including type and volume of solvent, extraction time and temperature [29,30]. According to Table 9, a comparison between the three different techniques used in the extraction of GMBL can be summarised. SXE and SVE have been recorded to be effective in extracting lipids and are widely used in many studies [34,35]. In extracting GMBLs, SVE (14.57–20.36%) and SXE (9.44–18.80%) were observed to have a higher yield compared with the use of UAE (4.30–7.50%).

In SXE and SVE, the transfer of heat between solid and liquid occurs from the outside to the inside of the sample membrane; meanwhile, the mass transfer happens and vice versa (from the inside to the outside) [27,36]. On the contrary, UAE utilises the frequency of electrostatic interactions arising from the development of high-intensity wave propagation. The UAE process involves a physical mechanism whereby the penetration of the cell walls and washing out of the cell material takes place before it becomes disrupted [36,37]. Low lipid yield in UAE may be caused by the lack of transmission consistency and being exposed to the sonic power which affects the biomass cell wall.

Table 9. Comparison between three lipid extraction techniques on GMBL yield [27,34,36–38].

Aspects	**Extraction Techniques**		
	SXE	**SVE**	**UAE**
Procedure	An appropriate size of cellulose thimble was chosen for the sample before being placed in the Soxhlet extractor. The solvent in a round bottom flask heated with a mantle	Sample soaked in the solvent with a magnetic stirrer in a flat bottom flask and then heated by a hot plate	Sample soaked in solvent, and placed in an ultrasonic water bath equipment
Solvent	hexane	hexane	hexane
Sample size	1–5 g	1–5 g	1–5 g
Temperature	60 °C	60 °C	60 °C
Solvent volume	50–150 mL	100–200 mL	100–300 mL
Extraction time	3–9 h	1–3 h	1–5 h
Advantages	No filtration is needed, easy to use	Short extraction duration	Moderate extraction duration, easy to use
Disadvantages	Long extraction duration	Filtration needed	High solvent amount, filtration needed
Lipid yield	9.44–18.80%	14.57–20.36%	4.30–7.50%
Quantity	Moderate	High	Low

4.4. Effect of Solvent Volume and Extraction Time

Based on the results obtained, a high solvent volume will help to accelerate the chemical reaction resulting in enhanced lipid production. Indeed, one study found a correlation between solvent volume and extraction time, in which a high solvent quantity was caused by the higher extraction time required [39]. Additionally, the author reported that a sufficient volume of the solvent is needed to guarantee the complete immersion of sample particles during the extraction process. Hexane has been selected for this analysis because it can be quickly recovered, is non-polar, has low latent vaporisation heat and has high solvent selectivity [29,40,41]. Alternatively, an optimal solvent volume and extraction time could be determined to enhance the yield of GMBLs.

In SXE, the lipid yield decreased when the time was set to the highest at 9 h (50 mL to 150 mL), while when the time was set up from 3 h to 6 h the lipid yields increased. This is because SXE reached an optimal time (3 h–6 h) and therefore after 6 h there was not much lipid to be obtained. Therefore, the lower duration is more efficient compared to the higher duration. In comparison to SXE and UAE, SVE had the shortest time duration (between 1 h to 3 h) and the smallest volume of solvents (100 mL to 200 mL) for the highest lipid production. The higher yield of GMBLs in SVE could be due to the use of a magnetic stirrer that continuously mixed the biomass sample with hexane solvent throughout the extraction process, while in SXE and UAE the sample was static [36,42].

4.5. Comparison of the Current Study with the Literature

4.5.1. Comparison of *G. lucidum* Optimization Using SmF

Table 10 presents the latest statistical optimisation in determining the best technique for achieving the efficient production of EPS, ENS and biomass from *Ganoderma lucidum* through shake-flask fermentation. As mentioned, it indicates that only three studies using *Ganoderma lucidum* in EPS, ENS and biomass production have previously been published using SmF.

As stated, the method of cultivation using RBF has been shown to doubly increase the productivity of EPS compared to the normal batch fermentation technique performed by Supramani, Ahmad [21]. They used the same ideal glucose concentration of 26.5 g/L and pH of 4, and were designed using RSM. Meanwhile, a low concentration of EPS was achieved when using an orthogonal matrix (1.723 g/L) and Taguchi orthogonal matrix (0.42 g/L), as studied by Chang, Tsai [43] and Yuan, Chi [44], and concentrations of just 0.42 g/L and 1.723 g/L EPS were reported, respectively.

Therefore, the study of Supramani, Ahmad [21] was the only one to produce ENS, but different cultivation methods and batch fermentation were applied in their research. Meanwhile, in the current study, RBF has been used and optimised by RSM for ENS production, and it showed the efficiency of this technique compared to their result. This can be seen in Table 10, where the present study generated the highest ENS with 2.44 g/L; meanwhile, 1.52 g/L was the result from Supramani. As a result, the current study suggests that the best productivity for ENS production could be achieved with numerous advantages.

In addition, referring to Table 10, Chang, Tsai [43] mentioned that under several optimum conditions, biomass production was significantly enhanced. Yuan, Chi [44] in their findings reported that the production of biomass was significantly increased by the optimal media. RSM was used in both the study of Supramani, Ahmad [21] and the present study. However, high biomass production was achieved in this study, demonstrating that the method proposed was effective compared to other works in the literature.

Table 10. Comparison of *Ganoderma lucidum* optimization using SmF with the previous studies.

Ganoderma lucidum Strains	Optimization Method	Cultivation Mode	Cultivation Method	Total Prep. Time (Days)	Initial pH	Glucose Concentration (g/L)	Agitation (rpm)	EPS (g/L)	ENS (g/L)	Biomass (g/L)	References
Ganoderma lucidum QRS 5120	Response surface methodology	Shake Flask	Repeated-batch fermentation	5 (up to 6 cycles)	4	26.5	100	4.34	2.44	34.31	Current study
Ganoderma lucidum QRS 5120	Response surface methodology	Shake Flask	Batch fermentation	10	4	26.5	100	2.64	1.52	5.19	[21]
Ganoderma lucidum CCRC 36124	Taguchi's orthogonal array	Shake Flask	Batch fermentation	7	6.5	12.1	160	0.420	NA [a]	18.70	[43]
Ganoderma lucidum CAU 5501	Orthogonal matrix	Shake Flask	Batch fermentation	4	-	50	150	1.723	NA [a]	7.235	[44]

[a] NA, Not Available.

4.5.2. Comparison of Ganoderma Lipid Profile

Lipids from *Ganoderma lucidum* were extracted with hexane as the extraction solvent using a Soxhlet apparatus based on a modified method [45]. Supernatants produced from the extract centrifugation for 20 min at 1500 rpm were then filtered and dried (sodium sulphate anhydrous). Therefore, using a rotary evaporator, supernatants were concentrated at 30 °C under a vacuum and the weight was recorded. Lipid profiling was analysed in accordance with ASTM E2881 [46].

The fatty acids in GMBLs primarily comprised 66.1 wt% oleic acids (C18:1), 16.2 wt% palmitic acid (C16:0) and 13.2 wt% linoleic acids (C18:2), as shown in Table 11. Therefore, a total of 95.5 wt% fatty acids contents were accounted for from all three fatty acids. In comparison, *Ganoderma austral* and *Ganoderma applanatum* mainly consisted of 49.5 wt% and 58.2 wt% linoleic acids (C18;2), respectively [47,48]. In this study, lipids produced from *G. lucidum* contained the highest saturated fatty acid, including palmitic acid (16:0) compared with lipids from *G. austral* and *G. applanatum*. According to Lin and Chiu [49], high saturated fatty acids cause a slow deterioration rate of lipids due to their high oxidative and thermal stability. Therefore, they are beneficial for long-term storage or surrounding high temperatures as they can delay the fuel degradation process.

Table 11. Comparison of Ganoderma lipid profile with previous studies.

Fatty Acid	Structure	*Ganoderma lucidum* (wt%)	*Ganoderma austral* (wt%)	*Ganoderma applanatum* (wt%)
Palmitic	16:0	16.2	10.7	10.2
Palmitoleic	16:1	2.2	1.9	2.0
Stearic	18:0	2.3	2.7	2.5
Oleic	18:1	66.1	30	15.5
Linoleic	18:2	13.2	49.5	58.2
References		This study	[47]	[48]

4.6. Biodiesel Properties of Ganoderma lucidum FAME

A comparison of biodiesel properties between GLFAMEs and international standards is shown in Table 12 [46]. Generally, GLFAMEs fulfilled the most necessary biodiesel properties according to the standard. By referring to the standards of the US (ASTM D6751-08) and the EU (EN 14214), our study found that GLFAMEs fell within the range including the ignition point (153.5 °C), kinematic viscosity (3.8 mm^2/s), water content (345 mg/kg), total glycerol content (0.15%), ester content (97.5%), iodine value (112 $g(I_2)/100$ g) and acid number (0.38 mg(KOH)/g). However, the oxidation stability was low (6 h) and this may be due to the 81.5% unsaturated fatty acids in *G. lucidum* oil which is considered to be high. This property does not satisfy the EU standard that requires more than 8 h of oxygen stability. However, it still passes the EU standard compliance countries, which is the US standard (>3 h), therefore the addition of antioxidants to GLFAMEs for longer storage capacity is recommended. On the other hand, the positive outcome of highly unsaturated fatty acids content in GLFAMEs could be seen from its expected excellent performance at low temperatures, represented by the good values of pour point (−3.0 °C), cold filter plugging points (−2.0 °C) and cloud point (−1.0 °C).

Table 12. FAME Properties of *Ganoderma lucidum* methyl esters in comparison with international guidelines [a].

Properties	Method	Unit	GLME (Biodiesel)	US (ASTM D6751-08)	EU (EN 14214)
Kinematic viscosity (40 °C)	ASTM D445	mm^2/s	3.8	1.9–6.0	3.5–5.0
Carbon residue	ASTM D4530	wt%	0.10	≤0.05	<0.30
Pour point	ASTM D2500	°C	−3.0	-	-
Cold filter plugging point	ASTM D6371	°C	−2.0	-	-
Ignition point	ASTM D93	°C	153.5	≥130	>101
Cloud point	ASTM D6749	°C	−1.0	-	-
Oxidation stability	EN 14112	h	6	-	8
Ester content	EN 14103	wt%	97.5	-	>96.5
Monoglyceride	EN 14105	wt%	0.05	-	<0.80
Diglyceride	EN 14105	wt%	0.03	-	<0.20
Triglyceride	EN 14105	wt%	n.d. [b]	-	<0.20
Total glycerol content	EN 14105	wt%	0.15	<0.24	<0.25
Water content	EN ISO12937	mg/kg	345	<500	<500
Acid number	EN 14104	mg(KOH)/g	0.38	<0.50	<0.50
Iodine value	EN 14111	$g(I_2)/100$ g	112	-	<120

[a] CEN (2019); ASTM (2008). [b] n.d., not detectable.

5. Conclusions

The implementation of CCD RSM to optimise EPS–ENS–biomass production from *Ganoderma lucidum* strain QRS 5120 was verified using SmF. The maximum production of EPS (4.34 g/L^{-1}), ENS (2.44 g/L^{-1}) and mycelial biomass (34.31 g/L^{-1}) was achieved under the optimised conditions of a 75% BRR, with day 13 and day 11 at 60% and 90%, respectively. All EPS, ENS and mycelial biomasses were accomplished within up to six cycles in a RBF strategy, as shown by continuous growth in the sixth cycle. Nevertheless, the RBF strategy offered high productivity of EPS, ENS and biomass, and reduced the fermentation period. Meanwhile, the total amount of GMBLs obtained from all three extraction methods was low. The SVE (20.36%) and SXE (18.8%) techniques extracted a higher amount of GMBL compared with the UAE method (7.5%). In SVE, the GMBL yield was enhanced as the amount of hexane (100–200 mL) and extraction time (1–3 h) increased. To maximise the lipid yield, the shortest time (3 h) and lowest volume of hexane solvent (200 mL) were needed in SVE. Among the extraction techniques used, SVE could be suggested as an effective extraction strategy to acquire lipids, as it requires a short extraction period with a moderately low solvent volume. The findings of this study will enhance the production of *Ganoderma lucidum* strain QRS 5120 and the capability to produce lipids as a new biodiesel feedstock. Therefore, the present study may be extended to different kinds of fungal fermentation for the effective production of EPS, ENS and biomass, and is ideal for the process of upscaling, allowing it to be introduced on an industrial scale to meet market demands. However, future research work should focus on other novel extraction techniques the optimisation of the extraction parameters in improving the lipid yield from *G. lucidum* mycelial biomass.

Supplementary Materials: The following supporting information can be downloaded at: https://www.mdpi.com/article/10.3390/su141710764/s1, Table S1: Raw data from *G. lucidum* QRS 5120.

Author Contributions: Conceptualization: B.M.M.J., W.A.A.Q.I.W.-M., Z.I. and N.R. Data curation: S.S., M.N.A.S. and M.F.I. Funding acquisition: W.A.A.Q.I.W.-M., Z.I., M.F.I. and S.A.-A. Investigation: B.M.M.J., W.A.A.Q.I.W.-M. and Z.I. Methodology: B.M.M.J., W.A.A.Q.I.W.-M., Z.I. and M.F.I. Project administration: Z.I., W.A.A.Q.I.W.-M., M.F.I. and S.A.-A. Resources: W.A.A.Q.I.W.-M., Z.I., M.F.I. and S.A.-A. Software: S.S. and M.N.A.S. Supervision: W.A.A.Q.I.W.-M., M.F.I. and S.A.-A. Visualization: B.M.M.J. Writing—original draft: B.M.M.J., W.A.A.Q.I.W.-M., Z.I. and N.R. Writing—review and editing: B.M.M.J., W.A.A.Q.I.W.-M., Z.I. and N.R. All authors have read and agreed to the published version of the manuscript.

Funding: This research was funded by Universiti Malaya International Collaboration Grant ST002-2022, ST007-2021, European Atlantic Area Neptunus (Project EAPA_576/2018) and Regional University Network European University (RUN-EU Project).

Institutional Review Board Statement: Not applicable.

Informed Consent Statement: Not applicable.

Data Availability Statement: Not applicable.

Acknowledgments: We would like to thank Universiti Malaya International Collaboration Grant ST002-2022 [W.A.A.Q.I.W.-M.], ST007-2021 [Z.I.], European Atlantic Area Neptunus (Project EAPA_576/2018) and Regional University Network European University (RUN-EU Project) [N.R.]. Z.I. is Fulbright Visiting Associate Professor at Cornell University during the completion of this work.

Conflicts of Interest: The authors declare no conflict of interest.

References

1. Bamisile, O.; Huang, Q.; Dagbasi, M.; Taiwo, W.; Adebayo, V. Energy, exergy and environmental analyses of a biomass driven multi-generation system. *Int. J. Exergy* **2020**, *31*, 249–267. [CrossRef]
2. Yang, M.; Yu, X. *Energy Efficiency: Benefits for Environment and Society*; Springer: London, UK, 2015.
3. Rowan, N.J.; Galanakis, C.M. Unlocking challenges and opportunities presented by COVID-19 pandemic for cross-cutting disruption in agri-food and green deal innovations: Quo Vadis? *Sci. Total Environ.* **2020**, *748*, 141362. [CrossRef] [PubMed]
4. Rowan, N.J.; Casey, O. Empower Eco multiactor HUB: A triple helix 'academia-industry-authority' approach to creating and sharing potentially disruptive tools for addressing novel and emerging new Green Deal opportunities under a United Nations Sustainable Development Goals framework. *Curr. Opin. Environ. Sci. Health* **2021**, *21*, 100254. [CrossRef]
5. Masinda, M.M.; Sun, L.; Wang, G.; Hu, T. Moisture content thresholds for ignition and rate of fire spread for various dead fuels in northeast forest ecosystems of China. *J. For. Res.* **2020**, *32*, 1147–1155. [CrossRef]

6. Goh, B.H.H.; Ong, H.C.; Cheah, M.Y.; Chen, W.-H.; Yu, K.L.; Mahlia, T.M.I. Sustainability of direct biodiesel synthesis from microalgae biomass: A critical review. *Renew. Sustain. Energy Rev.* **2019**, *107*, 59–74. [CrossRef]
7. Aburto, J.; Amezcua-Allieri, M.A. Biodiesel and Green Diesel Fuels: A Techno-Economic Analysis. In *Green Diesel: An Alternative to Biodiesel and Petrodiesel*; Springer: Singapore, 2022; pp. 309–324.
8. Quah, R.V.; Tan, Y.H.; Mubarak, N.; Khalid, M.; Abdullah, E.; Nolasco-Hipolito, C. An overview of biodiesel production using recyclable biomass and non-biomass derived magnetic catalysts. *J. Environ. Chem. Eng.* **2019**, *7*, 103219. [CrossRef]
9. Sukhadiya, M.L.; Thakur, N.S.; Patel, V.R.; Gunaga, R.P.; Kharadi, V.B.; Tyagi, K.K.; Singh, S. Provenance variations in proximate principles, mineral matter, total phenols and phytochemicals of Melia dubia drupes: An unexplored alternate livestock feed stock. *J. For. Res.* **2020**, *32*, 119–131. [CrossRef]
10. Nurul Amirah, H. Optimisation of Biomass and Exopolysaccharide Production in Wild Serbian *Ganoderma Lucidum* Strain BGF4A1 Using Response Surface Methodology/Nurul Amirah Hassan. Ph.D. Thesis, University of Malaya, Kuala Lumpur, Malaysia, 2019.
11. Supramani, S.; Jailani, N.; Ramarao, K.; Zain, N.A.M.; Klaus, A.; Ahmad, R.; Wan-Mohtar, W.A.A.Q.I. Pellet diameter and morphology of European Ganoderma pfeifferi in a repeated-batch fermentation for exopolysaccharide production. *Biocatal. Agric. Biotechnol.* **2019**, *19*, 101118. [CrossRef]
12. Du, Z.; Dong, C.-H.; Wang, K.; Yao, Y.-J. Classification, Biological Characteristics and Cultivations of Ganoderma. *Ganoderma Health* **2019**, *1181*, 15–58. [CrossRef]
13. Wu, D.-T.; Deng, Y.; Chen, L.-X.; Zhao, J.; Bzhelyansky, A.; Li, S.-P. Evaluation on quality consistency of *Ganoderma lucidum* dietary supplements collected in the United States. *Sci. Rep.* **2017**, *7*, 1–10. [CrossRef]
14. Chen, S. The Pharmacological Effects of Triterpenoids from *Ganoderma lucidum* and the Regulation of Its Biosynthesis. *Adv. Biol. Chem.* **2020**, *10*, 55–65. [CrossRef]
15. Smith, J.E.; Rowan, N.; Sullivan, R. Medicinal mushrooms: A rapidly developing area of biotechnology for cancer therapy and other bioactivities. *Biotechnol. Lett.* **2002**, *24*, 1839–1845. [CrossRef]
16. Sullivan, R.; Smith, J.E.; Rowan, N.J. Medicinal Mushrooms and Cancer Therapy: Translating a traditional practice into Western medicine. *Perspect. Biol. Med.* **2006**, *49*, 159–170. [CrossRef]
17. Feng, J.; Zhang, J.-S.; Feng, N.; Yan, M.-Q.; Yang, Y.; Jia, W.; Lin, C.-C. A novel *Ganoderma lucidum* G0119 fermentation strategy for enhanced triterpenes production by statistical process optimization and addition of oleic acid. *Eng. Life Sci.* **2016**, *17*, 430–439. [CrossRef]
18. Domingos, M.; de Souza-Cruz, P.B.; Ferraz, A.; Prata, A.M.R. A new bioreactor design for culturing basidiomycetes: Mycelial biomass production in submerged cultures of Ceriporiopsis subvermispora. *Chem. Eng. Sci.* **2017**, *170*, 670–676. [CrossRef]
19. Esmaeilzadeh, M.; Kianirad, M.; Sheykhinejad, A.; Khosravi, A.; Sharifzadeh, A. Ganoderic Acid and Exopolysaccharide Production by Ganoderma Lucidum from Semi-Solid-State and Submerged Fermentation. *Biol. J. Microorg.* **2018**, *7*, 63–75.
20. Hassan, N.A.; Supramani, S.; Sohedein, M.N.A.; Usuldin, S.R.A.; Klaus, A.; Ilham, Z.; Chen, W.-H.; Wan-Mohtar, W.A.A.Q.I. Efficient biomass-exopolysaccharide production from an identified wild-Serbian *Ganoderma lucidum* strain BGF4A1 mycelium in a controlled submerged fermentation. *Biocatal. Agric. Biotechnol.* **2019**, *21*, 101305. [CrossRef]
21. Supramani, S.; Ahmad, R.; Ilham, Z.; Annuar, M.S.M.; Klaus, A.; Wan-Mohtar, W.A.A.Q.I. Optimisation of biomass, exopolysaccharide and intracellular polysaccharide production from the mycelium of an identified *Ganoderma lucidum* strain QRS 5120 using response surface methodology. *AIMS Microbiol.* **2019**, *5*, 19–38. [CrossRef]
22. Birhanli, E.; Yesilada, O. Enhanced production of laccase in repeated-batch cultures of Funalia trogii and Trametes versicolor. *Biochem. Eng. J.* **2010**, *52*, 33–37. [CrossRef]
23. Wan-Mohtar, W.A.A.Q.I.; Ab Kadir, S.; Saari, N. The morphology of *Ganoderma lucidum* mycelium in a repeated-batch fermentation for exopolysaccharide production. *Biotechnol. Rep.* **2016**, *11*, 2–11. [CrossRef]
24. Fan, C.; Qi, K.; Xia, X.-X.; Zhong, J.-J. Efficient ethanol production from corncob residues by repeated fermentation of an adapted yeast. *Bioresour. Technol.* **2013**, *136*, 309–315. [CrossRef]
25. Ji, X.-J.; Zhang, A.-H.; Nie, Z.-K.; Wu, W.-J.; Ren, L.-J.; Huang, H. Efficient arachidonic acid-rich oil production by Mortierella alpina through a repeated fed-batch fermentation strategy. *Bioresour. Technol.* **2014**, *170*, 356–360. [CrossRef]
26. Naritomi, T.; Kouda, T.; Yano, H.; Yoshinaga, F.; Shigematsu, T.; Morimura, S.; Kida, K. Influence of broth exchange ratio on bacterial cellulose production by repeated-batch culture. *Process Biochem.* **2002**, *38*, 41–47. [CrossRef]
27. Hakimi, M.I.; Ilham, Z.; Kohar, R.A.A. Enhancement of Agro-Industrial Copra Residue Oil Yield Using Microwave-Assisted Extraction. *Waste Biomass- Valorization* **2018**, *10*, 2681–2688. [CrossRef]
28. Redzwan, G.; Amin, M.M.; Zulkarnain, N.N.; Abu Mansor, M.R.; Annuar, M.S.M.; Ilham, Z. Extrication of biodiesel feedstock from early stage of food waste liquefaction. *J. Mater. Cycles Waste Manag.* **2015**, *19*, 676–681. [CrossRef]
29. Perrier, A.; Delsart, C.; Boussetta, N.; Grimi, N.; Citeau, M.; Vorobiev, E. Effect of ultrasound and green solvents addition on the oil extraction efficiency from rapeseed flakes. *Ultrason. Sonochemistry* **2017**, *39*, 58–65. [CrossRef]
30. Tonato, D.; Luft, L.; Confortin, T.C.; Zabot, G.L.; Mazutti, M.A. Enhancement of fatty acids in the oil extracted from the fungus Nigrospora sp. by supercritical CO2 with ethanol as a cosolvent. *J. Supercrit. Fluids* **2019**, *146*, 180–188. [CrossRef]
31. Stanbury, P.F.; Whitaker, A.; Hall, S.J. *Principles of Fermentation Technology*; Elsevier: Amsterdam, The Netherlands, 2013.
32. *BS EN 14214:2012+A2:2019*; Fatty acid methyl esters (FAME) for use in diesel engines and heating applications: Requirements and test methods. Liquid Petroleum Products, British Standards Institutions (BSI): London, UK, 2019.

33. *ASTM D. D6751-08*; Standard Specification for Biodiesel Fuel Blend Stock (B100) for Middle Distillate Fuels. ASTM International: West Conshohocken, PA, USA, 2018. [CrossRef]
34. Asl, P.J.; Niazmand, R.; Yahyavi, F. Extraction of phytosterols and tocopherols from rapeseed oil waste by supercritical CO2 plus co-solvent: A comparison with conventional solvent extraction. *Heliyon* **2020**, *6*, e03592. [CrossRef]
35. Ma, Y.; Wu, X.; Zhao, L.; Wang, Y.; Liao, X. Comparison of the compounds and characteristics of pepper seed oil by pressure-assisted, ultrasound-assisted and conventional solvent extraction. *Innov. Food Sci. Emerg. Technol.* **2019**, *54*, 78–86. [CrossRef]
36. Oukebdane, K.; Portet-Koltalo, F.; Machour, N.; Dionnet, F.; Desbène, P. Comparison of hot Soxhlet and accelerated solvent extractions with microwave and supercritical fluid extractions for the determination of polycyclic aromatic hydrocarbons and nitrated derivatives strongly adsorbed on soot collected inside a diesel particulate filter. *Talanta* **2010**, *82*, 227–236. [CrossRef]
37. Rao, P.; Rathod, V. *Phytochemicals: An Insight to Modern Extraction Technologies and Their Applications*; Academic Press: Cambridge, MA, USA, 2017; pp. 495–521. [CrossRef]
38. Mohammadpour, H.; Sadrameli, S.M.; Eslami, F.; Asoodeh, A. Optimization of ultrasound-assisted extraction of Moringa peregrina oil with response surface methodology and comparison with Soxhlet method. *Ind. Crop. Prod.* **2019**, *131*, 106–116. [CrossRef]
39. Ibrahim, N.A.; Zaini, M.A.A. Microwave-assisted solvent extraction of castor oil from castor seeds. *Chin. J. Chem. Eng.* **2018**, *26*, 2516–2522. [CrossRef]
40. Efthymiopoulos, I.; Hellier, P.; Ladommatos, N.; Russo-Profili, A.; Eveleigh, A.; Aliev, A.; Kay, A.; Mills-Lamptey, B. Influence of solvent selection and extraction temperature on yield and composition of lipids extracted from spent coffee grounds. *Ind. Crop. Prod.* **2018**, *119*, 49–56. [CrossRef]
41. Kumar, S.P.J.; Prasad, S.R.; Banerjee, R.; Agarwal, D.K.; Kulkarni, K.S.; Ramesh, K.V. Green solvents and technologies for oil extraction from oilseeds. *Chem. Central J.* **2017**, *11*, 1–7. [CrossRef]
42. Menegazzo, M.L.; Fonseca, G.G. Biomass recovery and lipid extraction processes for microalgae biofuels production: A review. *Renew. Sustain. Energy Rev.* **2019**, *107*, 87–107. [CrossRef]
43. Chang, M.-Y.; Tsai, G.-J.; Houng, J.-Y. Optimization of the medium composition for the submerged culture of *Ganoderma lucidum* by Taguchi array design and steepest ascent method. *Enzym. Microb. Technol.* **2006**, *38*, 407–414. [CrossRef]
44. Yuan, B.; Chi, X.; Zhang, R. Optimization of exopolysaccharides production from a novel strain of *Ganoderma lucidum* CAU5501 in submerged culture. *Braz. J. Microbiol.* **2012**, *43*, 490–497. [CrossRef]
45. Bin Zakaria, M.; Vijayasekaran; Ilham, Z.; Muhamad, N.A. Anti-inflammatory Activity of Calophyllum Inophyllum Fruits Extracts. *Procedia Chem.* **2014**, *13*, 218–220. [CrossRef]
46. Sohedein, M.N.A.; Wan-Mohtar, W.A.A.Q.I.; Hui-Yin, Y.; Ilham, Z.; Chang, J.-S.; Supramani, S.; Siew-Moi, P. Optimisation of biomass and lipid production of a tropical thraustochytrid Aurantiochytrium sp. UMACC-T023 in submerged-liquid fermentation for large-scale biodiesel production. *Biocatal. Agric. Biotechnol.* **2020**, *23*, 101496. [CrossRef]
47. Papaspyridi, L.; Sinanoglou, V.; Strati, I.; Katapodis, P.; Christakopoulos, P. Fatty acid profile of Pleurotus ostreatus and Ganoderma australe grown naturally and in a batch bioreactor. *Acta Aliment.* **2013**, *42*, 328–337. [CrossRef]
48. Sande, D.; de Oliveira, G.P.; Moura, M.A.F.E.; Martins, B.D.A.; Lima, M.T.N.S.; Takahashi, J.A. Edible mushrooms as a ubiquitous source of essential fatty acids. *Food Res. Int.* **2019**, *125*, 108524. [CrossRef]
49. Lin, C.-Y.; Chiu, C.-C. Burning characteristics of palm-oil biodiesel under long-term storage conditions. *Energy Convers. Manag.* **2010**, *51*, 1464–1467. [CrossRef]

MDPI
St. Alban-Anlage 66
4052 Basel
Switzerland
www.mdpi.com

Sustainability Editorial Office
E-mail: sustainability@mdpi.com
www.mdpi.com/journal/sustainability

www.ingramcontent.com/pod-product-compliance
Lightning Source LLC
Chambersburg PA
CBHW061450210926
51850CB00008B/1671